Molecular Biology Intelligence Unit

Lipocalins

Bo Åkerström
Department of Clinical Sciences
University of Lund
Lund, Sweden

Niels Borregaard
Granulocyte Research Laboratory
Department of Hematology L, Rigshospitalet
Copenhagen, Denmark

Darren R. Flower
The Jenner Institute
University of Oxford
Compton, Berkshire, U.K.

Jean-Philippe Salier
Inserm, U519
University of Rouen, France
Institut Fédératif de Recherches Multidisciplinaires sur les Peptides
Rouen, France

CRC Press
Taylor & Francis Group
Boca Raton London New York

CRC Press is an imprint of the
Taylor & Francis Group, an informa business

Lipocalins

Molecular Biology Intelligence Unit

CRC Press
Taylor & Francis Group
6000 Broken Sound Parkway NW, Suite 300
Boca Raton, FL 33487-2742

First issued in hardback 2019

ISBN-13: 978-1-58706-297-1 (hbk)

Visit the Taylor & Francis Web site at
http://www.taylorandfrancis.com

and the CRC Press Web site at
http://www.crcpress.com

Library of Congress Cataloging-in-Publication Data

Lipocalins / Bo Åkerström ... [et al.].
p. ; cm. -- (Molecular biology intelligence unit)
Includes index.
ISBN 1-58706-297-6
1. Carrier proteins. I. Åkerström, Bo. II. Title. III. Series: Molecular biology intelligence unit (Unnumbered)
[DNLM: 1. Carrier Proteins. QU 55.2 L764 2006]
QP552.C34L57 2006
572'.696--dc22

2006014309

About the Editors...

BO ÅKERSTRÖM is a professor and group leader at the Department of Clinical Sciences, Section for Clinical and Experimental Infection Medicine, Lund University, Lund, Sweden. He earned his Ph.D. degree in Medical and Physiological Chemistry in 1982, M.D. degree in 1987, both at Lund University. His research is focussed on structure and function of the lipocalin α_1-microglobulin and bacterial immunoglobulin-binding proteins, anti-oxidant mechanisms and general aspects of the lipocalin protein family.

NIELS BORREGAARD is professor of hematology at the University of Copenhagen and head of the Department of Hematology, University Hospital, Rigshospitalet Denmark. His research interest is innate immunity with focus on antibiotic molecules and their expression in myeloid cells and epithelia. He received his M.D. from the University of Aarhus Denmark in 1978 and his Dr.med. Sci. degree from University of Aarhus in 1981.

DARREN R. FLOWER is a Jenner Fellow at the University of Oxford. He received his first degree in Chemistry and Biochemistry from the Imperial College of Science, Technology, and Medicine (1988) and a second degree in Molecular Biophysics from the University of Leeds (1992). After spending seven years in the Pharmaceutical Industry, Dr. Flower moved in 1999 to become a Group Leader at the Edward Jenner Institute for Vaccine Research, which became part of Oxford University at the end of 2005. Dr. Flower's research interests reflect his peripatetic multidisciplinary background and include: computational chemistry, immunoinformatics, drug design, and, of course, the lipocalin protein family. Dr. Flower is a Fellow of the Royal Society of Chemistry and chairs its Molecular Modelling Group. He is married with one child.

JEAN-PHILIPPE SALIER is a Senior Investigator and a group leader in the French *Institut de la Santé et de la Recherche Médicale* (Inserm). He received a Ph.D. (1975) and a D.Sci. (1984) in Immunogenetics from Rouen University (Normandy). Apart from his global interest in lipocalin genes, his current focus is on functional genomics with a major interest in the transcriptome-based analysis of gene transcription in health and disease in liver (acute phase response, cirrhosis, hepatocellular carcinoma) and in rheumatoid arthritis.

Dedication

This book is dedicated to our colleagues Syed Pervaiz and Keith Brew whose discovery of an unexpected structural relationship between seemingly unrelated proteins (Pervaiz S, Brew K. Homology of beta-lactoglobulin, serum retinol-binding protein, and protein HC. Science 1985; 228:335-337) is the cornerstone upon which the subsequent opening up of the lipocalin world is built.

CONTENTS

EDITORS

Bo Åkerström
Department of Clinical Sciences
University of Lund
Lund, Sweden
Email: bo.akerstrom@med.lu.se
Chapters 1, 10, 16

Niels Borregaard
Granulocyte Research Laboratory
Department of Hematology L, Rigshospitalet
Copenhagen, Denmark
Email: borregaard@rh.dk
Chapter 1

Darren R. Flower
The Jenner Institute
University of Oxford
Compton, Berkshire, U.K.
Email: Darren.flower@jenner.ac.uk
Chapters 1, 3, 17

Jean-Philippe Salier
Inserm, U519
University of Rouen, France
Institut Fédératif de Recherches Multidisciplinaires
sur les Peptides
Rouen, France
Email: Jean-Philippe.Salier@univ-rouen.fr
Chapters 1, 2

CONTRIBUTORS

Russell E. Bishop
Departments of Laboratory Medicine
and Pathobiology, and Biochemistry
Faculty of Medicine
University of Toronto
Toronto, Ontario, Canada
Email: russell.bishop@utoronto.ca
Chapter 4

Brian J. Burke
School of Biochemistry
and Microbiology
University of Leeds
Leeds, U.K.
Chapter 14

Christian Cambillau
Architecture et Fonction
des Macromolecules Biologiques
CNRS-Universités Aix-Marseille I and II
Marseille, France
Chapter 4

Andrea Cavaggioni
Dipartimento di Anatomia
e Fisiologia Umana
Università di Padova
Padova, Italy
Chapter 12

Jean-Benoit F. Charron
Université du Québec à Montréal
Département des Sciences Biologiques
Centre-ville, Montréal QC, Canada
Chapter 5

Philip C.N. Chiu
Department of Obstetrics
and Gynecology
University of Hong Kong
Queen Mary Hospital
Hong Kong, China
Chapter 11

Björn Dahlbäck
Department of Laboratory Medicine
and Clinical Chemistry
Wallenberg Laboratory
University Hospital
Lund University
Malmö, Sweden
Chapter 13

Sonia Do Carmo
Département des Sciences Biologiques
Université du Québec à Montréal
Montréal, Québec, Canada
Chapter 13

Stephen G. Edwards
Institute of Structural
and Molecular Biology
School of Biological Sciences
University of Edinburgh
Edinburgh, U.K.
Chapter 12

Naomi Eguchi
Department of Molecular
Behavioral Biology
Osaka Bioscience Institute
Osaka, Japan
Chapter 9

John B.C. Findlay
School of Biochemistry
and Microbiology
University of Leeds
Leeds, U.K.
Email: J.B.C.Findlay@leeds.ac.uk
Chapter 14

María D. Ganfornina
Departamento de Bioquímica
y Fisiología y Genética Molecular
Instituto de Biología
y Genética Molecular
Universidad de Valladolid-CSIC
Valladolid, Spain
Email: opabinia@ibgm.uva.es
Chapters 2, 3, 6

Anne-Christine Gauthier-Jauneau
Inserm, U519
University of Rouen, France
Institut Fédératif de Recherches Multidisciplinaires sur les Peptides
Rouen, France
Email: anne-christine.jauneau@univ-rouen.fr
Chapter 2

Lesley H. Greene
Department of Biochemistry and Molecular Biology
University College London
London, U.K.
Chapter 3

Gabriel Gutierrez
Departamento de Genética
Universidad de Sevilla
Sevilla, Spain
Email: ggpozo@us.es
Chapter 2

Osamu Hayaishi
Department of Molecular Behavioral Biology
Osaka Bioscience Institute
Osaka, Japan
Chapter 9

Derek Hsi
Departments of Laboratory Medicine and Pathobiology, and Biochemistry
Faculty of Medicine
University of Toronto
Toronto, Ontario, Canada
Chapter 4

Hartmut Kayser
Abteilung Biologie I (Allgemeine Zoologie und Endokrinologie)
Universität Ulm
Ulm, Germany
Chapter 6

Hannu Koistinen
Department of Clinical Chemistry
University Central Hospital, Helsinki
Helsinki, Finland
Chapter 11

Riitta Koistinen
Department of Clinical Chemistry, and Obstetrics and Gynecology
University Central Hospital, Helsinki
Helsinki, Finland
Chapter 11

Lennart Lögdberg
Department of Pathology and Laboratory Medicine
Emory University School of Medicine
Atlanta, Georgia, U.S.A.
Email: llogdbe@emory.edu
Chapters 10, 16

Rauno Mäntyjärvi
Department of Clinical Microbiology
University of Kuopio
Kuopio, Finland
Chapter 15

Marcia E. Newcomer
Department of Biological Sciences
Louisiana State University
Baton Rouge, Louisiana, U.S.A.
Chapter 7

David E. Ong
Department of Biochemistry
Vanderbilt University School of Medicine
Nashville, Tennessee, U.S.A.
Chapter 7

Paolo Pelosi
Dipartimento di Chimica e Biotecnologie Agrarie
Università di Pisa
Pisa, Italy
Chapter 12

Gilbert G. Privé
Division of Molecular
and Structural Biology
Ontario Cancer Institute
Toronto, Ontario, Canada
Chapter 4

Eric Rassart
Département des Sciences Biologiques
Université du Québec à Montréal
Montréal, Québec, Canada
Chapter 13

Bernhard Redl
Department of Molecular Biology
Innsbruck Medical University
Innsbruck, Austria
Chapter 14

Clara Redondo
School of Biochemistry
and Microbiology
University of Leeds
Leeds, U.K.
Chapter 14

Jean-Loup Risler
Laboratoire Génome et Informatique
UMR 8116 CNRS
Université d'Evry
Evry, France
Email: risler@genopole.cnrs.fr
Chapter 2

Diego Sanchez
Departamento de Bioquímica
y Fisiología y Genética Molecular
Instituto de Biología
y Genética Molecular
Universidad de Valladolid-CSIC
Valladolid, Spain
Email: lazarill@ibgm.uva.es
Chapters 2, 3, 6

Fathey Sarhan
Université du Québec à Montréal
Département des Sciences Biologiques
Centre-ville, Montréal QC, Canada
Email: sarhan.fathey@uqam.ca
Chapter 5

Lindsay Sawyer
Institute of Structural
and Molecular Biology
School of Biological Sciences
University of Edinburgh
Edinburgh, U.K.
Email: lindsay.sawyer@ed.ac.uk
Chapter 12

Markku Seppälä
Department of Clinical Chemistry
University of Helsinki
Biomedicum Helsinki
Helsinki, Finland
Email: mseppala@welho.com
Chapter 11

Arne Skerra
Lehrstuhl für Biologische Chemie
Technische Universität München
Freising-Weihenstephan, Germany
Chapter 17

James M. Sodetz
Department of Chemistry
and Biochemistry
University of South Carolina
Columbia, South Carolina, U.S.A.
Chapter 13

Roland K. Strong
Division of Basic Sciences
Fred Hutchinson Cancer
Research Center
Seattle, Washington, U.S.A.
Email: rstrong@fhcrc.org
Chapter 8

Elisabeth R.M. Tillier
Department of Medical Biophysics
University of Toronto
and
Ontario Cancer Institute
University Health Network
Toronto, Ontario, Canada
Chapter 4

Desiree Tillo
Department of Medical Biophysics
University of Toronto
and
Ontario Cancer Institute
University Health Network
Toronto, Ontario, Canada
Chapter 4

Yoshihiro Urade
Department of Molecular
Behavioral Biology
Osaka Bioscience Institute
Osaka, Japan
Email: uradey@obi.or.jp
Chapter 9

Willem van Dijk
Glycoimmunology Group
Department of Molecular Cell Biology
and Immunology
VU University Medical Center
Amsterdam, The Netherlands
Email: w.vandijk@vumc.nl
Chapter 13

Tuomas Virtanen
Department of Clinical Microbiology
University of Kuopio
Kuopio, Finland
Email: tuomas.virtanen@uku.fi
Chapter 15

William S.B. Yeung
Department of Obstetrics
and Gynecology
University of Hong Kong
Queen Mary Hospital
Hong Kong, China
Chapter 11

PREFACE

Six years have elapsed since the publication in 2000 of the first anthology devoted to lipocalins (Biochim Biophys Acta 1482, 2000), and only two years since the first Lipocalin International Symposium in Copenhagen in 2003 (Benzon Symposium no. 50 "The Lipocalin Protein Superfamily," Copenhagen, 2003) and the introduction of a public lipocalin website (http://www.jenner.ac.uk/lipocalins.htm). In spite of all these recent joint actions from the lipocalin community, we feel that it is time for another anthology. We have three major reasons for this. First, many new exciting publications have been issued during the past five years, partially outdating the 2000 BBA lipocalin anthology. Secondly, the three events mentioned above have undoubtedly had a positive effect upon lipocalin research and the exchange of research information, and thus the community of lipocalin researchers are highly motivated to continue such pan-lipocalin activities. Finally, we were invited by a very enthusiastic Ron Landes to edit this anthology; in fact, it was his idea.

Many new lipocalins have been discovered and it was not feasible to allow a full chapter to each lipocalin. Therefore, several of the chapters in this volume are reviews of groups of lipocalins with a similar phylogenetic or tissue distribution (Chapters 4-6, 12 and 13). In spite of this, a few lipocalins have not been covered at all in this book, and we apologize to all authors whom we have, regretfully, been unable to invite. Furthermore, two chapters discuss the evolutionary and structural relationships between the lipocalins (Chapters 2 and 3) and the penultimate three chapters are treatises on themes in lipocalin research: receptors, allergy, and clinical diagnosis (Chapters 14-16); the final chapter discusses how lipocalin research might go in future. Again, we feel that there are several new, exciting themes in the field that could have been covered in this issue (such as the dynamic folding of lipocalins) but which unfortunately have been omitted due to lack of space.

We are deeply indebted to all reviewers who have patiently read and commented on the manuscripts. We also wish to thank Cynthia Conomos and the staff at Landes Bioscience for their guidance in pulling through this project.

Bo Åkerström
Niels Borregaard
Darren R. Flower
Jean-Philippe Salier

Chapter 1

Lipocalins: An Introduction

Bo Åkerström,* Niels Borregaard, Darren R. Flower and Jean-Philippe Salier

The lipocalin protein family is discussed, in its totality, in Chapters 2 and 3 and most lipocalins are reviewed individually, or in groups, elsewhere in this volume. In this chapter, written after collecting and editing all other chapters, a few additional subjects related to the family as a whole, and that have not been treated elsewhere in the book, will be raised.

Growth of Lipocalin Research

Research on the lipocalin family has been characterized by the continuous discovery of new lipocalin members and, recently, by a dramatic increase in the interaction and transfer of knowledge among individual researchers in the field. Thus, the time-line of lipocalin research (Fig. 1) shows a steady increase in the number of reported individual lipocalins and published papers on the subject. To the best of our knowledge, the lipocalin protein family was first identified in 1981 when the amino acid sequence homology between rat urine alpha$_{2u}$-globulin and bovine beta-lactoglobulin was reported by Unterman et al.[1] Subsequently, the family has grown quickly, as is shown in Figure 1, and today encompasses more than 40 proteins from bacteria, plants, and animals. The exact number of members is difficult to determine; mainly because the information on several of the newly discovered lipocalins is too limited to allow researchers to determine properly whether they are in fact new lipocalins or merely orthologues or paralogues of existing lipocalins. Indeed, many of the new lipocalins have so far only been identified at the gene level.

Lipocalin Subgrouping

Structural Grouping

As proposed by Flower,[6] the lipocalins have been grouped into ***kernel and outlier*** members depending on the degree of amino acid sequence similarity. Thus, the kernel lipocalins share all three conserved sequence motifs, whereas the outlier lipocalins only have one or two in common. Based on nucleotide/amino acid sequence similarity and phylogenetic analyses, the lipocalins have been grouped into at least 14 *clades* (ref. 11; Chapters 2 and 3). This categorization of the lipocalins will probably be increasingly useful as the family grows.

Distribution

Almost all lipocalins that have been identified are *extracellular* proteins, exceptions including the bacterial lipocalins, plant epoxidases, some arthropodan lipocalins, mammalian probasin

*Corresponding Author: Bo Åkerström—Department of Clinical Sciences, Lund University, BMC, B14, 221 84 Lund, Sweden. Email: Bo.Akerstrom@med.lu.se

Lipocalins, edited by Bo Åkerström, Niels Borregaard, Darren R. Flower and Jean-Philippe Salier.

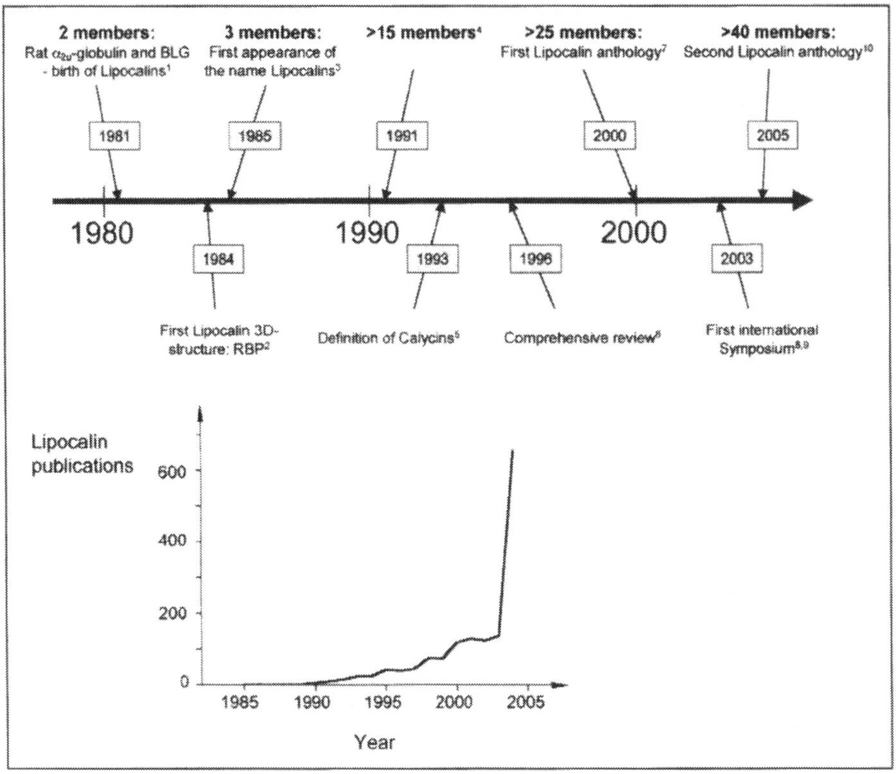

Figures 1. Upper panel) Time-line of lipocalin research. Lower panel) Time-development of the number of articles/year having the word "lipocalin" in the title or abstract, according to the search engine PubMed (http://www.ncbi.nlm.nih.gov/entrez/query.fcgi?DB=pubmed). Footnotes, see the list of references at the end of this Chapter.

and prostaglandin D synthase, which are intracellular proteins. Animal lipocalins are found in most tissue fluids, including ***blood***, and can therefore often be detected and isolated from ***urine***.

Functional Grouping

The prevalent view of the lipocalins is that its members have similar 3D-structures, but different functions, except perhaps a capacity to bind a small hydrophobic ligand within its internal binding site. However, it is possible to perceive groups of lipocalins with similar properties and functions among the characterized lipocalins. For example, ***allergen lipocalins*** (Chapter 15) are a group harbouring 15 or so family members from a variety of sources, most of them with unknown functions, but with a capacity to induce allergy in humans. The structural basis of this property is still unknown but it seems likely that the intrinsic lipocalin fold is crucial. A growing group is made up of proteins involved in ***defense and innate immunity***. Several of these lipocalins are scavengers of small molecules and some are regulators of cell-growth. A large group consists of ***odorant- and pheromone-binding*** lipocalins, of which most are regulators of animal behaviour. Very few lipocalins, no more than a handful, have been shown to be ***enzymes***. These include epoxidases in plants (Chapter 5), prostaglandin D synthase (Chapter 9) and α-microglobulin, which was recently shown to be a reductase.[12] Many more groups of lipocalins with related functions have been described (i.e., refs. 6,7), and it is difficult to say if they are the result of divergent evolution, convergent evolution or mere coincidence.

Lipocalin Nomenclature

Naming novel proteins is always difficult. Often names are given before the function is apparent, and may therefore be entirely descriptive, e.g., p25, which is a protein with a molecular mass of 25 kD, or NGAL, which stands for Neutrophil Gelatinase Associated Lipocalin. The use of more imagination creates names that are often easier to recall e.g., JAK (Janus kinase or just another kinase). As functions of novel proteins become apparent a need for renaming is often necessary and another name may be suggested, e.g., siderocalin for NGAL. The problem of naming after function is that the function which names the protein may not be the only or, indeed, the most important function of the protein. Likewise, such a function may also be shared by other proteins, which may then have the same claim to the name. Another frequent problem arises when more than one research groups claims priority to having discovered a protein. It happens frequently that discoveries are made independently and at almost the same time, and who then has the priority in naming the proteins? Finally, changing a name leads to the use of multiple names in databanks, etc and thus to difficulties in literature searches. For example, try to find all published articles on the neutrophil cytosol protein calgranulin A ***alias*** migration inhibitory factor-8 (MRP-8) ***alias*** cystic fibrosis antigen (CFAG) ***alias*** P8 ***alias*** leukocyte L1 complex light chain ***alias*** S100 calcium-binding protein A8!

It would be nice if we could all agree on one name for one lipocalin. Given the arguments above, it is safe to conclude that we can not. Certainly, one could try to come up with systematic names like lipocalin 1, lipocalin 2 etc; this is to a large extent how metalloproteinases are named, but not only is it difficult to associate such names with function, it will also be difficult to give names for homologous proteins between species, since the degree of homology between species may vary and doubt may arise as to which species should set the standard, not necessarily Man. We have therefore refrained from suggesting a unifying nomenclature for lipocalins.

A Family or a Superfamily? The Box-in-a-Box Dilemma

As mentioned above, lipocalins can be separated into several subgroups or considered as a whole. Whether the entire set of lipocalins is a family or a superfamily has been and remains a subject of some debate within the lipocalin community. This controversy obviously calls for a unified definition of a family versus superfamily, but this semantic and evolutionary issue remains unsettled. One view is that a family gathers proteins whose sequence similarity (by identical or conservative amino acid residues) reaches at least 35%, whereas any set of proteins that no longer retains this figure, despite a shared ancestry at the structural level, is a superfamily. Others do not rely so tightly on the similarity percentage. They consider that any series of proteins obviously resulting from gene duplication is a family[13] and two or more related, but strongly diverging, families make a super-family.[14] If we favour one of these definitions, say the lipocalins as a superfamily, then another problem arises immediately. As any box can fit into a bigger box, how should one name the calycins, a large protein group that includes the lipocalin superfamily? Are the calycins a super-superfamily or maybe a hyper-family? What box would they go in? A super-hyperfamily, perhaps? Conversely, if, instead, we describe the lipocalins as a family, then this classification does not obey the minimum 35% similarity in protein sequence noted above. It may still be too early to decide which precise container this whole set of proteins best fits in, and in this volume we have left it to individual authors and readers to choose.

References

1. Unterman RD, Lynch KR, Nakhashi HL et al. Cloning and sequence of several alpha2u-globulins. Proc Natl Acad Sci USA 1981; 78(6):3478-82.
2. Newcomer ME, Jones TA, Åqvist J et al. The three-dimensional structure of retinol-binding protein. EMBO J 1984; 3(7):1451-4.
3. Pervaiz S, Brew K. Homology of beta-lactoglobulin, serum retinol-binding protein, and protein HC. Science 1985; 228(4697):335-7.
4. Nagata A, Suzuki Y, Igarashi M et al. Human brain prostaglandin D synthase has been evolutionarily differentiated from lipophilic-ligand carrier proteins. Proc Natl Acad Sci USA 1991; 88:4020-4.

5. Flower DR, North ACT, Atwood TK. Structure and sequence relationships in the lipocalins and related proteins. Protein Sci 1993; 2(5):753-61.
6. Flower DR. The lipocalin protein family: Structure and function. Biochem J 1996; 318(Pt 1):1-14.
7. Åkerström B, Flower DR, Salier J-P. Lipocalins: unity in diversity. Biochim Biophys Acta 2000; 1482(1-2):1-8.
8. Benzon Symposium no. 50 "The Lipocalin Protein Superfamily". Copenhagen: 2003.
9. Salier JP, Åkerström B, Borregaard N et al. Lipocalins in bioscience: The first family gathering. Bioessays 2004; 26(4):456-8.
10. Present volume.
11. Ganfornina MD, Gutierrez G, Bastiani M et al. A phylogenetic analysis of the lipocalin protein family. Mol Biol Evol 2000; 17(1):114-26.
12. Allhorn M, Klapyta A, Åkerström B. Redox properties of the lipocalin α-microglobulin: Reduction of cytochrome c, hemoglobin and free iron. Free Rad Biol Med 2005; 38:557-67.
13. Creighton TE. Proteins. Structure and molecular properties. 2nd ed. New York: WH Freeman, 2002:507.
14. Hartl DL. A primer of population genetics. 3rd ed. Sunderland, USA: Sinauer Associates, 2000:180.

Chapter 2

Lipocalin Genes and Their Evolutionary History

Diego Sanchez,* María D. Ganfornina, Gabriel Gutierrez, Anne-Christine Gauthier-Jauneau, Jean-Loup Risler and Jean-Philippe Salier

Introduction

As extensively detailed elsewhere in this book, lipocalins exhibit three characteristic features, which include: (i) an unusually low amino acid sequence similarity (typically 15-25% between paralogs) (ii) a highly conserved protein tertiary structure, and (iii) a similar arrangement of exons and introns in the coding sequence of their genes. These shared protein and gene features are overwhelming arguments for the existence of a single lipocalin ancestral gene that once extended into a family.

The ancestral gene appears to have arisen in a group of bacteria, and possibly was inherited by eukaryotes as a result of genome fusion (see Chapter 4). Given this hypothetical beginning, lipocalins are expected to be found in all descendants of the eukaryotic common ancestor. Currently, and aside of prokaryotes, bona fide lipocalins have been recovered from a protoctist, a fungus, several plants, a nematode, several arthropods, a tunicate, a cephalochordate, and many examples of chordates.

This review will first focus on the structure of lipocalin genes in eukaryotes, and then on our current view of the evolutionary history of this family.

An Overview of Lipocalin Genes

Exon-Intron Organization of Eukaryote Lipocalins

Exon-intron arrangements in lipocalins were first deciphered mostly in human and rodents, but the virtually complete information on genome sequence in an ever growing number of organisms, as released over the last few years, has considerably extended our knowledge of gene organization in eukaryotes at large. Whether by direct sequencing of known genes or bioinformatics-aided identification of novel open reading frames (ORF), the lipocalin genes whose structure is now established has benefited from this flow of information. Figure 1, provides an overview of the exon-intron arrangement of lipocalin genes, from unicellular eukaryotes to human. Given the large number of lipocalin genes that arose from duplication and retained a similar organization (see below), not every lipocalin currently known is depicted in Figure 2. The latter intends to provide an overview of gene structure with major trends rather than a comprehensive list of genes. When the position of introns is marked in a protein sequence alignment of different lipocalins, the pattern that emerges immediately highlights a model gene arrangement with a maximum of five exons (e1-e5) and four introns (A-D) in

*Corresponding Author: Diego Sanchez—Departamento de Bioquímica y Biología Molecular y Fisiología-IBGM, Universidad de Valladolid-CSIC, Valladolid, Spain. Email: lazarill@ibgm.uva.es

Lipocalins, edited by Bo Åkerström, Niels Borregaard, Darren R. Flower and Jean-Philippe Salier. ©2006 Landes Bioscience.

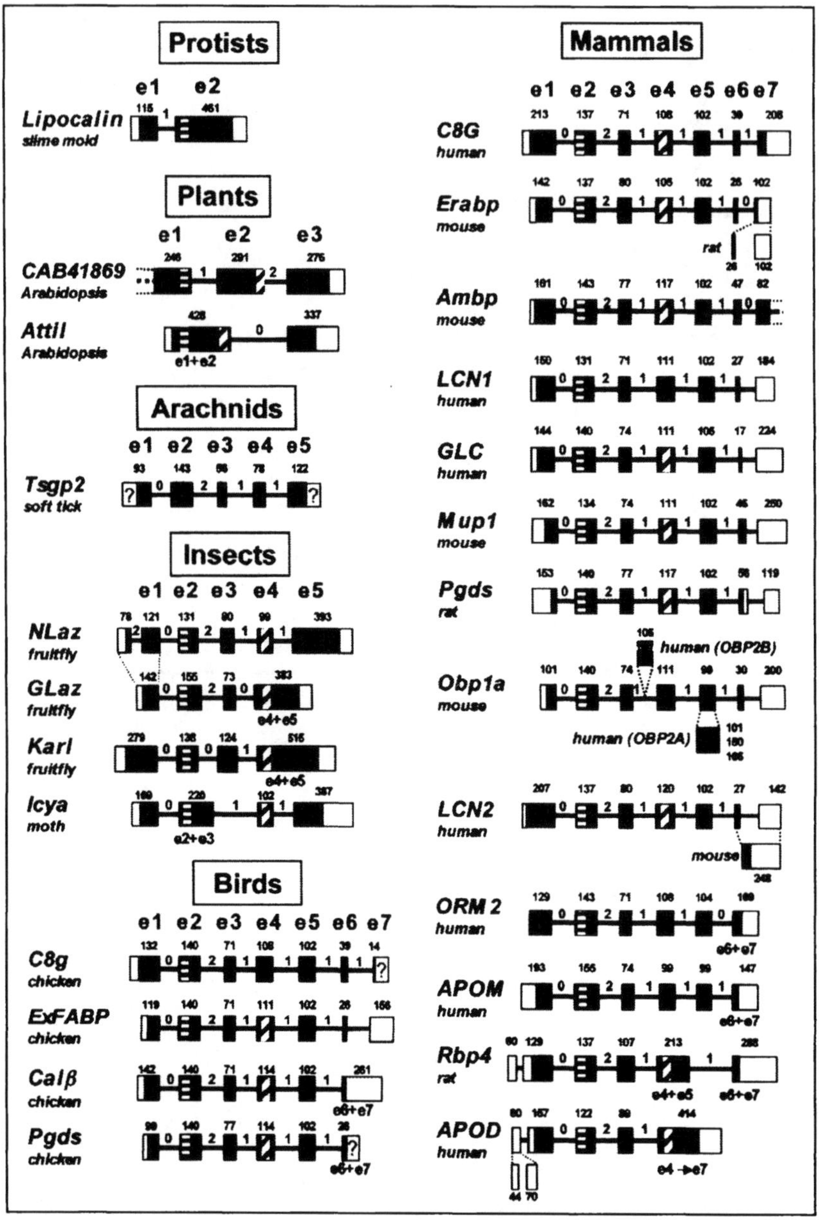

Figure 1. Intron-exon organization of lipocalin genes in eukaryotes. Only a limited series of informative genes is shown. Every gene name is identified on the left (capitals for human genes and lowercase letters for all others) and the species taken as an example is noted underneath. In a few instances of nonmammalian genes, the gene name of the mammalian ortholog is noted instead of the original name (e.g., *Pgds* instead of *Calγ* in chicken) for the sake of comparison. Introns are depicted as a solid line and they are not drawn to scale for the sake of exon comparison between genes. The intron phase is indicated between exons: 0 (intron inserted between codons) or 1 (intron after the first nucleotide in a codon) or 2 (intron in the second nucleotide in a codon). Every exon is depicted as a box with its size (bp) noted above and its coding part filled in. The figure legend is continued on the next page.

Figure 1, continued. When the 5' or 3' border of the utmost 5' or 3' exon is still undetermined, the noncoding part of the exon is given an arbitrary size and noted with a question mark, and only the size of the coding part is indicated above this exon. The archetypal lipocalin exons are noted e1-e7. Whenever possible, the exons of every gene are lined up with the archetypal exons. Note that in protoctists and plants the number of archetypal exons still remains highly speculative. A large exon noted with an arrow joining two exon numbers (e.g., ex→ez) suggest that this exon is the counterpart of the exons ex+ey+ez separately found in other genes. The genes are ordered by decreasing number of coding exons with respect to the maximal exon arrangement. When the exon arrangement of a given gene significantly varies among species, the arrangement closest to most lipocalin genes is shown in full, and some species differences are detailed underneath. The splice variants that occur within e5 of human *OBP2a* are shown, and the resulting exon sizes are noted in e5. When the 5' or 3' region of a fused gene does not code for a lipocalin-type polypeptide, this part is shown as a dotted line (e.g., *CAB44869, AMBP*). The canonical peptide motifs of lipocalins are shown as a stripped (GxW) or dotted (TDY) box within an exon. Absence of such a box in an exon means that the corresponding lipocalin lacks this motif. Modified from reference 1.

arthropods *vs* a maximum of seven exons (e1-e7) and six introns (A-F) in chordates, being the exon sizes and the pattern of intron phases of different lipocalins quite similar[1,2] (Figs. 1, 2). So far, too few lipocalin genes have been identified in other phyla to allow for rules of exon/intron arrangements to be inferred. Besides these features, a major tool for unambiguous exon identification is the presence in some of them of short peptide-encoding sequences. These are the well known common motifs of lipocalins (see Chapter 2), namely the GxW motif in e2, and the TDY motif in e4, that appear depicted in Figure 2, in the context of their preserved secondary structure. Because of the well aligned intron-exon boundaries in the ORF of all lipocalin family members, we can conclude that intron positions are homologous characters in this family. As an update of our previous work,[1,2] we will review structure properties of lipocalin genes in different organismal groups.

Unicellular Eukaryotes and Plants

As noted above very few lipocalin genes have been found in these groups, and therefore drawing any solid conclusion from the different exon/intron arrangement as currently found in two genes of a single plant species may be risky, nor is it possible to align the protist or plant genes with those found in higher eukaryotes. Again, the presence of the GxW and TDY motifs are the only reliable elements that allow one to propose that, for instance, the first exon in *Attil* may be a counterpart of the series of four e1 to e4 exons found in animals (see Fig. 1).

Arthropods

Exon alignment between lipocalin genes in arthropods highlights a gene arrangement with a maximum of five exons (e1-e5) / four introns (A-D), with the introns present in the 3'-end of chordate lipocalins being always absent (Fig. 1). Singular exons such as that appearing in the region coding for the signal peptide of *NLaz* are encountered as well. Also, *GLaz* and *Karl* lack the intron D that intervenes exons e4 and e5 of insects. In general, the lipocalin motifs GxW and TDY, encoded by e2 and e4 respectively, could help identify such exons. In the genes found so far in arachnids, the *TSGP1, -2,* and *-4* and the *HBP2* gene of the haematophagous ticks (*O. savignyi* and *R. appendiculatus*) both peptides motifs are absent.[3] However, the number, position and pattern of intron phases (0, 2, 1, 1) are shared with insect lipocalins (Fig. 1). This has permitted to align the tick proteins to other lipocalins[4] and include them in the phylogeny studies of the family (see below).

Vertebrates

Vertebrate lipocalins show a maximum of seven exons, usually including six coding exons (e1-e6), and the introns A-E show a fairly conserved pattern of intron phases (0, 2, 1, 1, 1) (Figs. 1, 2). Deviations from this scheme are seen in some genes which, for instance, lack intron F (*ORM2, APOM*), introns D and F (*RBP4*), or introns D to F (*APOD*). Yet, other more complex events have been noticed. For instance, extra noncoding 5' exons are found in *RBP4*

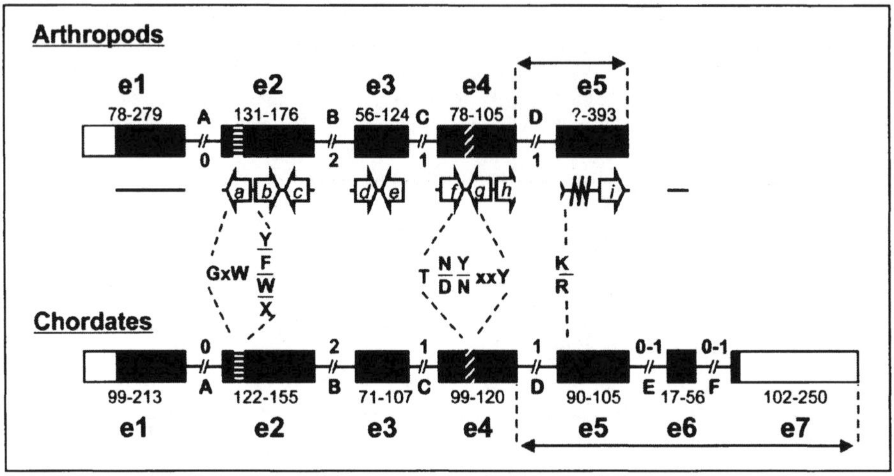

Figure 2. Lipocalin gene consensus and exon/peptide relationships in arthropods and chordates. Every exon e1 to e7 is depicted as a box whose coding part is filled in. The names (A-F) and phase (number) of introns intervening the ORF are indicated between the corresponding exons. The size range (bp) for e1-e7 is indicated (the range values are taken from Figure 1 and from other genes not shown, after excluding unusual cases of combined exons or alternative splicing). Areas where introns generate diversity in the 3' region of the lipocalin coding sequence are indicated with a double-headed arrow bordered by vertical dotted lines. An extra exon sometimes found at the 5' end of some genes as well as the variable location of the stop codon in e6 *vs* e7 in chordates (see text) are not depicted. The nine antiparallel β-strands (*a-i*) and the major C-terminal α-helix of lipocalins are indicated by head-to-tail arrows and by diagonal lines respectively. Other protein segments are depicted as horizontal lines. These elements are aligned with the corresponding gene exons. The three lipocalin motifs, namely two peptide (shown as a stripped or hatched bar within exons) and a basic amino acid residue are detailed below the appropriate β-strand or α-helix. Modified from reference 1.

and *APOD* in mammals. Also, in a limited number of cases (*ERABP* and *C8G*), the 3'-end of the ORF is intervened by intron F. Finally, a unique lipocalin gene (*AMBP*) exists (Fig. 1), which is part of a fusion of two unrelated genes whose 5' and 3' sides code for two in-frame proteins, namely the lipocalin A1m and the protease inhibitor bikunin, that is not a lipocalin. The gene region coding for A1m shows intron F at the 3'-end, and thus belongs to the group of lipocalins whose ORF is contained in 7 exons. Finally, other gene-specific events and arrangements have occurred as illustrated in Figure 1, (e.g., *Obp1a*), but they will not be commented any further herein.

Chromosomal Locations and Gene Clusters

The standard techniques of in situ hybridization and linkage studies, and the advent of information from several sequencing genome projects, have unveiled the physical mapping of lipocalins to particular chromosomes. In *Drosophila melanogaster*, three lipocalins are found in chromosome 2,[5] and one in chromosome X. The chromosomal arrangement in vertebrates displays a striking pattern, that is illustrated in Figure 2. With the exception of *APOD, APOM*, and *RBP4*, most human lipocalin genes are found on the long arm of chromosome 9 (HSA9q). Likewise, their orthologs in mouse and rat are clustered into two separate chromosomes that show syntenies with HSA9q.[1,6] Along these lines, a lipocalin gene cluster is also found in chicken.[7] This cluster locates onto chromosome 17 (Fig. 3), which is known to be a counterpart of HSA9.[8] Alike what is found in human, the *ApoD* and *Rbp* genes are isolated in different chromosomes in rodents and chicken. An intra-lineage duplication of *Rbp* that resulted in locations on two different chromosomes further appears in chicken.

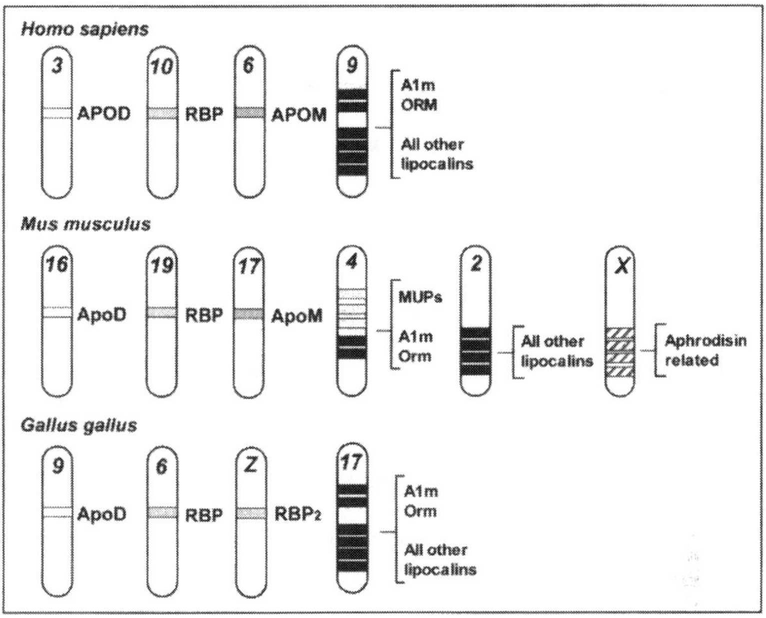

Figure 3. Chromosomal location of lipocalin genes in several chordate species. The areas of human chromosome 9, mouse chromosomes 2 and 4 and chicken chromosome 17 that harbor most lipocalin genes are known to be syntenic. The urinary proteins (*MUP* in mouse and *a2ug* in rats) of mouse chromosome 4, and the aphrodisin-related genes on mouse chromosome X are rodent-specific.

Not only are distantly related paralogs gathered in the above cluster(s) but, in many instances, more than one copy of a given lipocalin gene is further found within a cluster, which shows a trend for lipocalin genes to duplicate. This is illustrated, for example, by a duplication of the *LCN1* gene (human and rat), the *BLG-glycodelin* gene (in many mammals), a *PGDS*-like gene (chicken), the epididymal lipocalins (human and mouse), the *ORM* gene (human and mouse), and the cluster of over twenty genes in the *MUP* locus (mouse and rat). Remarkably, this trend to lipocalin gene duplication is also found in lower eukaryotes as illustrated with the *Icya* gene (in the tobacco hornworm) and the expansion of a series of lipocalins found in the saliva of hematophagous insects and chelicerates.[1,3,7,9,10] As stressed previously, these genes could not all be depicted in the limited setting of Figure 2.

Inferring the Evolution of the Lipocalin Family

Five properties can theoretically be used for inferring gene phylogenies: the gene/protein sequence, the protein three-dimensional structure, the exon-intron structure of the genes, their chromosomal position, and their organismal representation.

Only the protein tertiary structure remains to be used for reconstructing the evolutionary pathway followed by lipocalins, which is mainly due to the lack of a solid phylogenetic method that uses this information. We will now review the available phylogenetic information that can be extracted from all other properties,[2,11,12] and will update previous phylogenetic inferences with the addition of information from new lipocalins.

Lipocalin Evolution as Derived from Gene Structure Analysis

Since the gene architecture of lipocalins is well conserved (see above), the exon-intron boundaries can be assumed to be homologous, and therefore to contain traces of the evolutionary history of these genes. A new method to derive gene phylogenies from gene structure features has been developed and applied to the lipocalin family.[2] The method is based on a similarity measure of the intron-exon boundaries as mapped on a multiple protein sequence alignment of selected lipocalins with known gene structure. Three parameters are used to calculate a genetic distance: the number of introns intervening a lipocalin ORF, their phase, and the position of the exon-intron boundaries. A distance based method for phylogenetic reconstruction was then applied.[2]

Only the gene structure features present in the coding sequence are used for phylogenetic purposes. This restriction is based on the following grounds: (i) The difficulty to align the untranslated regions (UTR) of lipocalin genes, and therefore to assign homologous character to the exon-intron boundaries found in the UTR; (ii) the milder selective pressure on noncoding sequences that allows for more flexibility for independent intron gains or losses in these regions; and (iii) the presence of intron phase in the intron-exon boundaries located within the coding sequence, which is one of the phylogenetically informative characters used.

When this methodology was applied to the lipocalin family,[2] we obtained a gene tree that is congruent with our previous protein sequence-based phylogenies of the family,[11,12] adding support and helping us refine the evolutionary history of lipocalins. This new method has also allowed Mans and Neitz[4] to include histamine-binding proteins and related sequences from several arachnids in the lipocalin phylogeny. Given the extreme divergence of these protein sequences from the lipocalin shared sequence motifs, the set of characters derived from gene structure are the most reliable approach to ascertain their relationship to lipocalins.

An updated version of a gene structure-based phylogenetic tree of lipocalins[2] is shown in Figure 4, where we have included *ApoM* as a novel member of the lipocalin family, as well as an

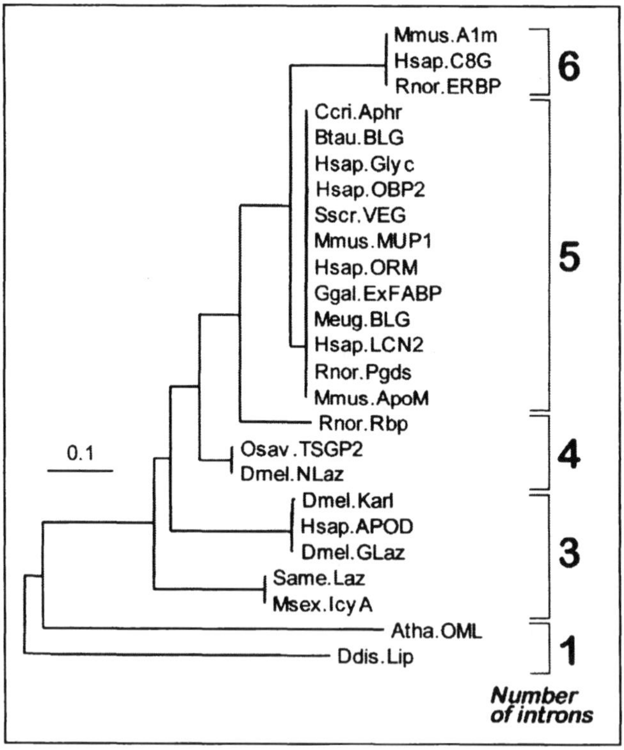

Figure 4. Lipocalin phylogeny (Neighbor-Joining) based on the exon-intron arrangement of selected genes, rooted with a protoctist lipocalin. Scale bar represents branch length (number of amino acid substitutions/site). The number of introns intervening the genes ORF are shown on the right.

example of the tick lipocalins. The tree, rooted with the lipocalin from the unicellular eukaryote Dictyostelium, shows agreement with the organismal phylogeny and sort arthropodan lipocalins in three groups, two of them with 3 introns and one group with 4 introns. The most basal chordate lipocalin in the gene tree is *ApoD*. The remaining chordate lipocalins assemble monophyletically and appear separated in three groups according to the 4-6 introns present in their ORF. The most important information that can be extracted from this tree is that lipocalins that have originated more recently contain more introns in their coding sequence. As mentioned above, introns E and F are missing in nonchordate lipocalins, while introns A-D display a broad phylogenetic distribution. In this sense, introns A and C are present in all metazoan lipocalins included in our study. Thus, the evolutionary history of metazoan lipocalins can be better traced from the distribution of introns B and D.

Lipocalin Evolution as Derived from Protein Sequence Analysis

The use of amino acid sequences and the multiple alignment of lipocalins allows us to explore their evolutionary history in more extent. Many more lipocalins can be studied in this way, given the larger dataset of mRNA and protein entries existing in molecular databases. This method also allows the addition of prokaryotic lipocalins, that could not be included in the phylogeny performed with the exon-intron arrangement.

We have aligned a set of 210 lipocalins using the same methods as in previous publications.[11,12] Namely, CLUSTAL X (1.8)[13] was used with a gap penalty mask to penalize the opening of gaps inside helices or strands, based on the secondary structure of lipocalins. Minor manual corrections were carried out based on the knowledge of lipocalin structure and function. With this alignment, we have performed a phylogenetic tree following a Bayesian approach,[14] as a newly developed method to be compared with our previous phylogenies in which a combination of neighbor-joining and maximum likelihood had been used (see refs. 11,12 for details). The use of different methodologies to reconstruct the phylogeny of lipocalins increases the confidence in the inferred relationships between extant lipocalins.

The parameters used in the Bayesian reconstruction are: 10^6 generations, a sample frequency of 10^2, a 'burn in' of 10^3, 4 chains, a mixed model, and a consensus tree representation following a 50% majority rule. The tree, rooted with the bacterial lipocalins is shown in Figure 2. Deep nodes supported by this and previous phylogeny reconstructions[11,12] are highlighted as open circles. As mentioned above, this tree includes new lipocalin members of paramount value to our phylogeny, such as those from the fungus ***Debaromyces hansenii***, the nematode ***Caenorhabditis elegans***, the tunicate ***Ciona intestinalis***, and the cephalochordate ***Branchiostoma belcheri***. Additionally, a group of Nitrophorins and of ApoM were included in our analysis.

Interestingly, the lipocalins of plants and fungi appear closely related to the bacterial lipocalins that root the tree. The protoctist dictyostelid lipocalin appears within a cluster of bacterial lipocalins. When comparing this phylogeny with our previous ones, the addition of lipocalins found in new organismal groups has resulted in the distribution of, for example, the bacterial and arthropodan lipocalins in several monophyletic groups, possibly reflecting functional relationships (see Chapter 6). Insect Nitrophorins appear grouped with a nematode lipocalin, but the tree position of this group is not well resolved possibly due to their strong sequence divergence, that could generate a long-branch attraction phylogenetic artifact.[15]

ApoD keeps being associated to a particular group of arthropodan lipocalins sharing gene expression in the nervous system, which suggest ApoD as the ancestral chordate lipocalin. This proposal gets further support from the finding that the tunicate lipocalin (Cint.Lip), which shows a basal tree position within the group of arthropodan lipocalins and ApoD, holds the highest pairwise sequence similarity with ApoD.

ApoM locates in our tree at a basal position of the chordate subtree with respect to all other chordate lipocalins. The rest of the tree does not significantly differ from our previous phylogenies, with Rbp, Blg and the Pgds-Ngal groups being most closely related to the ancestral ApoD.

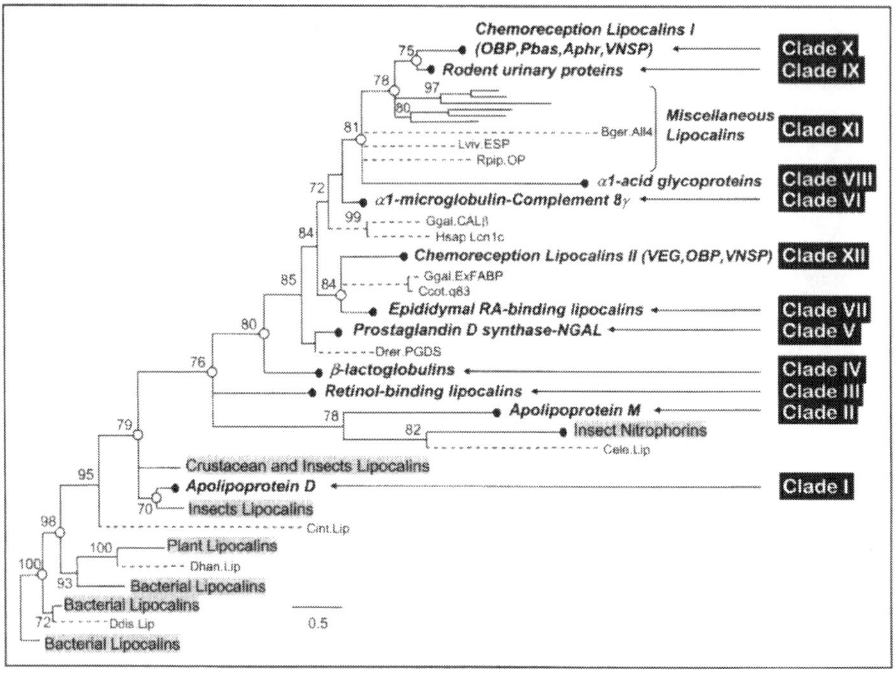

Figure 5. Phylogenetic tree of the lipocalin family derived from a multiple alignment of 210 lipocalin protein sequences, reconstructed by a Bayesian method, and shown as a consensus tree (50% majority rule) rooted with a group of bacterial lipocalins. Posterior clade probability values (>70) are shown at each node. The scale bar represents the branch length (number of amino acid substitutions/site). Lipocalin clades from chordates are resumed to the main node, marked with a black dot, and numbered on the right. The remaining groups of lipocalins are clustered and boxed in gray. Individual lipocalins without supported grouping are shown with a dashed line, and named with an abbreviated species name and a functional label. The deep nodes marked with open circles are the ones supported by different tree building methods previously applied to the lipocalin family (see text).

As part of an ongoing project, these phylogenetic representations are helping us classify lipocalins in evolutionarily related clades. By ascribing a particular lipocalin to a clade, our goal is also to assist researchers in their experimental designs to test lipocalin functions based on the knowledge gathered from other clade members.

The lipocalin clades, numbered by roman numerals in our previous publications,[11,12] are therefore in constant refinement. Based on our updated tree, we propose a new clade arrangement (Fig. 5). This classification will hereafter only cover the well represented lipocalins of the chordate phylum, where there are several organisms with fully sequenced genomes, and that has been subjected to exhaustive lipocalin searches (ref. 16 and our continuing work). Lipocalins from other phyla, like the growing group from arthropods (see Chapter 6), need further sampling and refinement of phylogenetic associations before we can assign meaningful and reliable clade memberships. The new clade classification aims at maintaining most clade numbers as in our previous phylogenies, while accommodating the addition of the ApoM clade. Future understanding of the function of the assorted clades of so-called chemoreception and miscellaneous chordate lipocalins, as well as the ungrouped ones (labeled with dashed lines in the tree), will definitely resolve the evolutionary organization of the chordate lipocalin tree.

Finally, the phylogenetic classification also intends to help unify lipocalin nomenclature for the potential lipocalin sequences emerging from the various genome sequencing projects.

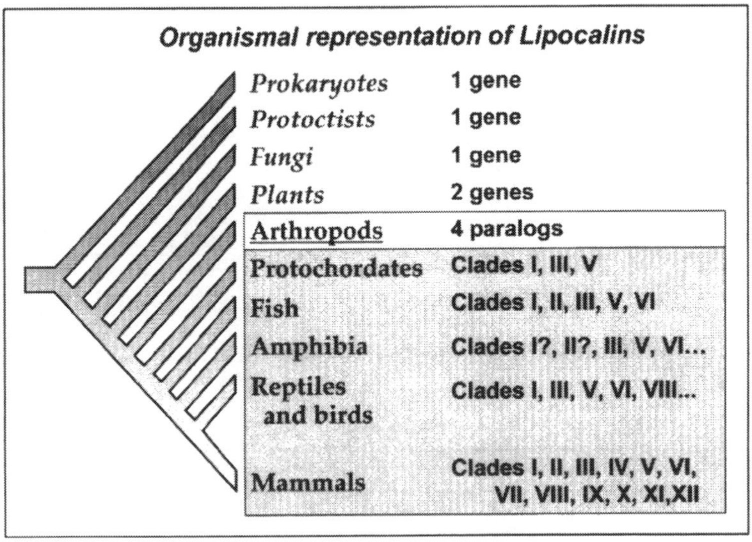

Figure 6. Representation of presumed lipocalin genes in the tree of life showing the five organismal kingdoms (metazoans are boxed). Lipocalin genes are grouped in clades in the chordate phylum, and we show the clades that have been found in each chordate group so far.

Lipocalin Evolution as Derived from Chromosomal Position

As described above, the chromosomal position of many lipocalin genes has been precisely defined in the insect Drosophila genome, and in a handful of vertebrate genomes. The study of chromosomal arrangement of lipocalins in selected vertebrates, such as chicken, mouse and human (see Fig. 3) has gathered information that let us to propose hypotheses about lipocalin evolution in this phylum. The fact that both the mammalian and chicken lipocalin genes of clades IV-XII are assembled in a single chromosome, suggests that the clustering of these clades was present in the common ancestor of reptiles and birds. Further tandem duplications of some of the genes located in that cluster have occurred, which is confirmed by the similarities they show both at the protein sequence and gene structure levels (see above), and in many cases by their similar expression and function. The same reasoning applies to the *APOD* and *RBP* genes, for which we suggest a location in separate chromosomes in the first terrestrial vertebrates.

Organismal Representation and Overview of Our Hypothesis of Lipocalin Evolution

In order to present a supported hypothesis on lipocalin evolution, besides reconstructing gene phylogenies, we need to assess their representation in the tree of life. Thus, we started by calculating the number of legitimate lipocalins present in extant taxa, without taking into account intra-species duplications. We then mapped this information onto a simplified organismal tree, which is shown in Figure 2. Using this scheme, and adding the information reported above, we can formulate a putative history of descent for lipocalins (illustrated in Fig. 7) since their appearance in the prokaryotic world.

The first thing that comes out of the analysis is that lipocalins are present in species of all five organismal kingdoms. However, a conservative number of 1 gene/species seems to be the catalog for prokaryotes, protoctists, and fungi. At least 2 lipocalin genes have been recovered from plants, but their singular exon-intron structure suggests their independent evolution. The animalia kingdom inherited the ancestral single lipocalin gene, but it subsequently duplicated giving rise to two genes, that evolved different gene structures with either 3 or 4 introns

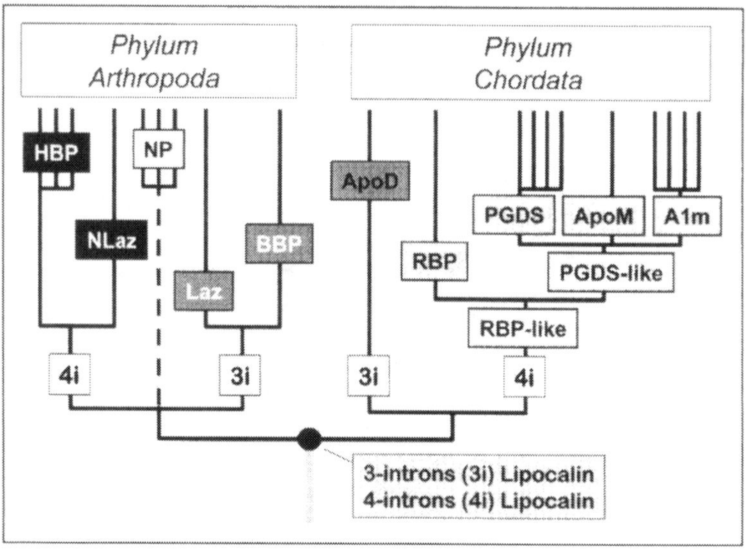

Figure 7. Hypothetical evolutionary pathway followed by arthropodan and chordate lipocalins after the split of these phyla from a common ancestor bearing two lipocalins with different gene structure.

intervening their ORFs. Based on the commonalities observed in terms of exon-intron arrangement between the two animal phyla better sampled today, arthropods and chordates, we propose that those two lipocalin genes were present in their common ancestor (see Fig. 7).

The two ancestral lipocalins followed a different history of change in these separate phyla, that were exposed to different adaptive landscapes. In arthropods, at least four paralogs can be assumed to be common to this phylum. However, as will be reviewed in Chapter 6, the diverse arthropodan lifestyles and physiology can impose different selective constraints to the divergence of lipocalins, giving rise to particular expansions of the lipocalins repertoire, such as the tick and insect saliva proteins that are specific of blood-sucking arthropods, or the milk proteins of the viviparous cockroaches. However, the gene catalog of this phylum is increasing, and more functional and expression data are being published that will help refine the independent evolution followed by lipocalins in these organisms.

As was discussed above, all evidences coincide in ApoD being the successor of one of the ancestral lipocalins present in the first chordate-like organism. The ancestral 5-exon lipocalin of chordates was probably an RBP-like lipocalin, given the basal position of RBP in all our phylogenies. This is also supported by the presence of an RBP ortholog in the cephalochordate Branchiostoma (A. Xu, personal communication). Coincidental with the whole-scale genome duplications that occurred early during chordate evolution, the ancestral *RBP* underwent duplications, giving rise to two new lipocalins that located in separate chromosomes (see Fig. 3). These two lipocalins were possibly the ancestors of nowadays PGDS and ApoM, a proposal supported by sequence and gene-structure phylogenies, as well as the presence of these two lipocalins in fishes, and of PGDS in Branchiostoma (A. Xu, personal communication).

Several arguments point to a PGDS-like protein as the originator of a series of tandem gene duplications along the evolution of chordates, that resulted in a number of lipocalins clustered in a single chromosome: (1) a basal position for PGDS in the sequence-derived tree, (2) a gene structure similar to the duplicants, (3) its presence in every sampled chordate, and (4) a site of expression similar to ApoD.[17] In the process of *PGDS* duplications, we propose A1m as the first descendant of PGDS, as suggested by the presence of A1m in fishes.[18] Subsequent rounds of tandem duplications of the genes coding for PGDS and A1m generated the remaining

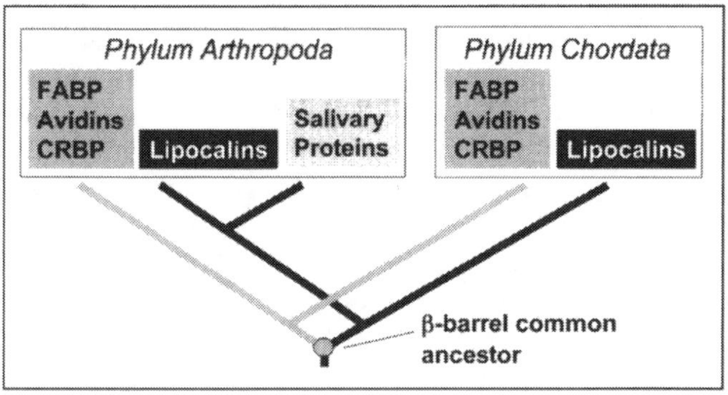

Figure 8. Evolutionary pathway of the members of calycin protein superfamily after duplication from a hypothetical β-barrel common ancestor.

lipocalins of the cluster, following a pattern not clearly known yet. Our proposed model of lipocalin gene evolution is depicted in Figure 7, but this working hypothesis keeps being refined by our ongoing studies including information from new genomes, and about the expression and function of lipocalin genes.

As an example of the 'evolving' character of our own work, a lipocalin sequence has been recovered from *Hydra magnipapillata* (UG # Hma.2173) in the process of writing this review. Although not included in the tree, this sequence bears strong similarities with the tunicate lipocalin and the chordate *ApoD*. This finding is very important, as it adds a new and fairly ancient metazoan phylum, the cnidaria, where lipocalins are found and can be studied.

Lipocalins and Their Sister Families in the Calycin Superfamily

In this review we have not included in depth the phylogeny of other families composing the calycin superfamily,[19] such as the FABP, the avidins, and the CRBP that show a compelling similarity to lipocalins in protein structure. Because of their marginal sequence (see Chapter 3) and gene structure[2] relationship, and because of their organismal representation, we propose that these families emerged from an eukaryotic common ancestral protein displaying a β-barrel structure. After that, they have diversified in different phyla (Fig. 8 illustrates this variety for the arthropodan and chordate lineages) by following divergent and independent evolutionary pathways, while folding constraints act as a selective pressure that maintains the protein structure.

Conclusions

Lipocalins are present in every organismal kingdom, and the current knowledge of their function (reviewed in other chapters of this book) suggests they have undergone striking functional diversification both after speciation and after gene duplication. Many cases are also known that point to lipocalins as moonlighter proteins,[20] able to perform more than one function at once, emphasizing the versatile nature of this protein folding.

The pathway proposed in this work for lipocalin evolution highlights the expansion of this gene family in metazoans, as well as the maintenance of the duplicated paralogous genes. However, the evolutionary mechanisms leading to the extant set of lipocalins in the two most sampled metazoan phyla (Arthropoda and Chordata) are quite distinct. A small number of lipocalins is present in most species of arthropods, while intra-lineage duplications multiply the number of lipocalins in some arthropod species adapting to new lifestyles (see Chapter 6).

Chordates also show intra-lineage gene duplications (e.g., the urinary proteins of rodents), but this phylum is characterized by a large number of paralogous lipocalins generated by large-scale duplications at the genomic level. In general these paralogs do not preserve the same protein function, as is the case for other families such as the globins or the Hox genes. The divergent protein sequences of the paralogs opened new avenues for molecular interactions (at the internal ligand-binding pocket and at the protein surface), while preserving the structural fold, and consequently increased the availability of functional pathways where to perform a novel task to be screened by natural selection.

References

1. Salier J-P. Chromosomal location, exon/intron organization and evolution of lipocalin genes. Biochim Biophys Acta 2000; 1482(1-2):25-34.
2. Sanchez D, Ganfornina MD, Gutierrez G et al. Exon-intron structure and evolution of the lipocalin gene family. Mol Biol Evol 2003; 20(5):775-783.
3. Mans BJ, Louw AI, Neitz AWH. The major tick salivary gland proteins and toxins from the soft tick, Ornithodoros savignyi, are part of the tick lipocalin family: Implications for the origins of tick toxicoses. Mol Biol Evol 2003; 20(7):1158-1167.
4. Mans BJ, Neitz AWH. Exon-intron structure of outlier tick lipocalins indicate a monophyletic origin within the larger lipocalin family. Insect Biochem Mol Biol 2004; 34(6):585-594.
5. Sanchez D, Ganfornina MD, Torres-Schumann S et al. Characterization of two novel lipocalins expressed in the Drosophila embryonic nervous system. Int J Dev Biol 2000; 44(4):349-359.
6. Chan P, Simon-Chazottes D, Mattei MG et al. Comparative mapping of lipocalin genes in human and mouse: The four genes for complement C8 gamma chain, prostaglandin-D-synthase, oncogene-24p3, and progestagen-associated endometrial protein map to HSA9 and MMU2. Genomics 1994; 23(1):145-150.
7. Pagano A, Giannoni P, Zambotti A et al. Phylogeny and regulation of four lipocalin genes clustered in the chicken genome: Evidence of a functional diversification after gene duplication. Gene 2004; 331:95-106.
8. Consortium ICGS. Sequence and comparative analysis of the chicken genome provide unique perspectives on vertebrate evolution. Nature 2004; 432:695-716.
9. Suzuki K, Lareyre J-J, Sanchez D et al. Molecular evolution of epididymal lipocalin genes localized on mouse chromosome 2. Gene 2004; 339:49-59.
10. Ribeiro JMC, Andersen J, Silva-Neto MAC et al. Exploring the sialome of the blood-sucking bug Rhodnius prolixus. Insect Biochem Mol Biol 2004; 34(1):61-79.
11. Ganfornina MD, Gutierrez G, Bastiani M et al. A phylogenetic analysis of the lipocalin protein family. Mol Biol Evol 2000; 17(1):114-126.
12. Gutierrez G, Ganfornina MD, Sanchez D. Evolution of the lipocalin family as inferred from a protein sequence phylogeny. Biochim Biophys Acta 2000; 1482(1-2):35-45.
13. Thompson JD, Gibson TJ, Plewniak F et al. The ClustalX windows interface: Flexible strategies for multiple sequence alignment aided by quality analysis tool. Nucleic Acids Res 1997; 24:4876-4882.
14. Ronquist F, Huelsenbeck JP. MrBayes 3: Bayesian phylogenetic inference under mixed models. Bioinformatics 2003; 19(12):1572-1574.
15. Philippe H, Laurent J. How good are deep phylogenetic trees? Curr Opin Genet Dev 1998; 8:616-623.
16. Salier J-P, Åkerström B, Borregaard N et al. Lipocalins in bioscience: The first family gathering. BioEssays 2004; 26(4):456-458.
17. Ganfornina MD, Sánchez D, Pagano A et al. Molecular characterization and developmental expression pattern of the chicken Apolipoprotein D gene. Implications for the evolution of vertebrate lipocalins. Dev Dyn 2005; 232:191–199.
18. Åkerström B, Logdberg L, Berggard T et al. alpha(1)-Microglobulin: A yellow-brown lipocalin. Biochim Biophys Acta 2000; 1482(1-2 SU -):172-184.
19. Flower DR. Structural relationship of streptavidin to the calycin protein superfamily. FEBS Lett 1993; 333(1-2):99-102.
20. Jeffery CJ. Moonlighting proteins. Trends Biochem Sci 1999; 24:8-11.

Chapter 3

The Lipocalin Protein Family: Protein Sequence, Structure and Relationship to the Calycin Superfamily

Maria D. Ganfornina, Diego Sanchez, Lesley H. Greene and Darren R. Flower*

Abstract

Lipocalins are remarkable in their diversity, as manifest at the levels of protein sequence and protein function. At the level of 3-dimensional structure, however, they are very similar. The lipocalins are also part of a larger protein superfamily: the calycins, which also includes the fatty acid binding proteins, avidins, a group of metalloproteinase inhibitors, and triabin. The superfamily is characterised by a similar structure (a repeated +1 topology β-barrel) and by the conservation of a remarkable structural signature. In this review, both of these aspects are explored.

Introduction

The lipocalin protein family is one of the most interesting and perplexing groups of homologous proteins. Diversity is the lipocalins' watchword. Lipocalins demonstrate extreme divergence at the sequence level: often beyond the ability of sequence analysis to readily recognize their relatedness. However, despite this sequence variability, the 3-dimensional structures of distinct lipocalins show remarkable similarity. Their structure comprises a slightly distended β-barrel, composed of eight β-strands, which forms an internal cavity lined by apolar residues ideally suited for the carriage of small hydrophobic residues. The barrel winds in a right-handed and conical manner around a central axis such that the first strand is hydrogen-bonded via its backbone to the last strand. The diversity of lipocalin sequences is matched by the variety of their function, their mechanisms of action, and their phyletic spread through a wide variety of species, which range from bacteria, through plants, to animals from ticks and arthropods to man. Physically, lipocalins are small (typically 150-250 residue) extracellular proteins, which share several common molecular recognition properties: the binding of small, predominantly hydrophobic molecules (such as retinol, long-chain lipids, or steroids); binding to specific cell-surface receptors;[1] and the formation of covalent and noncovalent complexes with other soluble macromolecules, such as human IgA.[2] Initially lipocalins were classified as transport proteins, their roles including carriage of retinol, odorants, and pheromones.[3] It is now clear, however, that lipocalins fulfill a wide variety of different functions: some act as tick anticoagulents, some show enzymatic activity, while others guide growing insect nerves or limit invading bacterial growth by iron sequestration.[4] The biological roles of other lipocalins are less clear-cut,

*Corresponding Author: Darren R. Flower—The Jenner Institute, University of Oxford, High Street, Compton, Berkshire RG20 7NN, U.K. Email: darren.flower@jenner.ac.uk

Lipocalins, edited by Bo Åkerström, Niels Borregaard, Darren R. Flower and Jean-Philippe Salier. ©2006 Landes Bioscience.

with their main biological interest being coincidental to their, as yet, unknown physiological functions. For example, α-1-acid glycoprotein (AGP) is implicated in drug binding in human blood and a set of lipocalins form a major group of mammalian aeroallergens. Lipocalins have also been shown experimentally to form partially unfolded states at low pH called molten globules.[5,6] The conversion to these nonnative conformers have been proposed to be an essential step in the mechanism of ligand release for vitamin A and lipids,[5,6] opening up another dimension to their sequence-structural relationships.

Despite many common themes and functions, membership of the lipocalin family has been defined primarily on the basis of sequence, or structural, similarity and now encompasses many different proteins with a wide phyletic spread. Within this the lipocalins display unusually low levels of overall sequence conservation, with pairwise comparisons routinely falling below 20%, the nominal cutoff for reliable alignment. However, almost all lipocalins share sufficient similarity, in the form of short characteristic conserved sequence motifs, for this to effectively define family membership.[7,8]

Known 3-dimensional lipocalin structures include a wide of variety of family members.[9,10] Recent additions include horse allergen Equ c 1,[11] Aphrodisin,[12] several anticalins,[13,14] various forms of crustacyanin,[15] Tear lipocalin,[16] and bacterial lipocalin.[17] Moreover, the structure of prostaglandin D synthase (PGDS) is imminent,[18] as well as those of several other anticalins. The common structure of the lipocalin protein fold is now well-described.[2,8,9] The lipocalin fold is a highly symmetrical all-β structure dominated by a single eight-stranded antiparallel β-sheet closed back on itself to form a continuously hydrogen-bonded β-barrel. Together with three other distinct protein families: the fatty-acid-binding proteins (FABPs), avidins, metalloproteinase inhibitors (MPIs) (and the presently enigmatic triabin), the lipocalin family forms part of a larger structural superfamily: the calycins.[2,8,19-21] This is an example of a "structural superfamily:" a set of proteins with closely related three-dimensional structures that show no significant overall similarity at the sequence level. In these pages, we shall review and update structure and sequence relations within the lipocalin protein family and also within the larger calycin protein superfamily.

Protein Sequence Relations within the Lipocalin Protein Family

Lipocalin genes are transcribed, with few exceptions, into mRNAs of 0.6-1 kb long. These, in turn, code for proteins of 160-230 amino acids. Most of these polypeptides have a signal peptide that would export the proteins into the extracellular environment. Exceptions to this include some bacterial lipocalins (see Chapter 4), a dictyostelid lipocalin, and the temperature-induced lipocalins of plants. Following the signal peptide cleavage, most eukaryotic lipocalins are secreted to the extracellular milieu, while those of prokaryotes are mainly attached by lipids to the bacterial membranes. However, some prokaryotic lipocalins appear as soluble proteins in the bacterial periplasmic space, and the grasshopper Lazarillo is GPI-linked to neuronal membranes (see Chapter 6). Moreover, the sub-cellular location of Rat probasin appears to be nuclear. Also, the chordate apolipoprotein M (ApoM) is unique in showing no cleavage site for the N-terminal signal peptide (see Chapter 13).

The mature polypeptides of lipocalins have an average predicted molecular weight (not counting postranslational modifications) of 19.4 KDa, ranging from the 17.7 KDa of ERBP to the 21.7 KDa of AGP. This variation in molecular weight results from the variable span of the N-terminal and C-terminal sides of the proteins revealed by a multiple protein sequence alignment that has been used for phylogenetic inference (see Chapter 2). Plant lipocalins, most arthropodan lipocalins, the group containing both α1-microglobulin (AMG) and C8γ, and also AGP show long C-terminal extensions. However, the C-terminal region of plant lipocalins and the grasshopper Lazarillo show cleavage sites for GPI-linkage. Similarly, the mature N-termini of lipocalins from Dictyostelium, the fungus Debaromyces, the ascidian ortholog of apolipoprotein D (ApoD), and the Drosophila Karl, are unusually long. Averaged values of the molecular weight of individual clades of lipocalins are shown in Table 1 (see Chapter 2 for details).

Table 1. Biochemical properties for lipocalin groups belonging to well established phylogenetic clades, as predicted from their protein primary sequence

Name	Clade	Length	MW (Da)	pI	N-Glycosyl	O-Glycosyl	S-S
Bacterial lipocalins	–	157	17915	8.1	0	0	0-1?
Plant lipocalins	–	190	21957	5.1	1	0	0
Arthropodan lipocalins	–	189	21391	6.6	0 / 1-5	0-1	2
ApoD	I	168	19275	5.9	2	0	2
RBP	III	179	20655	5.9	0	0	3
Blg	IV	162	18490	4.9	0-1	0	2
PGDS	V	169	19084	7.2	2	2-4	1
NGAL	V	177	20339	8.2	1-2	0-3	1
A1mg	VI	182	20735	6.6	1-2	0-1	1
C8GC	VI	181	20193	8.7	1	0	1
ERBP	VII	159	17792	7.6	0-2	0-1	1
A1GP	VIII	186	21723	5.5	4-5	0	2
RUPs	IX	164	18952	5.1	0-1	0	1
Chemoreception I	X	159	18388	5.9	0-1	0	0-1
Chemoreception II	XII	161	18262	6.5	0-1	0-2	0-1
ApoM	II	167	18668	6.1	0-1	0-2	3
Miscellaneous lipocalins	XI	162	18795	6.5	1-2	0-1	0-1

The alignment of lipocalins has always been a complicated and perplexing task, given the overall low conservation of their protein sequence. Pairwise sequence identities of 20-30% are common between family paralogs. This falls within the "twilight zone" for sequence assignment based on the protein primary structure. Alignments here are much less dependable than for higher levels of sequence identities. Certain pairs of lipocalins show less than 12% identity. This is generally apparent between the prokaryotic lipocalins and the odorant-binding lipocalins of chordates, while more specific examples include retinol-binding protein and the major urinary protein, aphrodisin and α-crustacyanin, the bilin-binding protein and lipocalin allergen Bos d 2. This value is deep within the "midnight zone", where sequence alignments lose almost all reliability and certainty. However, the presence of several conserved sequence motifs does allow the generation of accurate automated multiple alignments of most family members. Moreover, the structure of several lipocalins that has been solved experimentally and the conserved exon-intron arrangement of lipocalins (see Chapter 2) both offer additional guidance for the alignment process.

Three sequence motifs revealed in lipocalin multiple alignments, called SCRs (structurally conserved regions,[8] described in detail below), were initially proposed to be present in genuine lipocalins, and served to classify them as kernel (bearing all three SCRs) or outlier (missing one or more motifs). The discovery of new lipocalins, based on finding such motifs or a similar protein tertiary structure, has revealed that bona-fide lipocalins show a looser requirement for preserving sequence signatures. In a multiple alignment of 209 lipocalins: >90% show the SCR1 motif GxW and the SCR3 motif R/K, while >60% of the proteins show conservation in the SCR2 motif TDYxxY. Entire clades of orthologous lipocalins lack individual motifs (e.g., ApoM, AGP and odorant binding proteins lack SCR2), and individual proteins such as the marsupial late-lactation proteins, some AGPs, and some nitrophorins show significant nonconservation in SCR1.

The gaps generated in the lipocalin multiple alignment highlight important sequence characters since they represent atypical regions shared by groups of lipocalins or present only in individual proteins. Because of the secondary structure mask used to guide alignment, these gaps are all located in expected loop regions of the tertiary structure. Most gaps appear in loops placed at the open end of the β-barrel (L1, L3, L5 and L7). An expanded L1 occurs in AMG, ApoM, AGP, and some chemoreception lipocalins. The loop L5 appears elongated in retinol binding protein (RBP) and plant lipocalins, whereas L7 is extended in RBP, most arthropodan lipocalins, and a number of PGDS enzymes. The unique lipocalins Glaz of Drosophila and the BL baboon show a lengthened L3. Finally, the only elongated gap located in the closed end of the barrel is L2 in the Drosophila Glaz and Karl.

Other properties derived from the protein primary sequence are also shown in Table 1, and help us to catalog lipocalins in different family clades. The isoelectric point, calculated for the mature protein sequence, is an important factor for polypeptide solubility and folding. Most prokaryotic lipocalins, and the vertebrate neutrophil gelatinase associated lipocalins (NGALs) and C8γ show a basic p*I*, although the highest individual values (p*I*>10) are those for probasin (Clade X) and the epididymal Lcn10 (Clade VII). Lipocalin clades with predicted acidic p*I* are the Blg, AGP, RUP, RBP, Plant lipocalins, and ApoD, although in the later there is a subgroup of fish ApoD with a basic p*I*.

We have also explored the glycosylation potential for lipocalins, both for N-linked and O-linked oligosaccharides (see Table 1). The clade with the highest number of predicted glycosylation sites is a1GP. These sites are exclusively of the N-linked type. Other clades with high predicted glycosylation are PGDS (with both N-linked and O-linked residues), and the ApoD clade (only with N-linked residues). At the other extreme, RBP shows no glycosylation sites, except for some fish RBP and the functionally divergent Purpurin of chicken. The clade IV (Blg) shows a few members with one potential N-linked site, but the lipocalin Glycodelin presents three potential N-linked and one O-linked residues, highlighting its divergent function in humans (see Chapter 11). Also, the arthropodan lipocalins, show a high heterogeneity

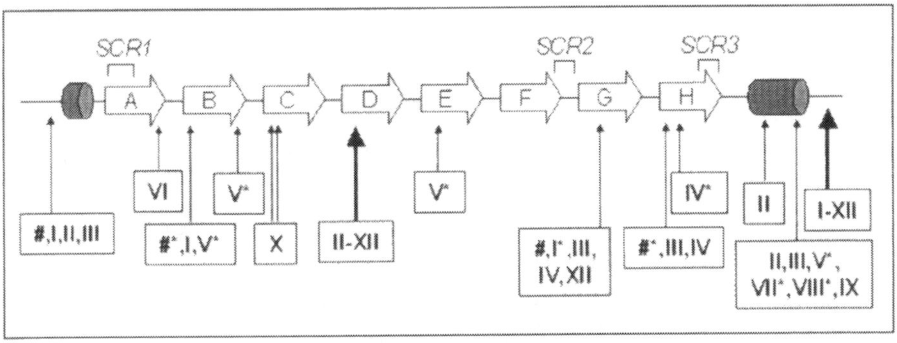

Figure 1. Cysteine locations in the lipocalin fold. Location of cysteine residues conserved in lipocalin clades in the context of the secondary structure of a prototypical lipocalin. β strands are represented by white arrows and α helices by gray cylinders. Clade numbers are defined in Chapter 2. The symbol # represents groups of lipocalins (outside of the chordate phylum) not assigned to clades. Cys residues marked with asterisks are present in subgroups of lipocalins belonging to a particular group or clade.

in terms of glycosylation potential (see Chapter 6). The remaining lipocalins display a low number of glycosylation sites.

Finally, most lipocalins have cysteine residues involved in intramolecular disulfide bonds. Figure 1 shows a schematic representation of the secondary structure of a model lipocalin, in which the position of conserved cysteines are indicated by arrows. Cysteines conserved in lipocalin clades (or groups of nonchordate lipocalins without clade assignment, shown as #) are depicted by clade number or #, while those conserved only in subgroups, within groups or clades, are labeled by the clade number with an asterisk. Table 1 shows the number of disulfide bonds present in lipocalin clades or groups. Plant, dictyostelid and fungal lipocalins lack disulfide bonds. However, the cysteines located in β-strand B and the protein C-terminus are conserved in most chordate lipocalins and form a disulfide bond. Six cysteines found in RBP and ApoM form three disulfide bridges. Finally, clades I, IV and VIII show a conserved pattern of two disulphides, while of the rest of the lipocalins, which constitute a heterogeneous group, have a single disulfide bond.

Structural Relationships in the Lipocalin Protein Family and Calycin Protein Superfamily

The folding pattern shared by members of the lipocalin protein family is that of a highly symmetrical all-β structure. Overall, it is dominated by a single antiparallel eight-stranded β-sheet. This is closed back on itself to form a continuously hydrogen-bonded β-barrel, which is slightly flattened in cross-section. The eight β-strands of the barrel, usually labeled A-H, are linked by a succession of + 1 connections, giving it the simplest possible β-sheet topology. The barrel winds in a conical and right-handed manner around the central axis so that strand A is hydrogen-bonded via its backbone to strand H. The seven loops, labeled L1 to L7, are all short β-hairpins, except loop L1: this is a large Ω loop. Loop L1 forms a lid folded back to close the internal ligand-binding site found at this end of the barrel. One end of the barrel (formed by loops L1, L3, and L5, and L7) is open, while the other end of the barrel is closed, with residues facing into the barrel form a tightly packed core. Beyond the eighth strand of the β-barrel is an α-helix. While this is a constant feature of all lipocalin structures, it is not conserved in its length, nor is it conserved in its position relative to the barrel. A simplified schematic is shown in Figures 2 and 3.

The β-barrel encloses a ligand-binding site composed of both an internal cavity and an external loop scaffold. It is the inherent structural and sequence diversity of this ensemble of cavity and scaffold that enables the lipocalin family to exhibit such a diversity of binding modes.

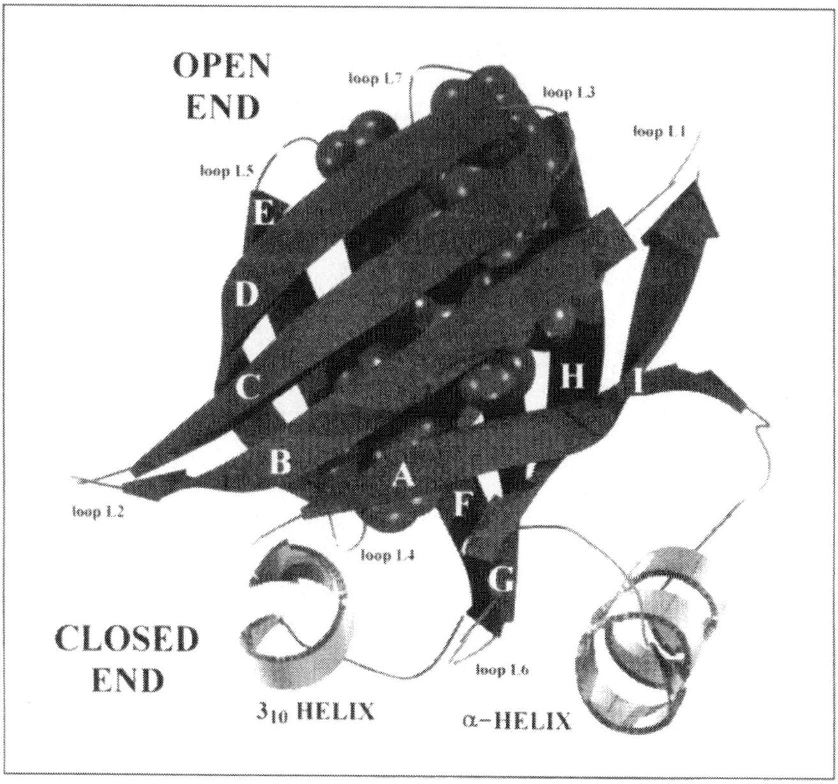

Figure 2. A schematic or ribbon-drawing of the lipocalin fold. The structure shown is a prototypical, not an actual, structure. The nine β-strands are shown labelled A-I. The N-terminal and C-terminal helices are shown and labelled. All loops (labelled L1-L7) are marked. The open end of the lipocalin β-barrel has four loops (loops L1, L3, L5, and L7). The closed end has three β-hairpin loops (L2, L4 and L6); the N-terminal polypeptide chain crosses this end of the barrel to enter strand A via a conserved 3_{10} helix closing this end of the barrel. The size of the ligand binding is shown by a collection of spheres. β-strands are depicted as curving arrows in grey, α-helices as spiral ribbons, and loops as thin cords. The figure was generated using ALTER[48] as an interface to POVRAY.

Each of these modes is capable of accommodating ligands with different chemotypes, sizes, and shapes. It is this diversity of structure in the ligand binding site which underlies much, but not all, of the functional diversity characteristic of the family. Contrasting with the overall highly conserved β-barrel topology, the loop region differs considerably between members of the family, both in amino acid composition, conformation, and length of the contributing polypeptide segments. This, in turn, gives rise to the particular ligand specificities displayed by individual lipocalins. Indeed, lipocalin binding sites can adopt very differing shapes. In the case of NGAL, it forms a wide, funnel-like opening to the solvent.[22] In mouse major urinary protein, the loops of the binding site close over the cavity fully encapsulating the ligand.[23] In RBP, the lumen of the binding site reaches down into the hydrophobic core of the barrel, deeply burying the α-ionone ring of retinal.[24] Finally, the binding site of human tear lipocalin (Tlc), forms an extended cavity with several lobes close to the base of the barrel.[16]

The common core characteristic of the lipocalin fold—which contains static features, like certain strands, rather than mobile features, such as the Helix—is dominated by three large structurally conserved regions (SCRs): SCR1 (strand A and the 3_{10}-like helix preceding it),

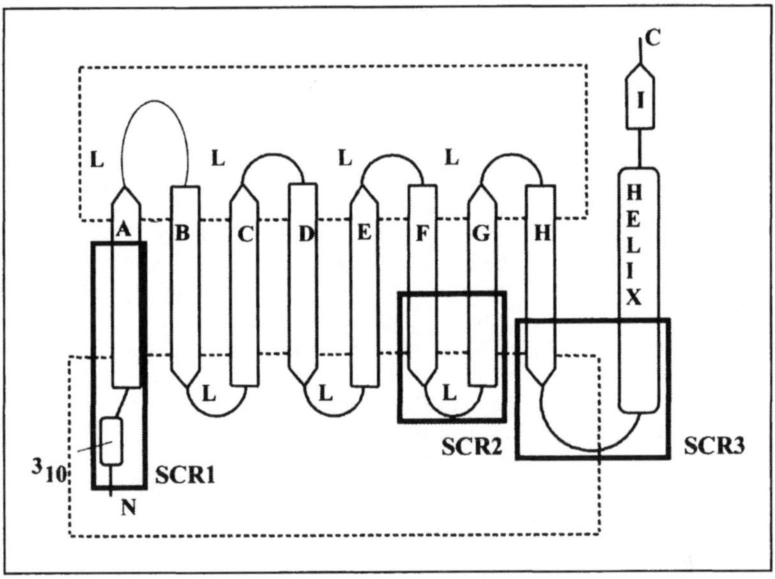

Figure 3. Schematic Structure of the lipocalin fold. An unwound view of the lipocalin fold orthogonal to the axis of the barrel. The nine β-strands of the antiparallel β-sheet are shown as arrows. The C-terminal α-helix A1 and N-terminal 3_{10} like helix are also marked. Loops are labelled L1-L7. A pair of dotted lines indicates the hydrogen-bonded connection of two strands. Those parts that form the three main conserved regions (SCRs) of the fold (SCRl, SCR2, and SCR3) are marked as heavy boxes.

SCR2 (strands F and G, and Loop L6 linking them), and SCR3 (strand H and adjoining residues). Recently, two other sequence motifs have been identified and implicated in the lipocalin folding process.[25] One of these motifs is located at the closed end of the β-barrel. The other is found at the beginning of the main C-terminal α-helix. However, the new motifs cannot be found in most nonchordate lipocalins, and thus cannot help to unequivocally ascribe new members to the family outside the phylum Chordata. As mentioned above, the three principal SCRs contain a sequence motif that is wholly, or partly, invariant. Interestingly, a complete absence of the three SCRs occurs in histamine-binding lipocalins of haematophagous insects. In this case, their membership of the lipocalin family is supported by a conserved overall protein structure and conserved exon-intron pattern at the level of gene structure (see Chapters 2 and 6).

The calycin protein superfamily, when originally identified,[8] was composed of a group of three families of ligand-binding proteins which includes the lipocalins, the fatty acid-binding proteins (FABPs), and the avidins. The avidins display an astonishing affinity for biotin, and have thus found key applications in biotechnology.[26] The FABPs are a family of predominantly intracellular proteins involved in lipid metabolism.[27] Calycins share related, if distinct, barrel structures. The structure of FABPs is built around a broken ten-stranded β-barrel structure. Avidins and related proteins, while eight-stranded, when compared to lipocalins lack a C-terminal helix or strand I and are more circular in cross-section. Others dispute the obvious and overwhelming structural similarity of the Avidins to other calycins,[19] primarily on the basis of differences in the hydrogen bonding pattern, which manifests itself as disparities of topology and shape.[28,29] Despite such differences, and the absence of global sequence similarity, these families are characterized by a similar folding pattern—an antiparallel β-barrel with a largely +1 topology—within which significant regions can be structurally equivalenced. Moreover, beyond structural propinquity, the calycins have a degree of functional similarity: many bind hydrophobic, or at least small, ligands and/or have key macromolecular interactions.

Over time, the size of the calycin superfamily has, on the basis of real or perceived structural similarity, grown to include a burgeoning group of other proteins: metalloprotease inhibitors,[30] triabin,[31] staphostatins,[32] domains from D-amino peptidase,[33] domains from quinohemoprotein amine dehydrogenase,[34] the exclusion domain from cathepsin C,[35] and various hypothetical proteins from bacteria. For some this is clearly correct and is borne out by detailed analysis. For other proteins, however, this is demonstrably not the case.[9] Photoactive yellow protein,[36] TCL-1,[37] MTCP-1,[37] catalase,[38] and cyclophilin,[39] have, for example, all been suggested as potential calycins. As, indeed, has the pleckstrin homology domain.[40] This has also been suggested as a structural homologue of other many proteins, including verotoxin and FK-506.

Superficially, calycins do resemble certain other all-β proteins with barrel-like structures,[41] and it is correct to say that β-barrels, in particular, are easily confused. This is contingent upon coincidental similarities that may arise as a product of the principles underlying barrel structure: only certain barrel geometries and shear numbers are possible. However, the relative resemblance at the level of gross structure is often insufficient to verify or refute convergent or divergent evolution, obliging us to look for synergistic indications, such as a greater or lesser degree of similarity in topology, size, and binding site composition, from both structure and sequence, to further characterise such similarities.

Beyond the similarities described above, members of the calycin superfamily have in common a characteristic structural pattern:[20] a lysine or arginine (from the terminal strand of the β-barrel) forms several potential hydrogen bonds to carbonyl groups of the main-chain of the short N-terminal 3_{10}-like helix while packing across a conserved tryptophan (from the initial strand of the barrel) in a structurally superimposible, nonrandom manner. Visual inspection of available lipocalin, avidin, and FABP structures all reveal a very similar arrangement of interacting residues. This signature corresponds to sequence determinants common to the calycin member families: a characteristic N-terminal sequence pattern centred on tryptophan, which displays preservation of key residues, and a weaker C-terminal motif centred on arginine or lysine. Although some conservation is apparent within these patterns, it is not of sufficient strength to allow the design of sequence discriminators able to identify all calycins with certainty.

For the newly identified calycins, several appear, on the basis of retained structural signatures, to be bone fide members of the superfamily. These include the MPIs and the relevant domains from D-amino peptidase, cathepsin C, and quinohemoprotein amine dehydrogenase. For other proteins, such as YodA[42] and the *Thermus thermophilus* HB8 polyprenyl pyrophosphate binding protein,[43] both globally and in terms of retained structural signatures, apparent structural propinquity may be simply coincidental. Many members of the calycin superfamily have variant signatures, which maintain some of the overall specificity of interaction but use different amino acids. Certain lipocalins—the late lactation proteins, for example—have a tyrosine instead of tryptophan and what appears to be a proline at the arginine position. A number of FABPs show a tryptophan to tyrosine, and even tryptophan to phenylalanine, substitution. Here an aromatic to arginine/lysine interaction is conserved, although structural data would suggest that it is not as strong as for tryptophan. Other calycins substitute lysine for arginine. Examples of this come from a group of highly diverged FABPs, typified by insect muscle FABPs, and also quinohemoprotein amine dehydrogenase.

Triabin and staphostatin remain enigmatic, in that they have similarity in terms of global conformation and the conservation of the family signature, yet have perturbed topologies. In triabin, Strands B and C are interchanged in position, altering the repeated +1 topology and antiparallel arrangement of adjacent strands. By allowing reverse matching of structural segments rather than exclusively ordered matching, large proportions of the triabin and other calycin structures can be equivalenced. At the level of protein sequence, triabin has some global similarity to nitrophorin and *Rhodnius prolixus* salivary platelet aggregation inhibitors. Staphostatin has a structural signature similar to other calycins but it is displaced with the sequence. The first three strands of the staphostatin barrel form a β-meander rather than two successive β-hairpins. As a consequence, the first strand interacts with the third and fourth strands (sequential numbering) so that the signature tryptophan, rather than coming from strand 1, instead comes from strand two (sequential numbering), which itself interacts with the third and eighth strands (sequential numbering).

It remains unclear whether these changes—variations in topology and conservation—generate alternative compensating interactions within these structures or are examples of drift in protein evolution. If certain proteins are tolerant to such alterations, then why, against a backdrop of significant sequence divergence across the whole family, are they required by most? In current structures, sequences giving rise to particular folding patterns may have evolved stability independent of this interaction of residues. Ignoring disordered proteins,[44] protein sequences in aqueous solvents generally fold into, essentially, a unique structure. However, distinct sequences can fold into similar structures. As Rost reports,[45] only 3-4% of amino acids appear to be crucial for protein structure and function. Residue identities for proteins which have evolved from the same (divergent) or different (convergent) ancestors are similar and it is problematic to differentiate them. Low sequence identity does not necessarily indicate a convergent route. What we see in the calycins is, perhaps, a distant evolutionary relic of the common calycin ancestor protein: still an important structural interaction but no longer essential. Nonetheless, conservation of a characteristic sequence signatures corresponding to an even more highly conserved structural signature supports the view that there is a common, if now very remote, origin from which the members of the calycin superfamily (lipocalins, FABPs, avidins, MPIs, triabin, etc.) have diverged.

Folding and Stability

The lipocalin protein family has continued to grow, both in terms of sequences and structures determined, but also in their diversity and interest. Members of the calycin protein superfamily share a β-barrel structure and certain member families share the ability to bind hydrophobic ligands, although many do not. As well as an overall structural similarity, the calycin proteins show conserved main chain conformations, amino acid side chains, and the interactions they make, which together forms a structural signature characteristic of the superfamily. In particular, an arginine or lysine, able to form a number of potential hydrogen bonds with the main chain carbonyls of a short 3_{10} helix, and which packs across a conserved tryptophan. Certainly, these conserved interactions act to "pin" together the two ends of the calycin β-barrel, but what other role does this structural signature play? Might its role be functional or might it be a protein-protein recognition site perhaps? Does it stabilize the structure maintaining the overall fold? Or is it involved in the folding pathway, perhaps guide the formation of the β-barrel?

A recent paper,[46] throws considerable light on this. Using RBP as their exemplar, the authors generated a series of mutant proteins. Changes to TRP24 or ARG139, both involved in the calycin signature, lead to similar significant losses in stability and decreased yields of protein as generated by folding in vitro. As a control they also mutated several other, more accessible trytophans and found that they did not affect stability or expression. These results, in concert with the nature of natural amino acid mutations at these positions, support the notion that conserved residues in homologous proteins act to increase the proportion of folded to misfolded proteins, thus stabilizing the native structure.

In a separate, more recent paper, these authors show that the main sequence motifs in the lipocalins, which form a superset that includes the main calycin motifs, are involved critically in the folding process of RBP.[25] This result synergises with results from Goto and coworkers, who have analysed the β-lactoglobulin (Blg) folding process.[47] In common with many other proteins (α-lactalbumin, lysozyme, plasminogen activator inhibitor type-1, annexin 1, etc.), studies on Blg indicate that there is a transient intermediate with significantly more α-helical content and less β-sheet than the native protein. They used ultra-rapid mixing techniques to monitor folding over a 100 μs to 10 s timescale. The folding intermediate detected in their experiments contains a well-ordered region formed from strands F, G and H, as well as the *C*-terminal helix and a region of the *N*-terminus (see Fig. 1). This last region normally adopts a β-strand conformation preceded by a rare, but conserved, 3_{10} helix. It is this region that adopts a nonnative α-helical conformation different to that in the fully folded protein. These results also help to explain the long-standing observation that polar solvents, such as water–alcohol mixtures, tend to increase the α-helical structure apparent in Blg, as observed by circular dichroism spectroscopy or NMR. Their observation fits well with an emerging consensus on the fundamental mechanisms underlying protein folding: a multi-dimensional energy landscape, sometimes described as a folding funnel, allows a large number of unequally populated alternative routes from the unfolded protein to the native state. For all but the simplest proteins, some of these routes will involve intermediates, or local minima, and may lead to kinetically trapped misfolded proteins.

Resolving the protein folding problem is one of the greatest challenges in science today. The lipocalins offer a promising system to investigate the determinants of topology, stability and the relationship between sequence conservation and folding. The great variation in their critical biological functions, their potential for specifically engineered drug transport, interactions with receptors, and their overwhelming pervasiveness throughout the eukaryotic kingdom have, biologically speaking, placed the lipocalins centre stage, with a thrilling new era of untold discoveries waiting to unfold.

References

1. Flower DR. Beyond the superfamily: the lipocalin receptors. Biochim Biophys Acta 2000; 1482:327-336.
2. Flower DR. The lipocalin protein family: Structure and function. Biochem J 1996; 318:1-14.
3. Pervaiz S, Brew K. Homology of beta-lactoglobulin, serum retinol-binding protein and protein HC. Science 1985; 228:335-337.
4. Flo TH, Smith KD, Sato S et al. Lipocalin 2 mediates an innate immune response to bacterial infection by sequestering iron. Nature 2004; 432:917-921.
5. Bychkova VE, Dujsekina AE, Fantuzzi A et al. Fold Des 1998; 3:285-291.
6. Gasymov OK, Abduragimov AR, Gasimov EO et al. Tear lipocalin: Potential for selective delivery of rifampin. Biochim Biophys Acta 2004; 1688:102-111.
7. Flower DR, North ACT, Attwood TK. Mouse oncogene protein-24p3 is a member of the lipocalin protein family. Biochem Biophys Res Commun 1991; 180:69-74.
8. Flower DR, North ACT, Attwood TK. Structure and sequence relationships in the lipocalins and related proteins. Protein Sci 1993; 2:753-761.
9. Flower DR, North AC, Sansom CE. The lipocalin protein family: Structural and sequence overview. Biochim Biophys Acta 2000; 1482:9-24.
10. Flower DR. Experimentally determined lipocalin structures. Biochim Biophys Acta 2000; 1482:46-56.
11. Lascombe MB, Gregoire C, Poncet P et al. Crystal structure of the allergen Equ c 1. A dimeric lipocalin with restricted IgE-reactive epitopes. J Biol Chem 2000; 275:21572-21577.
12. Vincent F, Lobel D, Brown K et al. Crystal structure of aphrodisin, a sex pheromone from female hamster. J Mol Biol 2001; 305:459-469.
13. Korndorfer IP, Beste G, Skerra A. Crystallographic analysis of an "anticalin" with tailored specificity for fluorescein reveals high structural plasticity of the lipocalin loop region. Proteins 2003; 53:121-129.
14. Korndorfer IP, Schlehuber S, Skerra A. Structural mechanism of specific ligand recognition by a lipocalin tailored for the complexation of digoxigenin. J Mol Biol 2003; 330:385-396.
15. Habash J, Helliwell JR, Raftery J et al. The structure and refinement of apocrustacyanin C2 to 1.3 A resolution and the search for differences between this protein and the homologous apoproteins A1 and C1. Acta Crystallogr D Biol Crystallogr 2004; 60:493-498.
16. Breustedt DA, Korndorfer IP, Redl B et al. The 1.8-A crystal structure of human tear lipocalin reveals an extended branched cavity with capacity for multiple ligands. J Biol Chem 2005; 280:484-493.
17. Campanacci V, Nurizzo D, Spinelli S et al. The crystal structure of the Escherichia coli lipocalin Blc suggests a possible role in phospholipid binding. FEBS Lett 2004; 562:183-138.
18. Irikura D, Kumasaka T, Yamamoto M et al. Cloning, expression, crystallization, and preliminary X-ray analysis of recombinant mouse lipocalin-type prostaglandin D synthase, a somnogen-producing enzyme. J Biochem (Tokyo) 2003; 133:29-32.
19. Flower DR. Structural relationship of streptavidin to the calycin protein superfamily. FEBS Letters 1993; 333:99-102.

20. Flower DR. A structural signature characteristic of the calycin protein superfamily. Protein Pept Lett 1995; 2:341-346.
21. Flower DR. The up-and-down beta-barrel proteins: Three of a kind. FASEB J 1995; 9:566-567.
22. Goetz DH, Willie ST, Armen RS et al. Ligand preference inferred from the structure of neutrophil gelatinase associated lipocalin. Biochemistry 2000; 39:1935-1941.
23. Bocskei Z, Groom CR, Flower DR et al. Pheromone binding to 2 rodent urinary proteins revealed by X-Ray crystallography. Nature 1992; 360:186-188.
24. Cowan SW, Newcomer ME, Jones TA. Crystallographic refinement of human serum retinol binding- Protein at 2 angstroms resolution. Proteins 1990; 8:44-61.
25. Greene LH, Hamada D, Eyles SJ et al. Conserved signature proposed for folding in the lipocalin superfamily. FEBS Lett 2003; 553:39-44.
26. Green NM. Avidin and streptavidin. Meth Enzymol 1990; 184:51-67.
27. Banaszak L, Winter N, Xu ZH et al. Lipid-Binding Proteins - A family of fatty-acid and retinoid transport proteins. Adv Protein Chem 1994; 45:89-151.
28. Murzin AG, Lesk AM, Chothia C. Principles determining the structure of beta-sheet barrels in proteins. I. A theoretical analysis. J Mol Biol 1994; 236:1369-1381.
29. Murzin AG, Lesk AM, Chothia C. Principles determining the structure of beta-sheet barrels in proteins. II. The observed structures. J Mol Biol 1994; 236:1382-1400.
30. Baumann U, Bauer M, Letoffe S et al. Crystal-Structure of a complex between Serratia-marcescens metalloprotease and an inhibitor from Erwinia-chrysanthemi. J Mol Biol 1995; 248:653-661.
31. Fuentesprior P, Noeskejungblut C, Donner P et al. Structure of the thrombin complex with triabin, a lipocalin-like Exosite-binding inhibitor derived from a triatomine bug. Proc Natl Acad Sci USA 1997; 94:11845-11850.
32. Rzychon M, Filipek R, Sabat A et al. Staphostatins resemble lipocalins, not cystatins in fold. Protein Sci 2003; 12:2252-2256.
33. Bompard-Gilles C, Remaut H, Villeret V et al. Crystal structure of a D-aminopeptidase from Ochrobactrum anthropi, a new member of the 'penicillin-recognizing enzyme' family. Structure Fold Des 2000; 8:971-980.
34. Satoh A, Kim JK, Miyahara I et al. Crystal structure of quinohemoprotein amine dehydrogenase from Pseudomonas putida. Identification of a novel quinone cofactor encaged by multiple thioether cross-bridges. J Biol Chem 2002; 277:2830-2834.
35. Turk D, Janjic V, Stern I et al. Structure of human dipeptidyl peptidase I (cathepsin C): Exclusion domain added to an endopeptidase framework creates the machine for activation of granular serine proteases. EMBO J 2001; 20:6570-6582.
36. Borgstahl GEO, Williams DR, Getzoff ED. 1.4 Angstrom structure of photoactive yellow protein, a cytosolic Photoreceptor - Unusual fold, Active-Site, and chromophore. Biochemistry 1995; 34:6278-6287.
37. Fu ZQ, Dubois GC, Song SP et al. Crystal structure of MTCP-1: Implications for role of TCL-1 and MTCP-1 in T cell malignancies. PNAS 1998; 95:3413-3418.
38. Russell RB, Sternberg MJE. A novel binding site in catalase is suggested by structural similarity to the calycin superfamily. Protein Eng 1997; 9:107-111.
39. Kallen J, Spitzfaden C, Zurini MGM et al. Structure of human cyclophilin and its Binding-Site for Cyclosporine-A determined by X-Ray crystallography and Nmr- Spectroscopy. Nature 1991; 353:276-279.
40. Orengo CA, Swindells MB, Michie AD et al. Structural similarity between the pleckstrin homology domain and Verotoxin - the problem of measuring and evaluating structural similarity. Protein Sci 1995; 4:1977-1983.
41. Efimov AV. A structural tree for proteins containing 3 Beta-Corners. FEBS Letters 1997; 407:37-41.
42. David G, Blondeau K, Schiltz M et al. YodA from Escherichia coli is a metal-binding, lipocalin-like protein. J Biol Chem 2003; 278:43728-43735.
43. Handa N, Terada T, Doi-Katayama Y et al. Crystal structure of a novel polyisoprenoid-binding protein from Thermus thermophilus HB8. Protein Sci 2005; 14:1004-1010.
44. Dyson HJ, Wright PE. Intrinsically unstructured proteins and their functions. Nat Rev Mol Cell Biol 2005; 6:197-208.
45. Rost B. Protein structures sustain evolutionary drift. Fold Des 1997; 2:S19-S24.
46. Greene LH, Chrysina ED, Irons LI et al. Role of conserved residues in structure and stability: Tryptophans of human serum retinol-binding protein, a model for the lipocalin superfamily. Protein Sci 2001; 10:2301-2316.
47. Kuwata K, Shastry R, Cheng H et al. Structural and kinetic characterization of early folding events in β-lactoglobulin. Nature Struct Biol 2001; 8:151-155.
48. Flower DR. ALTER: Eclectic management of molecular structure data. J Mol Graph Mod 1997; 15:161-169.

Chapter 4

Bacterial Lipocalins: Origin, Structure, and Function

Russell E. Bishop,* Christian Cambillau, Gilbert G. Privé, Derek Hsi, Desiree Tillo and Elisabeth R.M. Tillier

Abstract

The bacterial lipocalins were discovered in 1995 and first reviewed in the year 2000. In the subsequent 5 years, two important developments have been made. First, an explosion of molecular sequence information from microbial genome projects has uncovered more than 90 bacterial lipocalin sequences. The phylogenetic distribution indicates that lipocalins are absent from the archaebacteria and nonmycolic acid-producing Gram-positive bacteria, which lack a permeability barrier located extrinsic to the inner (cytoplasmic) membrane, a point of contrast with the lipocalin-encoding Gram-negative bacteria and mycolic acid-producing actinobacteria. This observation strongly supports the conclusion that bacterial lipocalins originated in association with the development of the bacterial outer membrane structure. Combined with the recent finding that the integral outer membrane enzyme PagP displays a compact lipocalin-like 8-stranded anti-parallel β-barrel fold, complete with an internal lipid-binding pocket, it becomes difficult to ignore the possibility that lipocalins might share a common origin with bacterial β-barrel membrane proteins. Second, the high resolution crystal structure of the prototypical bacterial lipocalin Blc, an outer membrane lipoprotein from *E. coli*, has now been solved. A recent reappraisal of the crystal packing contacts in this structure have led to the conclusion that Blc is a functional dimer with a lipid acyl-chain binding-site buried at the subunit interface. Tryptophan fluorescence-quenching experiments indicate micromolar affinities for various fatty acids and phospholipids, but nanomolar affinities for lysophospholipids. Outer membrane enzymes like PagP generate lysophospholipids as enzymatic products, suggesting that Blc might fulfil a role in outer membrane biogenesis or repair related to lysophospholipid metabolism.

Introduction

The lipocalin protein family consists mainly of small extracellular proteins that bind hydrophobic ligands and fulfill numerous biological functions including ligand transport, cryptic coloration, sensory transduction, the biosynthesis of prostaglandins, and the regulation of cellular homeostasis and immunity. The lipocalins can be structurally characterized as an 8-stranded antiparallel β-barrel followed by a C-terminal α-helix, as exemplified by serum retinol binding protein.[1] The lipocalin fold is widely distributed among vertebrates and a few have been isolated from invertebrates and plants. Until 1995, when the first bacterial lipocalins were identified,[2] the

*Corresponding Author: Russell E. Bishop—Departments of Laboratory Medicine and Pathobiology, and Biochemistry, Faculty of Medicine, University of Toronto, 6213 Medical Sciences Building, 1 King's College Circle, Toronto, Ontario M5S 1A8, Canada. Email: russell.bishop@utoronto.ca

Lipocalins, edited by Bo Åkerström, Niels Borregaard, Darren R. Flower and Jean-Philippe Salier. ©2006 Landes Bioscience.

family was thought to be restricted to eukaryotes. The recent crystal structure of a bacterial lipocalin suggests that phospholipid derivatives are likely ligands.[3] The existence of bacterial lipocalins provides insight into the origins of the lipocalin family and, combined with the powerful tools of bacterial genetics, provides fertile ground for the investigation of lipocalin structure and function.

The bacterial lipocalin Blc was first identified as an outer membrane lipoprotein of *E. coli* in 1995.[2] Approximately 100 lipoproteins with varied structures and functions are encoded within the *E. coli* genome.[4] Export of bacterial lipoproteins across the inner membrane is directed by a *sec*-dependent type-II signal peptide, which converts the amino terminal cysteine into *N*-acyl-*S*-*sn*-1,2-diacylglycerylcysteine. The lipidated protein remains anchored to the periplasmic surface of the inner membrane until it encounters the LolCDE ATP-binding cassette transport system, which can actively disengage the lipoprotein from the membrane and complex it with the periplasmic chaperone LolA. The LolA/lipoprotein complex moves across the periplasmic space and docks in the outer membrane with LolB, which is also a lipoprotein, to allow delivery of the lipoprotein from LolA into the periplasmic leaflet of the outer membrane.[5] The outer membrane β-barrel proteins are similarly exported across the inner membrane in a process that subsequently employs dedicated chaperones to move them through the periplasmic space before folding into the outer membrane.[6] Lipoproteins and integral membrane β-barrel proteins are the two major classes of proteins present in the bacterial outer membrane. Additionally, the outer membranes of Gram-negative bacteria are distinguished by an asymmetric lipid organization, with the inner and outer leaflets composed of phospholipids and lipopolysaccharides (LPS), respectively.

β-barrel membrane proteins are not normally found in the inner membrane, although there is no physical reason why they could not assemble in that location. The β-barrel structure is very effective at creating pores that would dissipate the proton motive force, suggesting that natural selection acts against their inner membrane localization. Indeed, several protein toxins act by forming β-barrel structures within energy-transducing membrane systems.[7] The β-barrel structure satisfies the requirement of providing continuous hydrogen bonding between the amide nitrogen proton and carbonyl oxygen of the peptide bonds, which are otherwise too polar to interact stably with the hydrocarbon chains in a lipid bilayer membrane. Roughly 8 amino acids are sufficient to span the membrane in the extended β-conformation, and half of these residues project into the mostly polar β-barrel interior region.[8] Only those residues whose registration projects them into the lipid milieu of the membrane need to possess hydrophobic side-chains. Consequently, β-barrel membrane proteins are only moderately hydrophobic, a fact that facilitates their coursing through the inner membrane and periplasmic space.[9] Nature's other solution to provide continuous peptide hydrogen bonding is the transmembrane α-helix, which demands roughly 20 hydrophobic amino acid side chains in order to span the membrane.[10] The greater hydrophobicity of transmembrane α-helical proteins dictates that their targeting to the inner membrane is essentially irreversible, a fact that explains why no transmembrane α-helical proteins are found in the outer membrane. The peptidoglycan exoskeleton that is sandwiched between the inner and outer membranes is believed to effectively block wholesale membrane exchange by vesicular trafficking. Consequently, outer membrane biogenesis likely depends on soluble periplasmic lipid-transfer proteins and/or localized contact bridges between inner and outer membranes.

The lipocalins represent a family of β-barrel proteins that are adapted to bind hydrophobic ligands, essentially chaperoning them through the aqueous compartments of the cell.[11] Accumulating evidence supports the hypothesis that lipocalins originated in association with the development of bacterial outer membrane systems, and that bacterial lipocalins exert a function in the maintenance of the outer membrane structure. Here we update the phylogenetic distribution of bacterial lipocalins and describe some interesting new relationships between their structure and function in the outer membrane. Some aspects of bacterial lipocalin structure-function relationships that have been reviewed previously are not addressed again here.[12] We suggest that interested readers consult this earlier article to gain a fuller picture of the bacterial lipocalin story.

Phylogenetic Distribution of Bacterial Lipocalins

Bacterial Lipocalins Are Codistributed with Outer Membrane Proteins

In the year 2000, 20 lipocalins were identified in Gram-negative bacteria, but that number now exceeds 90. Among the newly identified sequences are three that belong to Gram-positive bacteria, namely, *Nocardia* and two species of *Corynebacteria*. These organisms belong to the so called CMN branch (for *Corynebacteria, Mycobacteria, Nocardia*) of the actinomycetes. The CMN branch represents the most highly derived of the Gram-positive bacteria and do not actually display Gram-positive staining. Instead, the presence of mycolic acids in these organisms generates a permeability barrier extrinsic to the inner membrane that requires staining by the so-called "acid-fast" procedure. The CMN organisms are regarded by some as being intermediate between Gram-negative and Gram-positive bacteria, because the mycolic acid layer requires porin channels to facilitate nutrient uptake and these adopt a transmembrane β-barrel architecture[13]—a decidedly Gram-negative characteristic.[14] Despite this similarity, the 16s rRNA phylogeny places the CMN branch firmly within the Gram-positive actinomycetes. Therefore, the bacterial lipocalins are associated with organisms that produce either LPS or mycolic acids, the two classes of lipids associated with bacterial permeability barriers that harbor β-barrel membrane proteins. Although lipocalins do not appear among the *Mycobacteria*, the fatty acid binding-protein fold that resembles lipocalins has been identified in these organisms (Pedro Alzari, personal communication). The mycolic acids of *Mycobacteria* are unusually long and create a thicker permeability barrier compared to *Nocardia* and *Corynebacteria*. To date, no lipocalins have been identified among the Gram-positive or archaebacterial organisms that lack the defining characteristics of outer membranes.

Not All Bacterial Lipocalins Are Bacterial

BLAST searches using *E. coli* Blc as a probe can usually discern between bacterial and eukaryotic lipocalins, because the latter sequences display a number of obvious insertions/deletions. For example, apolipoprotein D (ApoD) and lazarillo, members of a group of eukaryotic lipocalins closely related to Blc, possess an insertion between β-strands G and H that is absent from Blc.[12] However, several eukaryotic lipocalins cannot easily be distinguished from the bacterial homologues (Table 1). Most striking among these are a group of six bacterial-like lipocalins from plants. Chromosome 5 of *Arabidopsis thaliana* encodes a lipocalin that appears to contain a single 84 nucleotide intron with the typical 5'-donor and 3'-acceptor splice sites. Removal of the intron yields a 186 amino acid polypeptide, which exhibits greatest similarity with the bacterial lipocalins. Plant genomes are unique among eukaryotes in that some encode most of the enzymes needed to make the hydrophobic anchor of Gram-negative LPS known as lipid A or endotoxin.[15] Green plants are thought to have acquired plastids by a recent endosymbiosis with Gram-negative cyanobacteria, and might consequently have retained some outer membrane components.[16] These same components were apparently lost from the mitochondrion after it was derived by an earlier endosymbiosis with a Gram-negative α-proteobacterium.[17] Indeed, at least one lipocalin is encoded in the genome of the cyanobacterium *G. violaceus*. The six bacterial-like lipocalins of plants lack any obvious signal sequence, suggesting they are soluble cytosolic proteins. The eukaryotic acquisition of bacterial genes that code for cytosolic functions is now thought to be an inevitable consequence of endosymbiosis.[18] Another bacterial-like lipocalin is derived from the marine yeast *D. hansenii*, and this protein exhibits a typical eukaryotic signal peptide that likely targets it to the endoplasmic reticulum.

Two Sub-Clades of Bacterial Lipocalins

The bacterial lipocalins are defined as belonging to Clade I,[19] but we can divide them further into two groups distinguished by the absence (Clade IA) or presence (Clade IB) of a lipocalin subgroup that displays a pair of cysteine residues flanking β-strands B and C in an apparent disulfide-bonding configuration (Fig. 1). The bacterial-like lipocalins from plants belong to a

Table 1. Listing of all Clade I lipocalin sequences (2005)[a]

Organisms	Signal Peptide[b]	βB-βC Cysteines	Accession Number
Bacteria			
Alpha-Proteobacteria (α)			
Agrobacterium tumefaciens	SPI	Yes	NC_003063
Caulobacter crescentus (1)	SPII	Yes	NC_002696
Caulobacter crescentus (2)	SPI	Yes	NC_002696
Hyphomonas neptunium	SPII	Yes	TIGR (incomplete)
Mesorhizobium loti (1)	None	Yes	NC_002678
Mesorhizobium loti (2)	None	Yes	NC_002678
Rhodobacter sphaeroides	SPII	Yes	AAAE01000158
Rhodospirillum rubrum	SPII	No	AAAG02000001
Silicibacter	SPI	Yes	NZ_AAFG01000010
Beta-Proteobacteria (β)			
Acidovorax (partial)	SPII	No	AB044565
Bordetella bronchiseptica	SPI	Yes	NC_002927
Bordetella parapertussis	SPI	Yes	NC_002928
Burkholderia cepacia (1)	SPII	Yes	AAEH01000003
Burkholderia cepacia (2)	SPII	No	AAEH01000003
Burkholderia fungorum (1)	SPII	No	NZ_AAAJ03000001
Burkholderia fungorum (2)	SPII	No	NZ_AAAJ03000001
Burkholderia fungorum (3)	SPII	Yes	NZ_AAAJ03000004
Chromobacterium violaceum	SPII	No	NC_005085
Dechloromonas aromatica	SPI	Yes	AADF01000001
Methylobacillus flagellatus	SPII	No	AADX01000001
Ralstonia eutropha	SPII	No	AADY01000001
Rubrivivax gelatinosus	SPI	Yes	AAEM01000005
Thiobacillus denitrificans	SPII	Yes	AAFH01000001
Gamma-Proteobacteria (γ)			
Acinetobacter (1)	SPI	Yes	CR543861
Acinetobacter (2)	SPII	Yes	CR543861
Azotobacter vinelandii	SPII	Yes	NZ_AAAU02000004
Citrobacter braakii (partial)	SPII	No	AF492447
Citrobacter freundii	SPII	No	U21727
Citrobacter murliniae	SPII	No	AJ607409
Citrobacter werkmanii (partial)	SPII	No	AF492448
Colwellia psychroerythraea	SPII	No	TIGR (Incomplete)
Enterobacter cancerogenus (partial)	SPII	No	AF492446
Enterobacter nimipressuralis	SPII	No	AJ487975
Erwinia carotovora	SPII	No	BX950851
Escherichia coli	SPII	No	P39281
Francisella tularensis	SPII	No	AY774926
Idiomarina loihiensis	SPII	No	NC_006512
Klebsiella oxytoca	SPII	No	Y17716
Klebsiella pneumoniae	SPII	No	Incomplete
Pseudomonas aeruginosa	SPII	Yes	AABQ07000004
Pseudomonas fluorescens	SPII	Yes	AAAT03000001
Pseudomonas putida (1)	SPII	Yes	NC_002947
Pseudomonas putida (2)	SPII	Yes	NC_002947

continued on next page

Table 1. Continued

Organisms	Signal Peptide[b]	βB-βC Cysteines	Accession Number
Gamma-Proteobacteria (γ)			
Pseudomonas syringae	SPI	Yes	AABP02000002
Salmonella paratyphi	SPII	No	Incomplete
Salmonella typhi	SPII	No	NC_006511
Salmonella typhimurium	SPII	No	AE008903
Shewanella oneidensis	SPII	No	AE015615
Shigella flexneri	SPII	No	NC_004741
Vibrio cholerae (1)	SPII	No	NC_002506
Vibrio cholerae (2)	SPII	No	NC_002506
Vibrio cholerae (3)	SPII	No	NC_002506
Vibrio cholerae (4)	SPII	No	NC_002506
Vibrio parahaemolyticus	SPII	No	NC_004605
Vibrio vulnificus	SPII	No	NC_004460
Xanthomonas axonopodis (1)	SPI	No	NC_003919
Xanthomonas axonopodis (2)	SPI	No	NC_003919
Xanthomonas axonopodis (3)	None	Yes	NC_003919
Xanthomonas axonopodis (4)	SPI	Yes	NC_003919
Xanthomonas campestris (1)	SPI	No	NC_003902
Xanthomonas campestris (2)	SPI	No	NC_003902
Xanthomonas campestris (3)	None	Yes	NC_003902
Xanthomonas campestris (4)	SPI	Yes	NC_003902
Yersinia enterocolitica	SPII	No	Sanger (Incomplete)
Delta-Proteobacteria (δ)			
Bdellovibrio bacteriovorus (1)	SPI	Yes	NC_005363
Bdellovibrio bacteriovorus (2)	SPI	Yes	NC_005363
Desulfotalea psychrophila	SPII	No	NC_006138
Geobacter sulfurreducens	SPII	Yes	NC_002939
Epsilon-Proteobacteria (ε)			
Campylobacter jejuni	None	Yes	AL139078
Campylobacter lari	None	Yes	NZ_AAFK01000002
Chlorobi			
Chlorobium tepidum (1)	SPII	No	NC_002932
Chlorobium tepidum (2)	None	Yes	NC_002932
Bacteroidetes			
Bacteroides fragilis	None	No	NC_006347
Bacteroides thetaiotaomicron	None	No	NC_004663
Cytophaga hutchinsonii	SPI	No	AABD03000002
Verrucomicrobia			
Verrucomicrobium spinosum (1)	SPII	No	TIGR (incomplete)
Verrucomicrobium spinosum (2)	SPII	No	TIGR (incomplete)
Chlamydiae			
Parachlamydia	SPI	Yes	NC_005861
Cyanobacteria			
Gloeobacter violaceus	SPII	Yes	NC_005125

continued on next page

Table 1. Continued

Organisms	Signal Peptide[b]	βB-βC Cysteines	Accession Number
Actinobacteria			
Corynebacterium efficiens	SPI	Yes	NC_004369
Corynebacterium glutamicum	SPI	Yes	NC_003450
Nocardia farcinica	SPI	Yes	NC_006361
Eukaryota			
Viridiplantae			
Arabidopsis thaliana	None	No	AY085685
Capsicum annuum	None	No	AY568589
Lycopersicon esculentum	None	No	BT013102
Oryza sativa	None	No	XM_466697
Triticum aestivum	None	No	AY077702
Zea mays	None	No	AY107141
Saccharomycetes			
Debaryomyces hansenii	ER	Yes	XM_460369
Metazoa (Outgroup)			
ApoD	ER	No	P05090
Lazarillo	ER/GPI	No	P49291

[a] A TBLASTN[44] search using the *E. coli* Blc protein sequence was performed on both the microbial genome databases and the non-redundant database at the NCBI website (http://www.ncbi.nlm.nih.gov). The genome sequence 500 bp upstream and downstream of each hit was retrieved. The six resulting translations were evaluated to identify both the full-length lipocalin sequence and the presence of the 16s ribosomal RNA binding site. The taxonomy of each organism was also retrieved using the NCBI TaxBrowser. [b] Each amino acid sequence was evaluated for the presence of signal peptides and these were assigned using SIGNALP.[45] SPI: signal peptidase I; SPII: signal peptidase II; ER: endoplasmic reticulum; GPI: glycosylphosphatidylinositol.

subgroup within Clade IB that appears to have lost this cysteine pair, whereas the *D. hansenii* lipocalin displays the apparent disulfide. These observations can be rationalized with the reducing cytosolic and oxidizing extracellular environments to which these lipocalins are targeted. Inexplicably, some cytosolic lipocalins within Clade IB still retain this pair of cysteine residues. The two nonconsecutive disulfides found in the closely related group of eukaryotic lipocalins that include ApoD and lazarillo each appear to have their own antecedents within the *B. fungorum2* and *G. violaceus* lipocalins from Clade IB (Fig. 2). Thus, it appears that Clade IB lipocalins are more closely related to the eukaryotic lipocalins than are those from Clade IA. An interesting feature within Clade IA is the *N*-terminal fusion with a short-chain dehydrogenase/reductase domain in two *Bacteroides* lipocalins. Similar domains are employed for carbohydrate epimerization during the synthesis of sugars needed for LPS biosynthesis.[20] Intriguingly, the *Chlorobium* genome reveals examples of both Clade IA and Clade IB bacterial lipocalins, suggesting that an early gene duplication event produced these two extant groups.

Vertical versus Horizontal Descent

In a previous analysis, we interpreted the absence of a homologue from an archaebacterium as indicating that lipocalins could not have existed in the last common ancestor of the three domains of life, and that they were probably acquired in eukaryotes horizontally by endosymbiosis.[12] Horizontal transfer is probably partly correct, especially in light of the bacterial-like lipocalins of plants and yeast described here. However, a recent study from Rivera and Lake

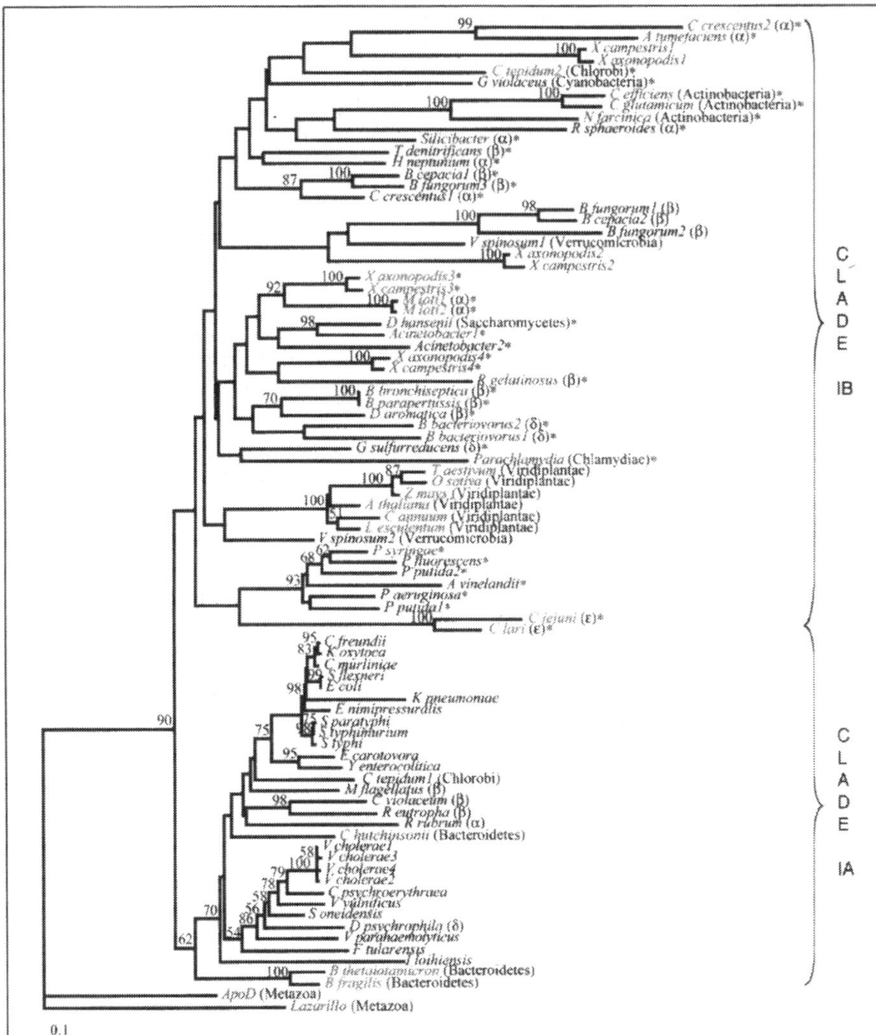

Figure 1. Evolutionary relationships between bacterial lipocalins and their closest eukaryotic descendants. Signal peptides were removed from the lipocalin sequences listed in Table 1. Partial sequences were excluded from the multiple sequence alignment, which was constructed using CLUSTAL W 1.83,[42] with default parameters (BLOSUM62 scoring matrix, gap opening penalty = -12, gap extension penalty = -2). In order to obtain an alignment that reflects lipocalin structure, a profile alignment and a gap penalty mask were constructed using the structural alignment and secondary structure information obtained from the published Blc structure.[3] Sequences were then added to the profile alignment one at a time, in increasing order of divergence from the *E. coli* Blc sequence. Once the multiple sequence alignment was complete, the profile alignment and gap penalty mask were removed. The neighbor-joining distance tree was then constructed using PROTDIST, with the Dayhoff option invoked, and NEIGHBOR, both from the PHYLIP 3.63 suite of programs (http://evolution.genetics.washington.edu/phylip.html). Confidence limits of branches were estimated from 100 bootstrap replications using SEQBOOT and CONSENSE. Bootstrap values greater than 50 are shown at the nodes. The tree was visualized using TREEVIEW[43] and rooted with ApoD and lazarillo. Asterisks mark those Clade IB lipocalins that possess conserved cysteines flanking β-strands B and C. Colors serve to identify secreted eukaryotic proteins (red), bacterial lipoproteins (black), secreted bacterial proteins (blue), and cytosolic proteins (green). Brackets identify taxonomic associations and the Greek symbols signify subdivisions of proteobacteria; those without brackets belong to the γ-proteobacteria.

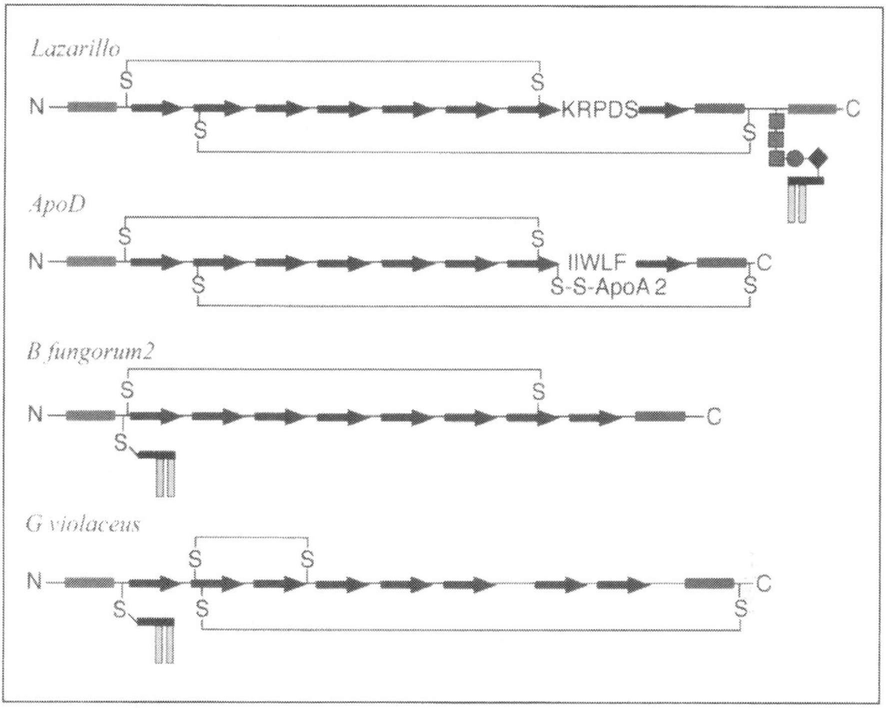

Figure 2. Relationships between apparent disulfide bonding patterns in lazarillo, ApoD, *B. fungorum*2, and *G. violaceus* lipocalins. Amino acid residues in the surface loop between β-strands G and H are shown for lazarillo, and as a consensus sequence for the ApoDs; however, the intermolecular disulfide bond with ApoA2 is unique to human ApoD. The apparent disulfide bond flanking β-strands B and C in the *G. violaceus* lipocalin is present in the majority of Clade IB lipocalins. The putative secondary structural elements are indicated as β-sheet (blue arrows), α-helix (red rectangles), or signal sequences (green rectangles). Diglyceride moieties of the lipid anchors in Blc and lazarillo are shown as black and yellow rectangles. The sugars of the GPI anchor of lazarillo are mannose (green squares), glucosamine (red circle), and inositol (blue diamond).

suggests that a genome fusion event occurred between an archaebacterium and a proteobacterium to create the first eukaryotic cell; this framework provides a more parsimonious explanation for the acquisition of most eukaryotic lipocalins by vertical descent from the proteobacterial component of the ancestral chimera.[21] This model builds on ideas from Gupta,[22] who proposes that monoderm Gram-positive bacteria are ancestral to both the monoderm archaebacteria and the diderm Gram-negative bacteria/CMN actinomycetes.[23] Our observations clearly fit with a model where the lipocalins represent a late development in bacterial evolution, having arisen in the diderm bacteria. However, this is not so late a development as to preclude ancestry to the first eukaryotic cell according to the archaebacterial-proteobacterial fusion model.[21] The earlier three domain model predicts that proteins acquired vertically by eukaryotes (Eukarya) will likely have ancestors in the archaebacteria (Archaea) if ancestors already exist in the eubacteria (Bacteria).[24]

Bacterial Lipocalin Structure and Function

Structural Similarities between PagP and Retinol Binding Protein

A striking similarity is seen between the structures of the β-barrel outer membrane protein PagP and the prototypic lipocalin serum retinol-binding protein (RBP).[25,26] Both PagP and RBP are 8-stranded antiparallel β-barrels with a deep lipid-binding pocket located inside one

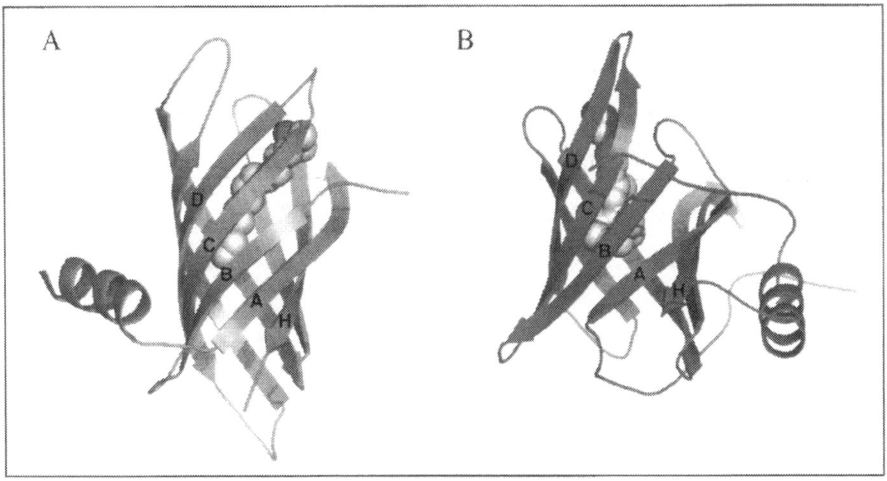

Figure 3. Comparison of PagP and serum retinol-binding protein. PagP (A) is shown in green (Protein Data Bank code 1THQ), and RBP (B) is shown in red (Protein Data Bank code 1AQB). The lauroyldimethylamine-*N*-oxide and retinol ligands are represented as spheres using CPK colors. PagP has an *N*-terminal α-helix, while most lipocalin structures, including RBP, have a single α-helix at the *C*-terminus. Both proteins are oriented so the β-strands correspond to each other; note that the first two β-strands of PagP are separated by the disordered L1 surface loop.

end of the barrel and a flanking α-helix at the other end (Fig. 3A,B). The structures superpose with the proper alignment of the β-strands (i.e., A-H of PagP superposes with A-H of RBP) with an r.m.s.d. of 2.9 Å over 93 Cα atoms. A structural similarity between the outer membrane protein OmpA and the lipocalins has also been noted, and several of the key differences between these proteins was pointed out, including the length of the barrels and the distinctly different protein interiors.[27] However, our case for similarity is greater because the PagP and RBP barrels have a higher percentage of alignable positions, and have similar inner lipid-binding pockets. Furthermore, in the lipocalins, a long loop spans strands A and B and typically forms a lid that partially closes the ligand-binding site.[11] It is intriguing that the long L1 loop from PagP is at an equivalent position and is strongly implicated in the function of this enzyme,[28,29] however, additional structural data will be required to test whether this loop interacts directly with ligand and/or the upper region of the barrel.

We cannot rule out the possibility that this structural similarity represents a case of convergent evolution. However, based on the absence of lipocalins from those bacteria that also lack β-barrel membrane proteins, we cannot currently exclude the hypothesis that PagP and lipocalins share a common ancestor. This hypothesis is supported by the observation that many bacterial lipocalins are anchored in the outer membrane as lipoproteins.[2] The lipocalin-lipoprotein might represent an intermediate state in the adaptation between membrane-bound and soluble globular domains. Outer membrane β-barrel domains should be suitably adapted to occupy both soluble and membrane-bound conformations because both environments are encountered during the outer membrane assembly process.[6]

Structure and Function of the Bacterial Lipocalin Blc

E. coli Blc was recently cocrystallized in the presence of the 18-carbon unsaturated fatty acid known as vaccenic acid, and data collected at 1.8 Å resolution (V. Campanacci, R.E. Bishop, L. Reese, S. Blangy, M. Tegoni, and C. Cambillau; submitted). The complex is isostructural with that of native Blc and two monomers are contained in the asymmetric unit. The Blc monomer has a typical lipocalin fold consisting of a β-barrel with eight anti-parallel strands and an α-helix

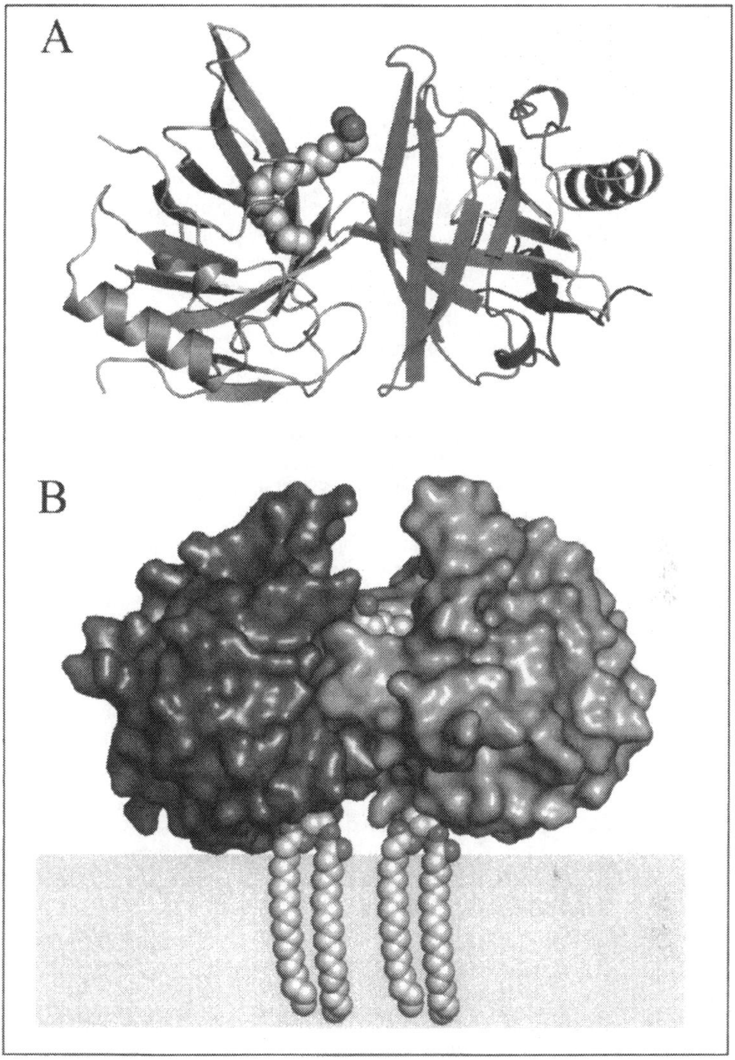

Figure 4. X-ray structure of Blc in complex with vaccenic acid. A) Ribbon view of Blc with vaccenic acid inside (spheres). Monomer B (left) is pink; monomer A (right) is rainbow colored from N- (blue) to C-terminus (red). B) Compact view of the Blc dimer attached to the membrane with anchor lipid *S-sn*-1,2-diacylglycerylcysteine attached to the first cysteine (computer model). Monomer B (right) is pink and monomer A (left) is blue; vaccenic acid and lipidic anchors are represented as spheres.

at the C-terminus. Inside the cavity of monomer B, a well defined elongated electron density could easily accommodate a vaccenic acid molecule (Fig. 4A). The molecule is bent at the position of the *cis*-double bond (Z11-12). Careful examination of the relationship between monomers A and B revealed that they interact tightly together: the interaction covers 786 Å^2 and 825 Å^2 of water accessible surface (WAS) area, on monomer A and B, respectively. Given that the total WAS area of each monomer is 7800 Å^2, the buried WAS area represents ~10% of the total, a value indicating that dimerization is not a crystallization artefact.[30] For comparison, these values of interacting surface area are comparable to those observed in immunoglobulin fragments/protein complexes.[31] The interaction between the two monomers was present and essentially identical in the original native structure, but it escaped earlier observation and was not described.[3]

The comparison of subunits A and B indicate significant conformational differences that might be necessary for dimerization. In order to fit within the other monomer, strands F and G, and particularly the loops between them, move up to 5 Å in one monomer relative to the other. The dimer interface involves in large part these loops, and other residues among which the aromatics Tyr113, Tyr137, and several surface-exposed phenylalanines including Phe53 A & B, Phe108 A & B, Phe109 A & B and Phe112 A & B together form an inter-dimer hydrophobic core. The presence of many exposed phenylalanines is an uncommon feature at the surface of a monomeric globular protein, and has to be regarded as a hallmark of protein-protein or protein-lipid interactions. Both subunits are involved in the binding site of Blc, accounting for the stoichiometry of one vaccenic acid molecule per dimer. The vaccenic acid interacts with both subunits and covers 89 $Å^2$ and 171 $Å^2$ of the WAS area of monomers A and B, respectively.

It should be stressed that Blc is anchored to membranes by a covalently attached lipid modification at the amino terminus of the protein. The manner in which the lipid-anchor of each monomer is arranged relative to the other is a critical issue. The Blc construct employed has the signal peptide and the first four residues (CSSP) removed and replaced by the Gateway ATTB1 sequence.[3] We have modelled into the dimeric structure the four original residues of Blc, as well as the major part of the anchoring lipid, *S-sn*-1,2-diacylglycerylcysteine, that we included at each *N*-terminal cysteine. This model reveals that the lipid-anchors of each of the 2 monomers are close to each other in the dimer, and are located on the same face in a geometry compatible with the insertion and stabilization of the dimer into the membrane (Fig. 4B). The binding site for vaccenic acid is located opposite to the membrane insertion site where it is expected to face the periplasmic space.

Tryptophan fluorescence quenching studies indicate that dimeric Blc binds fatty acids and phospholipids in a micromolar Kd range. An exposed and unfilled pocket seemingly suited to bind a polar group extending from the fatty acid prompted investigation of lysophospholipids (LPLs), which were found to bind in a nanomolar Kd range. Favorable steric and electrostatic interactions are observed when a glycerophosphoethanolamine moiety is modeled into this region of the structure. Given the high affinity of Blc for LPLs, it seems likely that Blc might fulfil a role in cell envelope LPL transport. Although LPLs are key inner membrane intermediates of phospholipid metabolism, we do not know of any evidence to indicate that they are exported to the outer membrane. However, exogenously supplied LPLs can be taken up by deep-rough LPS mutants and converted by reacylation into glycerophospholipids using inner membrane-associated enzymes.[32-34] To date, no accessory factors needed for LPL uptake have been identified. In wild-type cells, at least two enzymes can generate LPLs in the outer membrane, namely, the phospholipase OMPLA[35] and the lipid A palmitoyltransferase PagP.[36] Both enzymes preferentially generate the *sn*-1 LPL regioisomers, but these are known to spontaneously rearrange into the more stable *sn*-2 LPLs.[37] Only *sn*-2 LPLs were tested for binding to Blc.

The LPL products *sn*-2-lysophosphatidic acid and sphingosine-1-phosphate, together with the structurally related platelet activating factors, have been shown to function as potent biological mediators.[38,39] In eukaryotic cells, *sn*-2 LPLs can be generated by phospholipases that mobilize arachidonic acid for eicosanoid-mediated signal transduction. Additionally, *sn*-2 LPLs are enzymatic products of the lecithin:cholesterol acyltransferase LCAT, which is, together with ApoD, a key component of the plasma high density lipoprotein particle. Mammalian ApoD and *Drosophila* lazarillo are the closest eukaryotic homologues of Blc and the only eukaryotic lipocalins that, like Blc, are anchored to lipid membranes. The ligand for lazarillo is unknown, but ApoD binds a variety of ligands including arachidonic acid.[40] Perhaps the high affinity of Blc for LPLs might help shed some light on the enigmatic functions of lazarillo and ApoD.

Conclusions

The bacterial lipocalin Blc is distinguished from eukaryotic lipocalins by its dimeric organization with a lipid-binding pocket at the subunit interface and by its high affinity for LPLs. We have indicated that the structural and functional organization of Blc shares some common features with β-barrel membrane proteins. The structural similarity between PagP and lipocalins is illustrative of this point, but, in the absence of measurable amino acid sequence identity, structural similarity does not by any means indicate common ancestry. In fact, we would normally regard the similarity between PagP and lipocalins as an example of convergent evolution, and other similarities between lipocalins and a polyisoprenoid-binding protein, for example,[41] indicate that this structural organization has probably arisen independently on more than one occasion in the past; even the lipoprotein chaperones LolA and LolB are lipid-interactive β-barrel proteins.[5] However, the tight association of bacterial lipocalins with those organisms that encode outer membrane β-barrel proteins like PagP makes it difficult to pronounce a verdict of convergent evolution. Combined with the functional association of bacterial lipocalins with outer membranes, we cannot ignore the possibility that lipocalins could have either engendered β-barrel membrane proteins, or vice versa. Whatever the case may be, bacterial lipocalins promise to reveal molecular details of lipid membrane biogenesis relevant to both bacteria and higher organisms, and we hope our present insight into their plausible origins will serve to guide experiments aimed at addressing this subject.

Acknowledgements

Work in the laboratories of R.E. Bishop, G.G. Privé, and E.R.M. Tillier was supported by operating grants from the Canadian Institutes of Health Research. Work in the laboratory of C. Cambillau was supported by the French Ministry of Industry (grant ASG) and the Marseille-Nice Genopole.

References

1. Newcomer ME, Jones TA, Aqvist J et al. The three-dimensional structure of retinol-binding protein. Embo J 1984; 3:1451-1454.
2. Bishop RE, Penfold SS, Frost LS et al. Stationary phase expression of a novel Escherichia coli outer membrane lipoprotein and its relationship with mammalian apolipoprotein D. Implications for the origin of lipocalins. J Biol Chem 1995; 270:23097-23103.
3. Campanacci V, Nurizzo D, Spinelli S et al. The crystal structure of the Escherichia coli lipocalin Blc suggests a possible role in phospholipid binding. FEBS Lett 2004; 562:183-188.
4. Gonnet P, Rudd KE, Lisacek F. Fine-tuning the prediction of sequences cleaved by signal peptidase II: A curated set of proven and predicted lipoproteins of Escherichia coli K-12. Proteomics 2004; 4:1597-1613.
5. Tokuda H, Matsuyama S. Sorting of lipoproteins to the outer membrane in E. coli. Biochim Biophys Acta 2004; 1693:5-13.
6. Mogensen JE, Otzen DE. Interactions between folding factors and bacterial outer membrane proteins. Mol Microbiol 2005; 57:326-346.
7. Montoya M, Gouaux E. Beta-barrel membrane protein folding and structure viewed through the lens of alpha-hemolysin. Biochim Biophys Acta 2003; 1609:19-27.
8. Schulz GE. The structure of bacterial outer membrane proteins. Biochim Biophys Acta 2002; 1565:308-317.
9. Tamm LK, Hong H, Liang B. Folding and assembly of beta-barrel membrane proteins. Biochim Biophys Acta 2004; 1666:250-263.
10. White SH, Ladokhin AS, Jayasinghe S et al. How membranes shape protein structure. J Biol Chem 2001; 276:32395-32398.
11. Flower DR, North AC, Sansom CE. The lipocalin protein family: Structural and sequence overview. Biochim Biophys Acta 2000; 1482:9-24.
12. Bishop RE. The bacterial lipocalins. Biochim Biophys Acta 2000; 1482:73-83.
13. Faller M, Niederweis M, Schulz GE. The structure of a mycobacterial outer-membrane channel. Science 2004; 303:1189-1192.
14. Nikaido H. Molecular basis of bacterial outer membrane permeability revisited. Microbiol Mol Biol Rev 2003; 67:593-656.
15. Raetz CR, Whitfield C. Lipopolysaccharide endotoxins. Annu Rev Biochem 2002; 71:635-700.

16. Delwiche CF, Palmer JD. The origin of plastids and their spread via secondary symbiosis. Pl Syst Evol 1997; (Suppl)11:53-86.
17. Gray MW, Burger G, Lang BF. Mitochondrial evolution. Science 1999; 283:1476-1481.
18. Doolittle WF. You are what you eat: A gene transfer ratchet could account for bacterial genes in eukaryotic nuclear genomes. Trends Genet 1998; 14:307-311.
19. Gutierrez G, Ganfornina MD, Sanchez D. Evolution of the lipocalin family as inferred from a protein sequence phylogeny. Biochim Biophys Acta 2000; 1482:35-45.
20. Breazeale SD, Ribeiro AA, Raetz CR. Oxidative decarboxylation of UDP-glucuronic acid in extracts of polymyxin-resistant Escherichia coli. Origin of lipid A species modified with 4-amino-4-deoxy-L-arabinose. J Biol Chem 2002; 277:2886-2896.
21. Rivera MC, Lake JA. The ring of life provides evidence for a genome fusion origin of eukaryotes. Nature 2004; 431:152-155.
22. Golding GB, Gupta RS. Protein-based phylogenies support a chimeric origin for the eukaryotic genome. Mol Biol Evol 1995; 12:1-6.
23. Gupta RS. Protein phylogenies and signature sequences: A reappraisal of evolutionary relationships among archaebacteria, eubacteria, and eukaryotes. Microbiol Mol Biol Rev 1998; 62:1435-1491.
24. Woese CR. Bacterial evolution. Microbiol Rev 1987; 51:221-271.
25. Ahn VE, Lo EI, Engel CK et al. A hydrocarbon ruler measures palmitate in the enzymatic acylation of endotoxin. Embo J 2004; 23:2931-2941.
26. Bishop RE. The lipid A palmitoyltransferase PagP: Molecular mechanisms and role in bacterial pathogenesis. Mol Microbiol 2005; 57:900-912.
27. Pautsch A, Schulz GE. High-resolution structure of the OmpA membrane domain. J Mol Biol 2000; 298:273-282.
28. Hwang PM, Choy WY, Lo EI et al. Solution structure and dynamics of the outer membrane enzyme PagP by NMR. Proc Natl Acad Sci USA 2002; 99:13560-13565.
29. Hwang PM, Bishop RE, Kay LE. The integral membrane enzyme PagP alternates between two dynamically distinct states. Proc Natl Acad Sci USA 2004; 101:9618-9623.
30. Miller S, Lesk AM, Janin J et al. The accessible surface area and stability of oligomeric proteins. Nature 1987; 328:834-836.
31. Desmyter A, Spinelli S, Payan F et al. Three camelid VHH domains in complex with porcine pancreatic alpha-amylase. Inhibition and versatility of binding topology. J Biol Chem 2002; 277:23645-23650.
32. McIntyre TM, Bell RM. Escherichia coli mutants defective in membrane phospholipid synthesis: Binding and metabolism of 1-oleoylglycerol 3-phosphate by a plsB deep rough mutant. J Bacteriol 1978; 135:215-226.
33. Hsu L, Jackowski S, Rock CO. Uptake and acylation of 2-acyl-lysophospholipids by Escherichia coli. J Bacteriol 1989; 171:1203-1205.
34. Hsu L, Jackowski S, Rock CO. Isolation and characterization of Escherichia coli K-12 mutants lacking both 2-acyl-glycerophosphoethanolamine acyltransferase and acyl-acyl carrier protein synthetase activity. J Biol Chem 1991; 266:13783-13788.
35. Snijder HJ, Ubarretxena-Belandia I, Blaauw M et al. Structural evidence for dimerization-regulated activation of an integral membrane phospholipase. Nature 1999; 401:717-721.
36. Jia W, Zoeiby AE, Petruzziello TN et al Lipid trafficking controls endotoxin acylation in outer membranes of Escherichia coli. J Biol Chem 2004; 279:44966-44975.
37. Pluckthun A, Dennis EA. Acyl and phosphoryl migration in lysophospholipids: Importance in phospholipid synthesis and phospholipase specificity. Biochemistry 1982; 21:1743-1750.
38. Anliker B, Chun J. Lysophospholipid G protein-coupled receptors. J Biol Chem 2004; 279:20555-20558.
39. Prescott SM, Zimmerman GA, Stafforini DM et al. Platelet-activating factor and related lipid mediators. Annu Rev Biochem 2000; 69:419-445.
40. Morais Cabral JH, Atkins GL, Sanchez LM et al. Arachidonic acid binds to apolipoprotein D: Implications for the protein's function. FEBS Lett 1995; 366:53-56.
41. Handa N, Terada T, Doi-Katayama Y et al. Crystal structure of a novel polyisoprenoid-binding protein from Thermus thermophilus HB8. Protein Sci 2005; 14:1004-1010.
42. Thompson JD, Higgins DG, Gibson TJ. CLUSTAL W: Improving the sensitivity of progressive multiple sequence alignment through sequence weighting, position-specific gap penalties and weight matrix choice. Nucleic Acids Res 1994; 22:4673-4680.
43. Page RD. TreeView: An application to display phylogenetic trees on personal computers. Comput Appl Biosci 1996; 12:357-358.
44. Altschul SF, Gish W, Miller W et al. Basic local alignment search tool. J Mol Biol 1990; 215:403-410.
45. Nielsen H, Brunak S, von Heijne G. Machine learning approaches for the prediction of signal peptides and other protein sorting signals. Protein Eng 1999; 12:3-9.

Chapter 5

Plant Lipocalins

Jean-Benoit F. Charron and Fathey Sarhan*

Abstract

Lipocalins are widely distributed in animals, insect and bacteria but very little is known about plant lipocalins. The first lipocalin-like proteins reported in plants were the two key enzymes of the xanthophyll cycle, the violaxanthin de-epoxidases and the zeaxanthin epoxidases. However, the peculiar architecture of these proteins raised doubt as of their true belonging to the lipocalin family. We recently reported the identification and cloning of the first true plant lipocalins from wheat and *Arabidopsis*. The encoded proteins were named temperature-induced lipocalins and possess the three structurally-conserved regions that characterize lipocalins. Sequence analyses revealed that these plant lipocalins share significant homology with three evolutionarily-related lipocalins, the mammalian apolipoprotein D, the bacterial lipocalin Blc and the insect Lazarillo protein. Data mining of genomic databases and bioinformatic predictions revealed that plants possess two other lipocalin members: temperature-induced lipocalin-2 and chloroplastic lipocalin. Expression and regulation studies suggest that the plant lipocalins are associated with environmental stresses.

Introduction

Lipocalins are an ancient and functionally diverse family of mostly extracellular proteins.[1] This family has been studied in details in bacteria, invertebrates and vertebrates, and these studies have been summarized in several excellent reviews.[2-5] However, very little is known about plant lipocalins.[6-7] The rapidly expanding area of functional, structural and comparative genomics provides opportunities for the identification of lipocalin homologs in plants. Using an integrated approach of data mining of EST databases, bioinformatics predictions, phylogenetic studies, and structural, cellular localization and expression profiling analyses, we identified novel plant lipocalins. Here we describe the molecular characterization and evolution of plant lipocalins and discuss their putative function during plant development under environmental stresses.

Temperature-Induced Lipocalins

The first true plant lipocalins were recently identified from wheat and *Arabidopsis thaliana*.[7] A full length clone was first isolated from a cDNA library prepared from cold-acclimated wheat tissues and named *TaTIL* for *Triticum aestivum* temperature-induced lipocalin. This gene has since been renamed *TaTIL-1*. The open reading frame encodes a protein of 190 amino acids (aa) with a calculated molecular mass of 22 kDa and a theoretical pI of 5.5 (Table 1). A search in the GenBank ESTs database revealed homology (74% identity, 83% similarity) with a predicted putative protein from *Arabidopsis thaliana* that

*Corresponding Author: Fathey Sarhan—Université du Québec à Montréal, Département des Sciences Biologiques, C.P. 8888, Succ. Centre-ville, Montréal QC, H3C 3P8, Canada. Email: sarhan.fathey@uqam.ca

Lipocalins, edited by Bo Åkerström, Niels Borregaard, Darren R. Flower and Jean-Philippe Salier. ©2006 Landes Bioscience.

Table 1. Structural features of plant lipocalins and lipocalin-like proteins

Protein	Precursor/ Mature Molecular Mass (kDa)	Subcellular Localisation	Cleavage Site Position*	SCR1 GxWY	SCR2 TDY	SCR3 R	Conserved Cys Residues	Conserved N-glycosyl. Sites	Other Domains
*At*TIL-1	21 / 20	membrane	C-terminal	yes	D only	yes	0	1	no
*Os*TIL-1	22 / 20	membrane	C-terminal	yes	D only	yes	0	1	no
*Ta*TIL-1	22 / 20	membrane	C-terminal	yes	D only	yes	0	1	no
*Os*TIL-2	21 / 19	ND	C-terminal	yes	D only	yes	0	1	no
*At*CHL	39 / 26	chloroplast	N-terminal	yes	yes	yes	8	0	no
*Os*CHL	37 / 26	chloroplast	N-terminal	yes	yes	yes	8	0	no
*At*VDE	52 / 40	chloroplast	N-terminal	yes	no	yes	14	1	yes**
*Os*VDE	50 / 40	chloroplast	N-terminal	yes	no	yes	14	1	yes**
*Ta*VDE	52 / 40	chloroplast	N-terminal	yes	no	yes	14	0	yes**
*At*ZEP	74 / 68	chloroplast	N-terminal	yes	no	no	6	1	yes***
*Os*ZEP	68 / 63	chloroplast	N-terminal	yes	no	no	5	1	yes***

At, Arabidopsis thaliana; Ta, Triticum aestivum (wheat); *Os, Oryza sativa* (rice); Cys, Cysteine; ND, not determined. * C-terminal, GPI anchor site; N-terminal, signal peptide. ** N-terminal cyteine-rich region and C-terminal glutamic acid-rich region. *** N-terminal ADP-binding site and C-terminal FAD-binding site we named *At*TIL for *Arabidopsis thaliana* temperature-induced lipocalin. Sequence analysis of this *Arabidopsis* clone revealed that the cDNA encodes a 186 aa protein. The SCR1 region is located from aa 15 to 31 (GLDVARYMGRWYEIASF) in *Ta*TIL-1 and from aa 12 to 28 (GLNVERYMGRWYEIASF) in *At*TIL, and possesses the two conserved amino acids G and W (Table 1).[8-9] The SCR2 of *Ta*TIL-1 is found in the C-terminal portion of the protein from aa 105 to 119 (YWVLYVDDDYQYALV) while in *At*TIL it is found from aa 101 to 115 (YWVLYIDPDYQHALI). The SCR2 of animal and bacterial lipocalins generally contains a TDY triplet.[8-9] However, in *Ta*TIL-1 and *At*TIL, only the central D is present (Table 1). SCR3 is found in the C-terminal portion of both proteins, from aa 129 to 144 (ILCRKTHIEEEVNQL) in *Ta*TIL-1 and from aa 125 to 140 in *At*TIL (ILSRTAQMEEETYKQL). The conserved R residue that characterizes this fingerprint is present in both sequences (Table 1).[8-9] Further sequence analysis of *Ta*TIL-1 and *At*TIL indicated the presence of a putative N-glycosylation site (Table 1). Putative C-terminal cleavage sites are predicted by several targeting peptide prediction programs (DGPI, PSORT, and SignalP) to be at aa 172 in *Ta*TIL-1 and at aa 168 in *At*TIL.[10-11] Considering this putative cleavage site, the calculated molecular mass of the mature proteins in wheat and *Arabidopsis* is 20 kDa with a pI of 5.2 (Table 1).

The homology search revealed that *Ta*TIL-1 (accession no. AY077702) and its ortholog from *Arabidopsis* (accession no. AY062789) share significant similarity with three evolutionarily related lipocalins: the human apolipoprotein D (ApoD) precursor (accession no. P05090), the *Escherichia coli* outer membrane lipoprotein Blc precursor (accession no. P39281), and the American grasshopper Lazarillo precursor (accession no. P49291). These proteins respectively share 29%, 31%, and 23% identity, and 46%, 54% and 40% similarity with *Ta*TIL-1. Among all lipocalins, Blc, ApoD, and Lazarillo are the only ones known to be anchored to biological membranes.[3] The good similarity between these proteins and the plant TILs suggests that *Ta*TIL-1 and *At*TIL are also membrane-associated proteins. The sequence analysis also revealed that, like the *E. coli* Blc, *Ta*TIL-1 and *At*TIL differ from most lipocalins by the absence of intramolecular disulfide bonds. However, they are potentially N-glycosylated like human ApoD and Lazarillo. When the three SCRs of these five proteins are aligned, the start codons from *Ta*TIL-1 and *At*TIL are positioned at the cleavage sites of the N-terminal signal peptides of the three other proteins. This alignment suggests that *Ta*TIL-1 and *At*TIL do not possess an N-terminal signal peptide as is the case in Blc, ApoD and Lazarillo. The N-terminal portion of *Ta*TIL-1 is composed of hydrophilic residues followed by few hydrophobic residues. In *At*TIL, the hydrophobic section is even less accentuated. This profile does not fit the standard hydrophobic nature of the N-terminal signal peptide identified in ApoD, Blc and Lazarillo. Like Lazarillo, the *Ta*TIL-1 and *At*TIL proteins are longer than ApoD and Blc at their C-terminal end and possess a similar putative cleavage site. The hydrophobic C-terminal tail enables Lazarillo to receive a glycosylphosphatidylinositol (GPI) anchor.[12] This suggests that *Ta*TIL-1 and *At*TIL could also receive a GPI anchor. GPI anchoring is a post-translational addition of a lipid occurring in the endoplasmic reticulum lumen which links proteins to the external face of the plasma membrane. This type of modification has been reported in plants.[13] The fact that the N-glycosylation site is conserved between the wheat and *Arabidopsis* TIL orthologs supports the possibility that these proteins are processed in the endoplasmic reticulum lumen. Another type of attachment to the membrane can also be suggested for *Ta*TIL-1 and *At*TIL. It has been proposed that human ApoD is associated with the external face of the membrane by a hydrophobic loop.[3,14-15] *Ta*TIL-1 and *At*TIL also possess a hydrophobic stretch of seven amino acids that is inserted into a loop between two β-strands. This hydrophobic stretch is in the loop between β-strands 5 and 6 instead of being in the loop between strands 7 and 8, as is the case in the human ApoD (Fig. 1 B1,C2). It is nevertheless possible that this stretch favours the attachment of TILs to the plasma membrane. The loop scaffold in *Ta*TIL-1 and *At*TIL is two amino acids longer than in the human ApoD and there is a proline at positions 32 and 29 respectively. These modifications suggest that the plant TILs have a different binding specificity. A recent proteomic analysis of highly purified plasma membranes from *Arabidopsis* showed

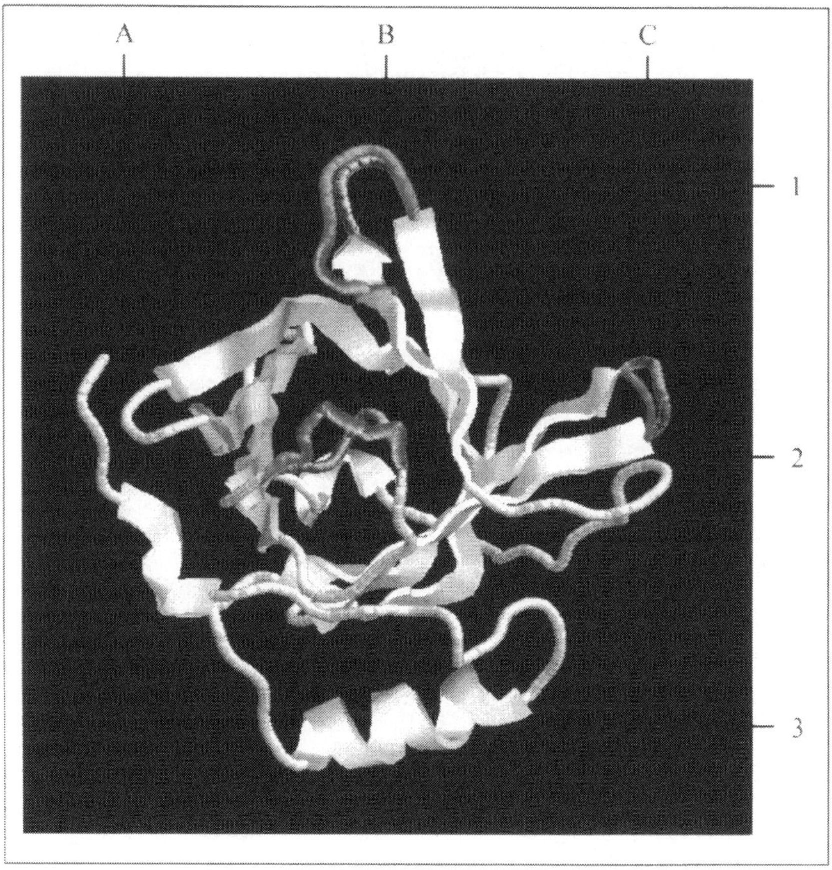

Figure 1. Structural models of human ApoD and wheat *Ta*TIL-1. Tertiary structure analyses were carried out using the Swiss-Model program.[16] The lower BLAST limit was set at 0.00001 and the human ApoD model (PDB ID: 2APD)[14] was used as template. The initial result was then resubmitted through the optimizing mode of the program. The final result was then visualized using the Swiss-Pdb Viewer and the model was adapted according to sequence comparison. Differences between the wheat and the human models were superimposed and colored. Grey sections are common to both models. The red (*Ta*TIL-1) and blue (ApoD) sections represent structural differences between the two proteins. Reprinted with permission from: Frenette Charron JB, Breton G et al. FEBS Lett 2002; 517(1-3):129-132. ©2002 Published by Elsevier Science B.V. on behalf of the Federation of European Biochemical Societies.

that *At*TIL is associated with this membrane fraction.[17] This result confirms the prediction that *Ta*TIL-1 and *At*TIL are membrane-associated proteins. However, the nature of the association or attachment is still unknown.

Northern blot analysis revealed that the *TaTIL-1* transcripts accumulate to high levels upon exposure to low temperature and heat-shock treatments (10-fold) and to a lesser extent after a water stress (3.5-fold).[7] Abscisic acid, high salt and wounding treatments have no measurable effect. The *TaTIL-1* transcripts accumulate gradually to a maximum level after 36 days of cold acclimation. Upon deacclimation, the level of transcripts returns to the level seen in the control nonacclimated plants. The accumulation of *TaTIL-1* transcripts in wheat was found to be tissue-specific, as they were detected only in cold-acclimated leaves. The expression analyses revealed that the dicot ortholog *AtTIL* is also induced by low temperature (6-fold) and heat-shock treatments (9-fold). RNA blot hybridization studies also demonstrated that cold acclimation induces the accumulation of *TaTIL-1* transcripts in both less tolerant and cold hardy wheat. However, this increase is greater in the hardy winter cultivars. Low levels of expression are also found in oat and barley, two less cold tolerant species. This difference in accumulation indicates that the *TaTIL-1* expression is correlated with the plant's capacity to develop freezing tolerance.

Analysis of the promoter regions of *AtTIL*, *TaTIL-1*, *OsTIL-1* and *OsTIL-2* revealed the presence of several low temperature response elements (LTREs), dehydration response elements (DREs) and heat shock elements (HSEs). *TaTIL-1* and *AtTIL* promoter sequences contain more LTREs than the *OsTIL-1* promoter sequence. On the other hand, the *OsTIL-1* promoter contains more HSEs than the *AtTIL* and *TaTIL-1* promoters. This situation is not unexpected since rice does not have the ability to cold acclimate but possesses a higher thermotolerance than wheat and *Arabidopsis*. The fact that TIL promoters possess several light-responsive elements supports the specific expression of the corresponding genes in green photosynthetic leaves.

Temperature stresses are known to induce membrane injuries.[17] The membrane-anchored lipocalins (Blc, ApoD, Lazarillo, and possibly *Ta*TIL-1 and *At*TIL) all appear to be expressed in response to conditions that cause membrane stresses, which suggests a biological role in membrane biogenesis and/or repair under severe stress conditions.[3] The plant *Ta*TIL-1 and *At*TIL proteins, like the human ApoD, may bind a wide variety of potential ligands of varying structures and functions. The mammalian ApoD is reported to bind arachidonic acid, bilirubin, steroid hormones (progesterone and pregnenolone) and cholesterol.[4] It is interesting to mention that plants also synthesize a wide variety of steroid hormones called brassinosteroids. A treatment with 24-epibrassinolide, a brassinosteroid, increases the tolerance of plants to heat and cold stresses.[18] The enhanced resistance to temperature stress is attributed to increased membrane stability and osmoregulation. It is known that sterol insertion in the plasma membrane increases its fluidity at low temperature and maintains the phospholipids order at high temperature.[19] *Ta*TIL-1 may be involved in the transport of these sterol molecules to the membrane in response to stress conditions.

Other Plant Lipocalins

Since plant lipocalins were last reviewed, the sequencing of the *Arabidopsis thaliana* and *Oryza sativa* (rice) genomes has been completed.[6,20-21] The newly identified *Ta*TIL-1 and *At*TIL proteins were used to search the proteins predicted from the DNA sequence information of these two genomes using the BLAST program. The search revealed that rice possesses two other lipocalin members, TIL-2 and CHL. Sequence analysis revealed the presence of two different genes in rice encoding TIL lipocalins: *OsTIL-1* and *OsTIL-2* on chromosomes 2 and 8, respectively, whereas *Arabidopsis thaliana* has only *AtTIL* on chromosome 5. The *Os*TIL-1 and *Os*TIL-2 proteins share 65% identity and 80% similarity. *Os*TIL-2 is a protein of 179 aa with a calculated molecular mass of 21 kDa (Table 1). The absence of a N-terminal target peptide suggests that the *Os*TIL-2 protein would, like *Os*TIL-1, accumulate in the cytosol. Further sequence analysis of the wheat and rice TIL-2 proteins indicated the presence of a conserved putative N-glycosylation site. In addition, a putative C-terminal cleavage site is predicted by several target peptide prediction programs: DGPI, PSORT,[10] and SignalP.[11] Considering this putative cleavage site, the calculated molecular mass of the mature *Os*TIL-2 protein is 19 kDa.

The second new member identified from *Arabidopsis* and rice was named CHL (for chloroplastic lipocalin). This protein was identified in *Arabidopsis* as a putative lipocalin (CAB41869).[6] An homology search revealed that *At*CHL shares only 23% identity and 40% overall similarity with *At*TIL. However, a region of 16 amino acids corresponding to SCR1 shows a high similarity with TIL lipocalins. The encoded mature proteins in *Arabidopsis* and rice are respectively 314 aa and 322 aa long with calculated molecular masses of 35 and 36 kDa (Table 1). SignalP and ChloroP predict N-terminal chloroplastic targeting peptides with high scores in both proteins (Table 1).[11,22] However, the exact length of the chloroplast transit peptide and the location of the proteins within the chloroplast is still unknown. A pairwise sequence alignment predicts chloroplast transit peptide cleavage sites near the beginning of SCR1 in both *At*CHL and *Os*CHL sequences. The mature CHL proteins would have a molecular mass of 26 kDa, which is approximately the usual lipocalin size (Table 1). CHL proteins also possess 8 conserved cysteine residues probably involved in the three-dimensional structure of the protein by forming disulfide bridges. Motif searches against the PROSITE database,[23] after exclusion of patterns with a high probability of occurrence, revealed that *Arabidopsis* and rice CHL proteins possess the SCR1 lipocalin signature (Table 1). This signature perfectly fits the SCR1 consensus used by the ScanProsite software and exhibits the two invariant amino acids G and W that are key features of SCR1.[8-9,24] As in most lipocalins, CHL SCR2 is found in the C-terminal half of the protein and bears the conserved TDY triplet (Table .1).[8-9] SCR3 is also found in the C-terminal portion of both proteins and the conserved R residue that characterizes this fingerprint is present (Table 1).[8-9]

Violaxanthin De-Epoxidases and Zeaxanthin Epoxidases

Violaxanthin de-epoxidases (VDEs) and zeaxanthin epoxidases (ZEPs) are the most puzzling members with regards to their classification as plant lipocalins. The size and the exon–intron architecture of the genes encoding these enzymes show no significant similarity to the genomic organization of bacterial and animal lipocalin genes and for these reasons, they were not considered as true lipocalins in most studies.[25-26] These enzymes are involved in photoprotection of the photosynthetic apparatus, and are first synthesized as precursor proteins that bear the transit peptide needed for translocation to the thylakoid space of chloroplasts.[6,27] They share the common substrate antheraxanthin and are believed to exhibit similar tertiary structure.[6] VDEs are predicted to be proteins with a central barrel structure flanked by a cysteine-rich N-terminal domain and a glutamate-rich C-terminal domain (Table 1).[28] ZEPs possess ADP-binding and FAD-binding domains and fit the description of a lipocalin based on SCR1 homology (Table 1). Functional analyses of the different domains of VDEs demonstrated that the deletion of any of the cysteine residues in the N-terminal region resulted in a total loss of activity.[28] This is likely because cysteine residues allow the formation of disulfide bridges, which are important determinants of protein conformation. It thus appears that the conformation of the mature protein in the N-terminal portion of VDEs is essential to retain their activities. Deletion analysis of the C-terminal region demonstrated that 71 out of 98 aa could be removed without any loss of activity.[28] However, removal of another 12 aa resulted in a 90% loss of activity and an important reduction of the binding of VDEs to the thylakoid membrane.[28]

Given the feature of VDEs and ZEPs and the strict definition of lipocalins, it is difficult to unequivocally consider these two proteins as true lipocalins. They are at best lipocalin-like proteins that could have arisen from the fusion of an ancestral plant lipocalin to proteins with enzymatic functions.[26,29] Thus, VDEs and ZEPs may represent the first examples of lipocalins evolution towards the acquisition of novel functions.

Evolutionary Origin of Plant Lipocalins and Lipocalin-Like Proteins

To help elucidate the evolutionary origin of plant lipocalins, we investigated the presence of lipocalins and lipocalin-like proteins in algae and cyanobacteria. Algae are considered primitive photosynthetic eukaryotes while cyanobacteria carry a complete set of oxygenic photosynthetic genes. The chloroplast is believed to have evolved from the endosymbiosis of a cyanobacterial ancestor with a eukaryotic host cell. An homology search performed with the *Ta*TIL-1 protein sequence revealed several ESTs from red algae. The search also revealed that cyanobacteria possess a lipocalin gene.

Phylogenetic analyses suggest that TIL lipocalin members were probably inherited from a bacterial gene present in the original host cell, the common ancestor of plants and animals.[1] In some plant species, the TIL-2 lipocalin may have arisen from the duplication of the gene encoding the TIL-1 lipocalin. However, the remaining plant lipocalin and lipocalin-like members CHLs, VDEs and ZEPs might have evolved from a series of duplication of the cyanobacterial ancestor gene after cyanobacteria endosymbiosis from which the chloroplast originated. VDE and ZEP sequences subsequently diverged and acquired new cellular function as xanthophylls cycle enzymes.

Conclusion

The identification and characterization of plant lipocalins and lipocalin-like proteins will help in designing experiments aimed at the understanding of their cellular function in plants and their role in modulating the responses to temperature and oxidative stresses. Using forward and reverse genetics in the model system *Arabidopsis* should provide the information needed to elucidate the function of each protein in the plant metabolism. In addition, microarray analyses will help in the identification of the target genes associated with over / under expression of the different proteins. The ease with which plants can be manipulated and the availability of mutants are tremendous tools that should enable us to understand the cellular function of lipocalins and lipocalin-like proteins in plants. This information could even help understand the cellular function of lipocalins in mammals.

Acknowledgements

This work was supported by a Natural Sciences and Engineering Research Council of Canada discovery grant, and by Genome Canada, Genome Québec, and Genome Prairie grants to F. Sarhan. We thank Dr. François Ouellet for helpful discussions and editorial help.

References

1. Sánchez D, Ganfornina MD, Gutiérrez G et al. Exon–intron structure and evolution of the lipocalin gene family. Mol Biol Evol 2003; 20(5):775-783.
2. Åkerström B, Flower DR, Salier JP. Lipocalins: Unity in diversity. Biochim Biophys Acta 2000; 1482(1-2):1-8.
3. Bishop RE. The bacterial lipocalins. Biochim Biophys Acta 2000; 1482(1-2):73-83.
4. Rassart É, Bedirian A, Do Carmo S et al. Apolipoprotein D. Biochim Biophys Acta 2000; 1482(1-2):185-198.
5. Sánchez D, Ganfornina MD, Bastiani MJ. Lazarillo, a neuronal lipocalin in grasshoppers with a role in axon guidance. Biochim Biophys Acta 2000; 1482(1-2):102-109.
6. Hieber AD, Bugos RC, Yamamoto HY. Plant lipocalins: Violaxanthin de-epoxidase and zeaxanthin epoxidase. Biochim Biophys Acta 2000; 1482(1-2):84-91.
7. Frenette Charron JB, Breton G, Badawi M et al. Molecular and structural analyses of a novel temperature stress-induced lipocalin from wheat and Arabidopsis. FEBS Lett 2002; 517(1-3):129-132.
8. Flower DR, North AC, Attwood TK. Structure and sequence relationships in the lipocalins and related proteins. Protein Sci 1993; 2(5):753-761.
9. Flower DR. The lipocalin protein family: Structure and function. Biochem J 1996; 318(Pt1):1-14.
10. Nakai K, Horton P. PSORT: A program for detecting sorting signals in proteins and predicting their subcellular localization. Trends Biochem Sci 1999; 24(1):34-36.
11. Nielsen H, Engelbrecht J, Brunak S et al. A neural network method for identification of prokaryotic and eukaryotic signal peptides and prediction of their cleavage sites. Int J Neural Syst 1997; 8(5-6):581-599.
12. Ganfornina MD, Sánchez D, Bastiani MJ. Lazarillo, a new GPI-linked surface lipocalin, is restricted to a subset of neurons in the grasshopper embryo. Development 1995; 121(1):123-134.
13. Morita N, Nakazato H, Okuyama H et al. Evidence for a glycosylinositolphospholipid-anchored alkaline phosphatase in the aquatic plant Spirodela oligorrhiza. Biochim Biophys Acta 1996; 1290(1):53-62.
14. Peitsch MC, Boguski MS. Is apolipoprotein D a mammalian bilin-binding protein? New Biol 1990; 2(2):197-206.

15. Bishop RE, Penfold SS, Frost LS et al. Stationary phase expression of a novel Escherichia coli outer membrane lipoprotein and its relationship with mammalian apolipoprotein D. Implications for the origin of lipocalins. J Biol Chem 1995; 270(39):23097-23103.
16. Peitsch MC. ProMod and Swiss-Model: Internet-based tools for automated comparative protein modelling. Biochem Soc Trans 1996; 24(1):274-279.
17. Kawamura Y, Uemura M. Mass spectrometric approach for identifying putative plasma membrane proteins of Arabidopsis leaves associated with cold acclimation. Plant J 2003; 36(2):141-154.
18. Clouse SD, Sasse JM. Brassinosteroids: Essential regulators of plant growth and development. Annu Rev Plant Physiol Plant Mol Biol 1998; 49:427-451.
19. Demel RA, De Kruyff B. The function of sterols in membranes. Biochim Biophys Acta 1976; 457(2):109-132
20. The Arabidopsis Initiative 2000. Analysis of the genome sequence of the flowering plant Arabidopsis thaliana. Nature 2000; 408:796-815.
21. Yu J, Hu S, Wang J et al. A draft sequence of the rice genome (Oryza sativa L. ssp. indica). Science 2002; 296(5565):79-92.
22. Emanuelsson O, Nielsen H, von Heijne G. ChloroP, a neural network-based method for predicting chloroplast transit peptides and their cleavage sites. Protein Sci 1999; 8(5):978-984.
23. Sigrist CJ, Cerutti L, Hulo N et al. PROSITE: A documented database using patterns and profiles as motif descriptors. Brief Bioinform 2002; 3(3):265-274.
24. Gattiker A, Gasteiger E, Bairoch A. ScanProsite: A reference implementation of a PROSITE scanning tool. Appl Bioinformatics 2002; 1(2):107-108.
25. Gutiérrez G, Ganfornina MD, Sánchez D. Evolution of the lipocalin family as inferred from a protein sequence phylogeny. Biochim Biophys Acta 2000; 1482(1-2):35-45.
26. Salier JP. Chromosomal location, exon/intron organization and evolution of lipocalin genes. Biochim Biophys Acta 2000; 1482(1-2):25-34.
27. Bugos RC, Hieber AD, Yamamoto HY. Xanthophyll cycle enzymes are members of the lipocalin family, the first identified from plants. J Biol Chem 1998; 273(25):15321-15324.
28. Hieber AD, Bugos RC, Verhoeven AS et al. Overexpression of violaxanthin de-epoxidase: Properties of C-terminal deletions on activity and pH-dependent lipid binding. Planta 2000; 214(3):476-483.
29. Ganfornina MD, Gutiérrez G, Bastiani M et al. A phylogenetic analysis of the lipocalin protein family. Mol Biol Evol 2000; 17(1):114-126.

Chapter 6

Lipocalins in Arthropoda:
Diversification and Functional Explorations

María D. Ganfornina,* Hartmut Kayser and Diego Sanchez

Introduction

The number of sequenced arthropodan lipocalins adds up to over eighty (see Table1). From our currently fragmented knowledge of arthropodan genomes, the last common ancestor of this phylum is proposed to possess two lipocalins (see Chapter 2). Intra-lineage duplications enlarged the number of lipocalins, with some large family expansions occurring independently in blood-feeding insects and arachnids.

Most arthropodan lipocalins share the common signature and structural properties with the rest of the family. They are single modular proteins of around 200 amino acids that fold tightly in a β-barrel with potential for binding small hydrophobic molecules in a central pocket. They usually have two α-helices at the N- and C-terminal sides, packed against the outer surface of the barrel. Nonetheless, we will encounter in this phylum variations around this main theme. While maintaining the lipocalin fold, the arthropodan lipocalins call our attention because of the panoply of interesting structural and functional explorations that have arisen throughout evolution. Because of the recent extensive review on arthropodan lipocalins by Kayser,[1] with special emphasis in their structure and function, it is our aim to offer a review focusing on the peculiarities of these lipocalins while framing them in an evolutionary context.

As a first step of analysis, we can simply look at an alignment of amino acid sequences of arthropodan lipocalins. A schematic representation of such an alignment is shown in Figure 1. The location of all important gaps is coincident with the loops (L) of the β-barrel. While L1, L4 and L6 are relatively well conserved, all other loops accommodate significant variations in size. Particularly large extensions in L5 are present in one of the Pallidipin sequences, and in other lipocalins related to Nitrophorins (see below). Other expanded loops (L3 and L7) are also expected in these lipocalins. These three loops (L3, L5 and L7) face the entrance of the binding pocket ('the open side of lipocalins', see Chapter 3), and their variations could condition the ligand binding properties of these lipocalins. Other lipocalins such as the Drosophila lipocalins GLaz and Karl show unique extensions in L2. Because this loop faces the closed side of the β-barrel, it is most probably involved in protein-protein interactions. Loop L3 of GLaz is also long, but there is no sequence similarity with the L3 loop of *R. prolixus* Pallidipin.

Variations are also observed in the length of the N- or C- terminal segments. Most lipocalins can be classified roughly in tree groups, which differ in the length of their C-termini after the last cysteine residue. The longest of these C-terminal extensions have a size similar to the hydrophobic GPI anchoring signal unique to the grasshopper Lazarillo (shown as a zigzag line in Fig. 1). However, the sequence signatures for GPI anchorage are absent in all other cases. In

*Corresponding Author: María D. Ganfornina—Departamento de Bioquímica y Biología Molecular y Fisiología-IBGM, Universidad de Valladolid-CSIC, 47005 Valladolid, Spain. Email: opabinia@ibgm.uva.es

Lipocalins, edited by Bo Åkerström, Niels Borregaard, Darren R. Flower and Jean-Philippe Salier. ©2006 Landes Bioscience.

Table 1. Summary of properties of all known arthropodan lipocalins

Taxon	Protein Name (*)	Species	Expression/ Occurrence	Function	Ligand	Protein Interactors and Other Properties	Refs.
Crustacea-Decapoda							
	Crustacyanins (2)						
	α-Crustacyanin: two genes: Crustacyanin 1 Crustacyanin 2	*Homarus gammarus* (lobster)	Epidermis (carapace)	Camouflage, carapace colour regulation	Astaxanthin (2 molecules per dimer); strong bathochromic shift	Homophilic interactions: 16-mer (α-crustacyanin) composed of 8 dimers (β-crustacyanin). Crystal structures of four dimers known.	16
Hexapoda-Insects							
	Biliproteins and related lipocalins (9)						
	Bilin binding protein	*Pieris brassicae* (butterfly)	Larval epidermis; dorsal fat body sheet; wings; hemolymph.	Camouflage? CO-signaling? Prevention of cellular damage by reactive oxygen and nitrogen species?	Biliverdin IXγ	Homophilic interactions (dimer). Crystal structure known.	25,26, 29
	Insecticyanin, two genes: IcyA and IcyB	*Manduca sexta* (moth)	Larval epidermis; fat body; hemolymph; oocytes	Camouflage in larva	Biliverdin IXγ	Homophilic interactions (tetramer). Crystal structure known.	38,41
	Biliverdin-binding proteins, BBP-I and BBP-II	*Samia cynthia ricini* (silkmoth)	Larval epidermis	Camouflage in larva	Biliverdin IXγ	Homophilic interactions (dimer of BBP-II)	46
	Gallerin	*Galleria mellonella* (wax moth)	Central nervous system; also in fat body	Unknown	Unknown		47
	Bombyrin	*Bombyx mori* (silkmoth)	Brain	Unknown	Unknown		48

continued on next page

Table 1. Continued

Taxon	Protein Name (*)	Species	Expression/ Occurrence	Function	Ligand	Protein Interactors and Other Properties	Refs.
	Lipocalins 1 and 4	*Lonomia oblique* (silkmoth)	Spicule; integument	Lip1:Hemorrhagic syndrome Lip4:Prothrombin activator with serine-protease activity.	Unknown		49,50
Hexapoda-Insects							
	Lazarillo/ApoD related proteins (10)						
Orthoptera	Lazarillo	*Schistocerca americana*	Subset of neurons and neuroblasts; Malpighian tubules; sessile mesodermal cells; nephrocytes.	Axonal growth and guidance	Unknown	Homophilic interactions. Membrane bound: GPI-linked	7,8
Diptera	Glial Lazarillo	*Drosophila melanogaster*	Subset of glia and glioblasts; developing gut; salivary glands.	Oxidative stress balance and aging	Unknown		14
	Neural Lazarillo	*Drosophila melanogaster*	Subset of neurons; amnioserosa; fat body; developing gut.	Oxidative stress balance and aging	Unknown		14
	Karl	*Drosophila melanogaster*	Hemocytes (plasmatocytes); lymph glands.	Unknown	Unknown		15
	Dmel.Lip4	*Drosophila melanogaster*	Unknown	Unknown	Unknown	Similar to Lazarillo	

continued on next page

Table 1. Continued

Taxon	Protein Name (*)	Species	Expression/ Occurrence	Function	Ligand	Protein Interactors and Other Properties	Refs.
	Agam.Lip1 and Lip2	*Anopheles gambiae* (mosquito)	Unknown	Unknown	Unknown	Similar to Lazarillo Lip2 similar to Nlaz.	
Hymenoptera	Amel.Lip1 and 2	*Apis mellifera* (honeybee)	Unknown	Unknown	Unknown	Lip1 similar to Karl. Lip2 similar to Glaz.	
Lepidoptera	Hyphantrin	*Hyphantria cunea* (silkmoth)	Epidermis (pupa)	Unknown	Unknown	Similar to Glaz	
Hexapoda-Insects							
	Insect saliva lipocalins (up to 31)						
Hemiptera	Nitrophorins (up to 11 described)	*Rhodnius prolixus* (blood sucking insect)	Salivary glands	Enhancement of vasodilation. Reduction of immune and inflammation response. Interference with wound healing and blood coagulation cascades. Thiol-oxidase activity. Heme-peroxidase activity.	Heme-NO Heme-Histamine	Temporal regulation through larval and adult stages	51,65, 66
				NP2: Interference with blood coagulation.		Interaction with Factor Xase	61
				NP7: Interference with prothrombinase by binding to activated platelet cell membrane.		Interaction with cell membranes containing phosphatidylserine	73

continued on next page

Table 1. Continued

Taxon	Protein Name (*)	Species	Expression/ Occurrence	Function	Ligand	Protein Interactors and Other Properties	Refs.
	BABP (Biogenic amine binding protein)	*Rhodnius prolixus*	Salivary glands	Vasodilation and platelet aggregation inhibitor.	Serotonin, epinephrine, norepinephrine.		75
	RPAIs (up to 7)	*Rhodnius prolixus*	Salivary glands	Anti-coagulation factors: blocking collagen- and ADP-induced platelet aggregation.	ADP (High affinity)	Binding of two ADP molecules in a single site	65, 76
	Pallidipins (3)	*Triatoma pallidipennis, T. brasiliensis, Rhodnius prolixus*	Salivary glands	Inhibition of collagen-induced platelet aggregation.			81
	Triabins (6)	*Triatoma pallidipennis, T. brasiliensis, Rhodnius prolixus*	Salivary glands	Inhibition of thrombin		Interaction with thrombin	78
	Procalin (3)	*Triatoma pallidipennis, T. brasiliensis, Rhodnius prolixus*	Salivary glands	Allergen			83
Hexapoda-Insects							
	Other lipocalins related to insect saliva lipocalins (3)						
Diptera	Dyak.Lip	*Drosophila yakuba*					
	Aaeg.Lip	*Aedes aegipti* (mosquito)	Hemocytes		Unknown		54

continued on next page

Table 1. Continued

Taxon	Protein Name (*)	Species	Expression/ Occurrence	Function	Ligand	Protein Interactors and Other Properties	Refs.
Dictyoptera	Bger.All4	*Blatella germanica* (cockroach)		Allergen	Unknown		97
	Dpun.Milk	*Diploptera punctata*	Brood sack epithelium	Nutrition for embryos	Cholesterol?		96
	Lmad.Lip	*Leucophaea maderae*	Male tergal glands	Aphrodisiac			98
Chelicerata, Arachnida							
	Arachnida saliva lipocalins (up to 20)						
Acari	HBP1-3 (Histamine binding protein)	*Rhipicephalus appendiculatus* (hard tick)	Salivary glands. Stage and sex specific.	Supression of inflammation during blood-feeding	Histamine. Two hydrophilic binding sites.	Major reconstruction of the closed end of lipocalins.	86
	SHBP (Serotonin and histamine binding protein)	*Dermacentor reticularis* (hard tick)	Salivary glands		Serotonin + Histamine. Two ligand binding sites.		88
	Isca.Lips (up to 11)	*Ixodes scapularis* (hard tick)	Salivary glands				85
	TSGP1-4 (tick salivary gland protein)	*Ornithodoros savignyi* (soft tick)	Salivary glands	Salivary gland granule biogenesis. TSGP-2 and 4: Toxins (host cardiovascular system)			85, 91
	Moubatin	*Ornithodoros moubata* (soft tick)	Saliva	Inhibition of collagen-induced platelet aggregation			89, 90

* Number of full-lengh proteins

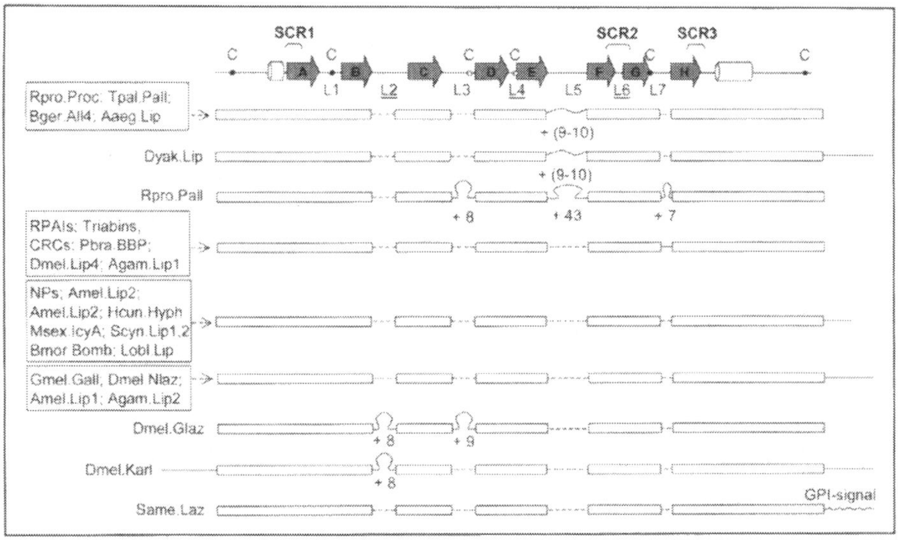

Figure 1. Amino acid sequence alignment of arthropodan lipocalins. The alignment was performed with Clustal X (1.8)[99] using a Gonnet series scoring matrix. We used a gap penalty mask based on the resolved secondary structure of some lipocalins, represented schematically by arrows (β strands) and cylinders (α helices). The known N-terminal signal peptides were not included. The sections aligned without gaps or with small gaps (one to two residues) are shown as continuous boxes. Large gaps are represented by dashed lines, and the sequence fragments unique to one or a group of lipocalins are depicted as continuous lines. Cysteine residues conserved in all arthropodan lipocalins are represented by filled dots. Open dots represent cysteines only present in some saliva lipocalins of hematophagous insects.

addition, Karl shows a uniquely long N-terminal extension, while some of the insect saliva lipocalins are shorter than average (not shown).

Therefore, we find in arthropods a wide range of variation in particular loops, and in the N- and C-terminal regions of the proteins. These changes are not expected to alter significantly the β-barrel core, and thus provide a substrate of variation for potential functional diversification and specialization.

Using the amino acid sequences, we have reconstructed a phylogeny of arthropodan lipocalins (Fig. 2) including the group of chordate lipocalins most related to them (Apolipoprotein D, ApoD), and using lipocalins from other kingdoms (plants, fungi and protoctists lipocalins) as the outgroup. However, the arachnid saliva lipocalins were not included. Because of their highly divergent sequences, the reliability of the alignment and the tree-reconstruction methods decreases due to long-branch attraction artifacts.[2] Instead, we have superimposed the branch of arachnid saliva lipocalins (dashed lines) as deduced from the analysis of their intron-exon structure[3] (see Chapter 2).

Several groups of arthropodan lipocalins are supported in the tree, and these relationships will guide us through the following sections of the chapter.

At the most basal position stands the group of Lazarillo (Laz) related lipocalins, which, in turn, closely relates to the chordate ApoDs currently found in many species ranging from tunicates to humans. The Lazarillo clade also shows, so far, the widest species representation within insects, with one lipocalin found in Orthoptera and Lepidoptera, and two in several Diptera species (fruitflies and mosquitoes) and hymenoptera (bees). Some of the new sequences retrieved from mosquitoes and bees can be ascribed as orthologous to one of the four lipocalins found in the fruitfly *Drosophila melanogaster*. However, divergence rates are either unequal or very high, so that the sequence of an older Drosophila species, *D. yakuba*, is set apart in the phylogeny.

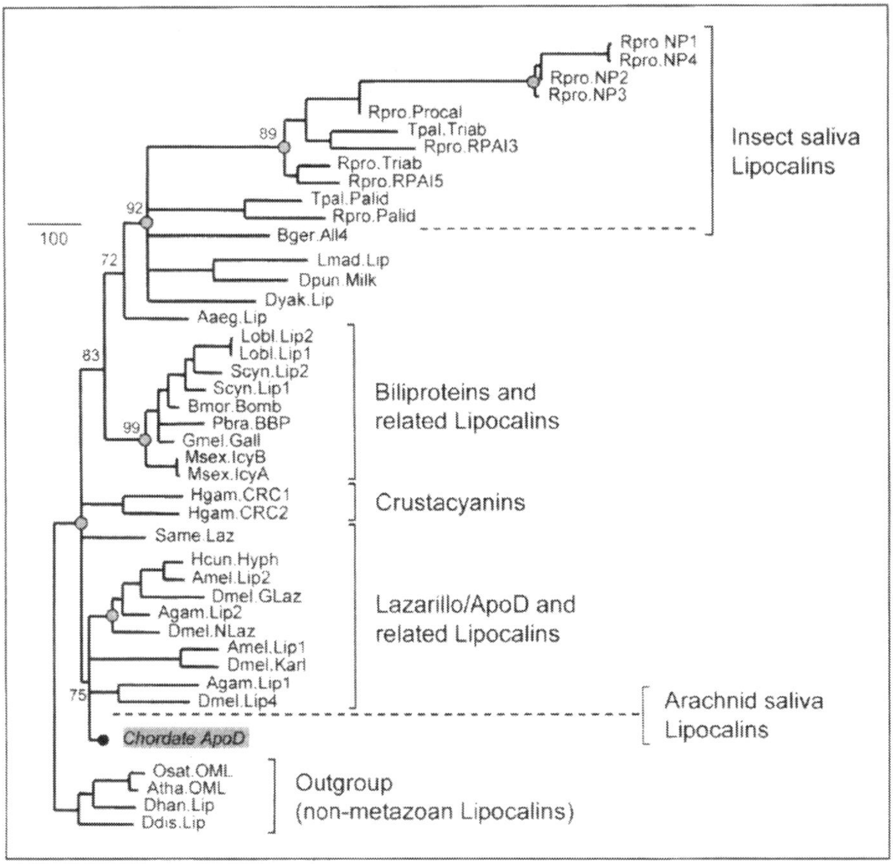

Figure 2. Protein sequence based phylogenetic tree of arthropodan lipocalins and their most related vertebrate group (the Apolipoprotein D) resumed to a single node (black dot). The tree was built by a maximum likelihood-based method that has been used for lipocalin phylogenies.[5] Main grouping nodes with local bootstrap proportion values >70 are shown. Unsupported nodes are excluded, and the branches joined as a polytomy. The scale bar represents the branch length (number of amino acid substitutions/100 residues). The tree was rooted with an outgroup of nonmetazoan lipocalins. The main grouping nodes marked with gray dots are also supported by a tree reconstructed with Mr Bayes with a 50% majority rule consensus. Proteins included are named with abbreviated species names and functional labels.

The group of tick saliva lipocalins might have stemmed out of an ancestor similar to these Lazarillo lipocalins, since their gene structure is most similar to the Drosophila Neural Lazarillo (NLaz).[3] This group represents a big expansion of the lipocalin family through gene duplication and divergence, probably associated to the adaptation to blood-feeding habits. They are all expressed in the salivary glands. Whether more "conventional" lipocalins are also present in other tissues of ticks, remains to be investigated.

A clade relates the only fully sequenced lipocalins reported in crustaceans: the Crustacyanins (CRCs). They have been well characterized in the lobster *Homarus gammarus*, but proteins with very similar biochemical properties are now found in more species of lobsters, crayfish and crabs. Two CRCs per species are found at the most (only the fully sequenced CRCs have been included in the tree).

The biliproteins (BPs) and their relatives are monophyletically related. All biliproteins so far established as lipocalins are found in Lepidoptera, a relatively young and very diverse group of insects. BPs have also been isolated from other insect orders but their sequences are not yet known and hence their lipocalin nature is still open (reviewed in ref. 1). A few of the lepidopteran lipocalins from this clade are well characterized by having a bilin bound as a specific ligand resulting in a blue protein. The ligands, if any, of the related members of this clade are not known; we therefore refer to them as biliprotein-related proteins (BPRPs). As it is the case for CRCs, a maximum of two lipocalins have been found in each species sampled.

The insect saliva lipocalins relate to the biliproteins clade. This group represents another independent expansion of lipocalins associated with hematophagy adaptation in insects. Interestingly, among them are the Nitrophorins (NPs) that are heme-containing lipocalins. The branches of the insect saliva lipocalins clade are very long, reflecting their high sequence divergence. Caution should thus be taken, since mistaken relationships due to long branch attraction are not ruled out: sequences that are more dissimilar tend to group together. Finding more lipocalins in the same species of blood-sucking insects, as well as the use of other phylogenetically informative characters, should help to establish their relationship to the rest of the family.

In the next sections we will review each of the five groups of lipocalins. We have found big contrasts in the level of knowledge about arthropodan lipocalins. From very fine details on protein structure, ligand-protein interactions and physiological functions, to the vastly unexplored functional and biochemical properties of new sequenced lipocalins. In any case, arthropods are an excellent arena in which to scrutinize and admire the enormous versatility of such an apparently simple protein fold.

The Lazarillo/Apolipoprotein D Related Lipocalins

According to the consensus pattern of gene structure and protein sequence motifs of the family, Lazarillo is a conventional lipocalin with the three conserved protein motifs of the family (see Chapter 2) and its gene ORF intervened by three introns,[4] two of which are conserved in every metazoan lipocalin found so far (see Chapter 3). As shown in Figure 2, phylogenetic analyses support the orthologous relationship of Laz and the vertebrate ApoD lipocalin.[4-6] In addition, Laz molecular features, expression pattern and function closely resemble those of ApoD.

Lazarillo in Grasshopper (Schistocerca americana)

The grasshopper Lazarillo (Laz) was identified by two of us thanks to a monoclonal antibody (mAb) raised against embryonic neural proteins. This mAb allowed the study of the expression profile and of a number of biochemical properties, and the protein purification that led to the Laz transcript by standard cloning techniques.[7,8] Laz gene is transcribed as a long (3 kb) mRNA, with a fairly long (> 2 Kb) 3'-UTR.

Grasshopper Laz shows some striking biochemical features, since it is, so far, the only lipocalin bound to the plasma membrane via a glycosylphosphatidylinositol (GPI) anchor. However, the functional relevance of this singular type of membrane attachment remains unknown. The protein is also heavily glycosylated: Cleavage with PNGase F demonstrates that around 40% of Laz molecular weight is due to N-linked oligosaccharides. Interestingly, the five Asn residues predicted to be glycosylated appear to be polarized to one side of the protein by homology-modeling of the tertiary structure of Laz.[9] The shift in electrophoretic mobility of the protein under reducing vs. nonreducing conditions, and Laz homology-modeling suggest that the four cysteine residues of Laz form two alternating disulfide bonds and contribute to its folding.

Although we are in the process of analyzing the ligand capabilities of the grasshopper Laz and other relatives, the study of the Laz 3D-model and preliminary experiments (Ganfornina, Ortega, Åkerström, and Sanchez unpublished observations) suggest that Laz can bind small and rod-shaped hydrophobic ligands, but not the flat-shaped bilins bound by other insect lipocalins (see above).

Due to its GPI tail, Laz is expected to interact homophilically, or with other proteins, in order to transduce signals from the cell surface to the intracellular compartment. We have evidence that various recombinant forms of Laz can oligomerize in solution (Ganfornina et al, unpublished observations), although no information is available for interactions with other proteins such as potential receptors.

In terms of the spatial and temporal pattern of expression, the grasshopper Laz appears on the surface of a subset of neural precursors, the neuroblasts, as well as in many postmitotic neurons along the embryonic central, peripheral, and enteric nervous system.[8,10,11] An interesting observation is the transient expression observed in the neuroblasts, and in some neurons that express Laz during the period of axon outgrowth. The restricted expression of Laz within the developing nervous system has made the anti-Laz mAb an excellent tool for the characterization of the NS in economically important species like *Locusta migratoria* or *Schistocerca gregaria.*[11-13]

Outside the nervous system, Laz is expressed by the undifferentiated tips of the excretory Malpighian tubules, by sessile mesodermal cells that originate at the tip of the invaginating proctodeum and then migrate anteriorly to the amnioserosal membrane, and by nephrocytes of the subesophageal body.

*Lazarillo in the Fruitfly (*Drosophila melanogaster*)*

The Drosophila genome sequencing project recovered two regions in the second chromosome of the fruitfly with sequence similarity to the lipocalins. The transcription of the two putative genes was confirmed by RT-PCR from Drosophila embryos,[14] and their deduced amino acid sequence showed two conventional lipocalins (Neural Laz, NLaz and Glial Laz, GLaz) with the three lipocalin motifs, and the four cysteines involved in forming alternate disulfide bonds (filled dots in Fig. 1).

Both Drosophila lipocalins show the highest sequence similarity to grasshopper Laz. However, they do not show the GPI tail. They are secreted to the extracellular environment as predicted by their signal peptide and suggested by the difficulty of recognizing the NLaz protein by whole-mount immunohistochemistry unless the secretory machinery is inhibited by the use of monensin.[14] No further details of their biochemical properties are known at present.

The Drosophila Laz genes pose a typical problem of orthology ascription. Two character sets closely relate GLaz to grasshopper Laz: amino acid sequence similarity[14] and exon-intron arrangement[4] (see Chapter 2). However, other protein properties support NLaz as the grasshopper Laz ortholog:

1. Sequence alignment (Fig. 1, see also Chapter 3) suggests the existence of two unique loops in GLaz that could be involved in specific functional interactions with a ligand or other proteins. These features make GLaz more dissimilar to Laz, while the alignment of NLaz and Laz has only minor gaps (1-2 residues).
2. NLaz is more glycosylated (five predicted sites) than GLaz (only one site).
3. NLaz is predicted to have an acidic *pI* (4.5), similar to the grasshopper protein (4.8) and distinct from the strongly basic GLaz (*pI* 9.0).
4. NLaz, but not GLaz, has a long C-terminal segment similar in size to the GPI anchoring signal of Laz (Fig. 1).

Neither of the other Drosophila lipocalins is more related to grasshopper Laz. Thus, the Laz ancestor gene probably gave rise to NLaz and GLaz by duplication in the fruitfly, the traces of which have been subsequently blurred by their divergence and acquisition of new functions.

Looking at the spatio-temporal expression pattern of these lipocalins, we have collected more data to ascertain their relationships to the grasshopper Laz. Time wise, the transcripts of both lipocalins are abundant at late stages of embryogenesis, mainly when the nervous system is being formed, in the pupal stage, and in adulthood. In addition, GLaz mRNA is present all along the embryonic period, with a maternal contribution at early stages. The low abundance during larval development and the burst observed after pupariation for both lipocalins is correlated with the profound tissue reorganization that occurs at this critical period of metamorphosis.

In terms of the tissue expression profile, NLaz holds its name because of its expression in a subset of CNS neuroblasts and neurons, while GLaz is expressed by the longitudinal glia of the ventral nerve cord and a specific glial precursor. Besides, a subset of neurons expresses NLaz in larval and adult CNS (Sanchez et al, unpublished observations). The pattern of NLaz transcription is particularly interesting, as it changes drastically when considering different neuromeres and even hemisegments. This indicates a very dynamic transcription for NLaz and suggests that different neurons need this lipocalin for particular time periods. NLaz protein is observed in axons, which suggests that the protein is transported along them and is possibly secreted to the extracellular environment by the growth cone. Outside the nervous system, only the embryonic expression of these lipocalins has been explored. NLaz appears in the amnioserosa, the fat body and the developing gut, while GLaz is mainly expressed in the developing gut and salivary glands.

Lazarillo Function, or Functions?

Laz perturbation by in vivo incubation of grasshopper embryos with the anti-Laz mAb resulted in axonal growth delay and growth cone misrouting.[8] Thus, the grasshopper Laz is functionally related to the modulation of growth and guidance of developing axons.

While functional analyses of the Drosophila Laz lipocalins are underway, a hypothesis based on their molecular properties had been proposed.[14] They differ mostly in surface regions of the proteins (see above), while an analysis by homology-modeling predict that the binding pocket environment of NLaz and GLaz is very similar. Because they are secreted and expressed in nearby embryonic cells, it is reasonable to propose that their functional specificity is due to protein-protein interactions. The binding of a ligand could serve as a modulator for those interactions.

We are currently analyzing the effect of loss-of-function mutations for the genes NLaz and GLaz (Ganfornina, Sanchez et al, unpublished observations). Surprisingly, no drastic effects are observed after abolishing the function of these two lipocalins during development, thus reducing support for an essential role in axonal pathfinding and growth during normal development. However, during adulthood, the lack of NLaz or GLaz causes a decrease of life span and more sensitivity to the oxidative stress produced by paraquat, a generator of free radicals. Therefore, in terms of neural development, it remains as a possibility that these lipocalins are needed under particular developmental conditions that pose a threat to the proliferation and survival of neurons and/or to their axonal growth. In this sense, ApoD, the chordate ortholog of Laz, is functionally involved in the cell-proliferation/cell-death balance, and in cell migration upon injury (reviewed in Chapter 13). Most data support that Laz/ApoD is a multifunctional protein used in different cellular or physiological contexts.

Other Lipocalins Related to Lazarillo

The lipocalin Karl was found when screening for genes specifically expressed in the blood cells[15] (and Bailey's unpublished observations). The first expression is observed in plasmatocytes during late embryonic development. Later, during larval stages, Karl is strongly expressed in the lymph glands (the larval hematopoietic organ) and in circulating hemocytes. Blood cells are important for the immune system of insects, since they act as macrophages and as producers of antimicrobial peptides. A potential role of Karl in the immune system of insects is being investigated.

The EST and genome sequencing projects are helping to identify more Laz related lipocalins. In addition to Karl, GLaz and NLaz, we have found a fourth lipocalin in the genome of *D. melanogaster*, yielding four as the maximum number of lipocalins in this well studied species. An equal number of lipocalins have been found in the hymenoptera *Apis mellifera*, the honeybee (although the full sequence is not available yet for all of them). It is worth noting that at least one of them is also expressed in the nervous system.

Several mosquito genome projects are underway and at least two lipocalins in *Anopheles gambiae* show sequence similarity to Laz. However, other mosquito sequences become associated with the more divergent group of insect saliva lipocalins in our protein phylogeny reconstruction (Fig. 2).

Finally, in the fall webworm moth *Hyphantria cunea*, we found the only example of a Laz related lipocalin within Lepidoptera. Interestingly, Hyphantrin was initially regarded as a biliprotein due to its (distant) similarity to the BPRPs clade of lipocalins and to its expression in the epidermis of pupae. However, the Hyphantrin sequence is more related to the Laz-like lipocalins (Fig. 2). In contrast, other BPRPs (like Gallerin or Bombyrin) are neuronal lipocalins (see below), reflecting that nervous system expression is not an exclusive character of the Laz-related lipocalins.

Crustacyanins: Blue Lipocalins in Crustaceans

As the name suggests, Crustacyanin (CRC) denotes a blue protein from a crustacean. Proteins with biochemical properties of CRC have been isolated from several species of lobsters, crabs and crayfishes, although the CRCs from *Homarus gammarus* are by far the best known. By contrast to the insect biliproteins discussed below, the blue color of this protein is not due to a bilin, but to a carotenoid. Hence, CRC represents a blue carotenoprotein. Over the past 50 years, many attempts have been undertaken to uncover the building principle and the physical basis of the blue color of CRC, but only recent work succeeded to solve some key features of its structure. The many facets of this long-standing endeavor has been well reviewed recently.[16]

CRC represents a protein complex with astaxanthin (3,3'-dihydroxy-β,β-carotene-4,4'-dione), a red carotenoid derived from β-carotene by metabolic oxidation that takes place in many marine arthropods. In the native protein complex, however, the absorption spectrum of the carotenoids is red-shifted by about 150 nm, resulting in an absorption maximum of 632 nm and hence in a blue carotenoprotein. The red color of the free ligand comes back upon protein denaturation, what happens when the lobster is cooked.

The native CRC complex with a molecular mass of ~320 kDa is called α-Crustacyanin. It is composed of eight β-Crustacyanin subcomplexes that are dimers of 20-kDa monomers, the ultimate protein subunits. As each subunit contains one molecule of astaxanthin, the 16-mer α-Crustacyanin complex contains 16 molecules of astaxanthin. Unusually, the dissociation into the β-Crustacyanin dimers is irreversible. Even more surprising is the shift of the absorption maximum of astaxanthin from 472 nm (in hexane) in the unbound form to about 585 nm in β-Crustacyanin and further to 632 nm in α-Crustacyanin. While the resolved structure of β-Crustacyanin provides a reasonable basis to explain the spectral properties of the carotenoid in the dimer (a 100-nm shift), the further shift of about 50 nm awaits the crystal analysis of the α-complex.[16,17] According to a very recent study, the tight packing of the subunits (yet unknown in detail) allowing exciton coupling between excited states of astaxanthin molecules is the likely cause for this impressive bathochromic effect.[18]

When Crustacyanin protein is separated electrophoretically, it appears as five distinct monomers. However, they are encoded by only two genes (CRC1 and CRC2). Recent results suggest that the nonencoded differences result from nonphysiological alterations during purification and storage of α-crustacyanin.[19] The protein sequences of CRC1 and 2 predict a typical lipocalin folding with two conserved cysteine bridges. The structures of four different β-Crustacyanin complexes have been solved,[19-22] with two bound astaxanthin per dimer. In fact, the crystal structures demonstrate a β-barrel made up of eight antiparallel strands and the presence two helices. In the dimers, the monomers are oriented in such a way that their cavity openings are facing each other. Even more surprisingly, each of the two astaxanthin ligands is bound to both cavities by bridging both protein subunits thus sharing each cavity. This unusual mode of ligand binding is a consequence of the large extended conformation of the carotenoid that does not fit into a single lipocalin cavity. The specific arrangement of the two carotenoid chromophores positioned fairly in parallel and close together, explains the high stability of the dimeric pigment complex. The structures further reveal a conformational change of the bound carotenoid that, together with protein contacts, is thought to result in the observed shift of the absorption spectrum of the dimeric β-Crustacyanin versus free astaxanthin.

It is obvious, that the presence of the blue CRC in the carapace of the lobster significantly contributes to camouflage coloration in its marine habitat that is dominated by blue to blue-green light. This is why other marine arthropods (and also other animals) have developed a similar protective coloration in their exoskeleton or underlying tissues. Representative species are the brine shrimp Artemia, some other shrimps and the water flea Daphnia, from which a blue high-molecular weight glycoprotein, called artemocyanin, has been purified.[23,24] In contrast to CRC, however, the blue color of artemocyanin is not due to astaxanthin or any other carotenoid but due to a bilin, reported to be similar to biliverdin IXα. An N-terminal sequence obtained from one of these biliproteins did not reveal any relationship to insect biliproteins that have been identified as lipocalins. So, these crustacean biliproteins seem to have evolved independently as functional equivalent means of protection.

The Bilin-Binding Lipocalins and Their Relatives

The term biliproteins denotes proteins that are specifically associated with a bilin as ligand. They are widespread among insects though mostly known from lepidopteran species. A number of these typically blue proteins have been purified and characterized in more or less detail, but crystal structures have been obtained only for the two proteins discussed below in detail (for an overview, see the review by Kayser in ref. 1). However, other lipocalins, for which the ligand binding properties are not yet known, appear grouped with the biliproteins in a strongly supported clade (Fig. 2). Whether they represent cases of functional divergence without much sequence variations will be discussed below.

Bilin-Binding Protein of Pieris brassicae

The name 'bilin-binding protein' (BBP) specifically refers to the biliprotein from *Pieris brassicae*, the large white cabbage butterfly. BBP is mainly found in the hemolymph of the last larval instar and in the wings of the adult stage, while it is also present in other tissues such as fat body and epidermis. The blue protein in the wings is masked by the large amounts of white pterins in the overlaying scales.

BBP was isolated from whole butterflies by one of us, crystallized and unexpectedly identified as a prototypic lipocalin, the first from an invertebrate source.[25,26] This contrasts to the well known cyanobacterial biliproteins, which are of mainly helical structure and function in photosynthetic light harvesting.[27,28] BBP is found predominantly as a ~19.8 kDa monomer composed of 174 amino acids,[29,30] is not associated with sugar or lipid, and has one molecule of biliverdin IXγ noncovalently bound.

BBP crystallizes as a tetramer, i.e., as a dimer of dimers, with the monomers showing the typical lipocalin folding of an eight-stranded antiparallel β-barrel with two cysteine bridges at positions known to be conserved in lipocalins (filled dots in Fig. 1).[25,26] Besides of two short helical structures before the first β-strand and close to the C-terminus, respectively, there is one long α-helix attached to a side of the barrel (Fig. 3). The bilin ligand of BBP is bound in the cavity of the barrel and displays the cyclic helical, porphyrin-like structure that is normally adopted also by open-chain tetrapyrroles (for a review, see ref. 31). The orientation of the bilin is such that the two carboxyl groups at the outer pyrrole rings point to the opening of the cavity, i.e., to the solvent. Most remarkably, the bilin, whose two enantiomeric cyclic conformations are rapidly interconverted in solution and therefore optically inactive, is fixed by the protein in one enantiomeric form (Fig. 3) resulting in an optically active bound state.[32] This specific mode of binding affects also the visible absorption spectrum of the bilin, which shows a red shift of about 25 nm to a plateau around 670 nm in the bound state; the UV-absorption band of the bilin in BBP is at 383 nm.

According to studies employing radiolabeled precursors for the bilin and the apoprotein, respectively,[33,34] there are two major developmental periods of BBP synthesis, one in the last

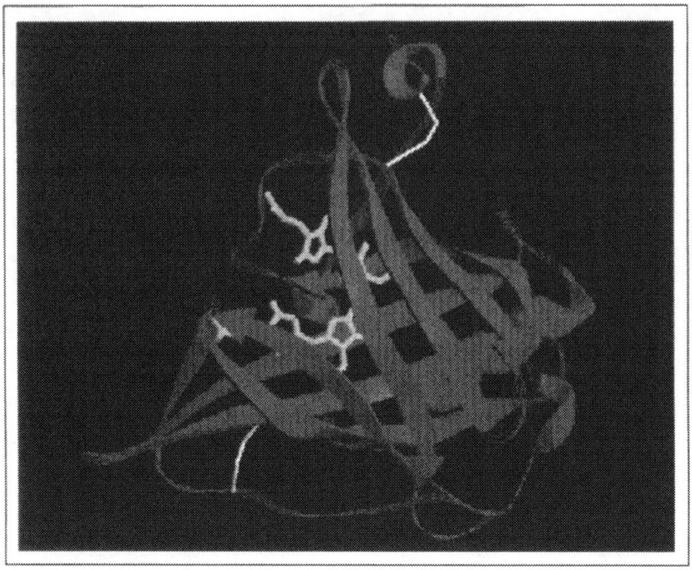

Figure 3. Crystal structure of *Pieris brassicae* bilin-binding protein (BBP), a prototypic lipocalin. Ribbon drawing, in magenta, of the eight antiparallel β-strands forming the core barrel structure. The bound ligand biliverdin IXγ is shown in pale blue and the two conserved disulfide bridges are labeled yellow. (Structure from Swiss-Prot, Expasy website www.expasy.org; accession code P09464). A color version of this figure is available online at http://www.Eurekah.com.

larval instar during the feeding period, and another one during development of the adult insect starting at mid-pupal stage. These studies further revealed that the bilin and the protein are made in parallel and that opening of the porphyrin ring must immediately follow its synthesis. This developmental pattern of biosynthesis has been confirmed by Northern analysis of BBP messenger RNA, which also identified the fat body as the major site of BBP synthesis in the larva.[29] Interestingly, holo-BBP in the developing butterfly is produced in the wings, as revealed in isolated wings in culture (Sehringer and Kayser, unpublished results).

The fact that BBP biosynthesis is under strict developmental and tissue control and, furthermore, that most of the tetrapyrrole precursor 5-aminolevulinate is used for bilin synthesis[34] suggest a vital role of this lipocalin in this insect. However, the functional implications of this program remain to be discovered. While the blue biliproteins of insects are generally said to play a role in camouflage coloration, which is certainly true in many species (see below for Insecticyanin), this is not obvious in *P. brassicae*, for example, where the coloration of larvae and adults is dominated by other pigments (for details, see ref. 1). A hypothetical role of BBP based on light absorption, as in cyanobacterial biliproteins,[28] can be excluded as it lacks the required photochemical properties.[35] This is likely true for all insect biliproteins. More realistic, though unproved roles of insect biliproteins related to metabolic regulation are discussed in a recent lipocalin review (see ref. 1), and they include prevention of cellular damage by reactive oxygen and nitrogen species, CO signaling cascades, and regulation of soluble guanylyl cyclase by biliverdin.

Insecticyanin of Manduca sexta

Insecticyanin (INS) is the counterpart of the butterfly BBP in the tobacco hornworm moth, *Manduca sexta*. This insect biliprotein was the first one to be purified and physicochemically characterized as a major hemolymph constituent.[36] INS occurs in larval hemolymph as a tetrameric complex of ~23-kDa subunits and, like BBP, has noncovalently bound biliverdin IXγ at a stoichiometry of 1:1, and is also free of sugar and lipid. Sequencing confirmed the larger size of the monomer, compared to BBP, as it comprises 189 amino acids residues making up a molecular mass of ~21.4 kDa.[37]

In the crystallized state, INS, again like BBP, is obtained as a tetramer with the subunits exhibiting the characteristic lipocalin folding[38] that first became known for vertebrate retinol-binding protein (see Chapter 3). While the barrel with the two sheets of orthogonally arranged up-and-down β-strands is almost identical to that of BBP, the moth protein shows additional helical structures, not only before and after the eight strands (for structural details, see ref. 1). The bilin of INS is bound to the cavity of the barrel in a way comparable to that of BBP, i.e., it displays the same 'frozen' cyclic helical conformation with the carboxyl groups directed to the solvent. As generally observed in members of the lipocalin family, the high similarity in the crystal structures of INS and BBP contrasts with the low similarity in their amino acid sequences, which are about 40% identical. This discrepancy is even higher when vertebrate lipocalins with the same 3D structure, e.g., retinol-binding protein, are included.

INS occurs in two isoelectric forms, a more acidic form (INS-a) and a more basic form (INS-b). Both molecular variants are present in the larval integument, while INS-b is the only form in the hemolymph.[39] The isoforms are encoded by two distinct INS genes that are both expressed mainly in the epidermis and to a lesser degree in the fat body.[40,41] The two gene products, which differ in 13 amino acid residues, are differentially exported into the two compartments: cuticle and hemolymph. More detailed developmental studies revealed that the two INS genes are differentially regulated in epidermis and fat body, show different temporal expression patterns and undergo significant cyclic changes in their messenger RNAs during larval development.[39] INS is synthesized from the second larval instar up to end of last instar, when its messenger RNA is lost as a consequence of the secretion of 20-hydroxyecdysone, which follows a decline of juvenile hormone marking the transition to the pupal stage.[42] In accordance with this observation, expression of the two INS genes is enhanced in the presence of juvenile hormone.[43] No expression takes place in the pupal and adult stages. Despite the cease of synthesis at pupation, INS is present in the hemolymph of the pupa and even the adult insect. In the young female moth, the larval biliprotein is sequestered into developing oocytes via a membrane receptor that specifically binds INS with high affinity.[44,45]

The uptake of INS into the eggs of *M. sexta* suggests a significant role of the biliprotein in embryogenesis or related processes. However, any role in the egg is open to speculation. During the larval stages, on the other hand, the presence of the blue INS together with the yellow carotenoids in hemolymph and epidermis create a green coloration that conceivably provides a protective camouflage effect (see also ref. 1). This is in contrast to the situation in larvae of *P. brassicae*, as discussed above.

The Biliverdin-Binding Proteins of Samia cynthia ricini

Two biliproteins have been purified from larval hemolymph of the moth *Samia cynthia ricini* and described as biliverdin-binding proteins BBP-I and BBP-II.[46] In contrast to the two Insecticyanins from *M. sexta*, the two proteins from *S. cynthia* share only 43% sequence identity and the length of their branches in the phylogenetic tree (Fig. 2) suggests that more divergence has taken place after the underlying gene duplication in this silkworm than in the tobacco hornworm. The subsequent independent evolution resulted in a significant difference in molecular size and in oligomeric state: while BBP-I occurs as a monomer of 20.5 kDa, BBP-II behaves as a dimer of 22.7-kDa subunits. Moreover, these two proteins are immunologically distinct from each other. BBP-II is most likely identical to the biliverdin-binding protein that is present in the molting fluid during larval-pupal ecdysis in this silkmoth. Both BBPs seem to be synthesized in the epidermis and secreted into the hemolymph and molting fluid, respectively. Each monomer of BBP is associated with one molecule of biliverdin tentatively identified as the IXγ isomer, which is also bound in the biliproteins of *P. brassicae* and *M. sexta*.

Biliproteins of about 20 kDa, the size of typical lipocalins, have also been described from at least four other lepidopteran species, as reviewed in.[1] As their N-terminal sequences only are known to date, their definite identification as lipocalins is still open.

Biliprotein-Related Lipocalins with Unknown Ligand

Some insect proteins, whose amino acid sequence identifies them as lipocalins, are monophyletically related to the known biliproteins described above (see Fig. 2). The one first described was Gallerin, which is mainly expressed in the nervous system of *Galleria mellonella*, the wax moth,[47] particularly in glial cells of the nerve cord. The sequence of the Gallerin precursor, comprising 203 amino acid residues, shows a close relationship to BBP and INS (see Fig. 2).

The clade of biliprotein-related proteins (BPRPs) also includes Bombyrin, a lipocalin from the silkworm *Bombyx mori*.[48] As Gallerin, it is also expressed in the nervous system (pupae brain). Whether Gallerin or Bombyrin have the potential to bind a bilin or if they occur as true biliproteins in the insect is not known. Their nervous system expression also links them, at least functionally, with the Laz related lipocalins, expressed in glial and neuronal cells.

Quite recently, lipocalin sequences from the moth *Lonomia oblique* became known (Acc. No. AY829833,[49] AY829809,[49] and AY908986[50]). Two of them have been included in our phylogenetic tree (Lip1 and Lip4, see Fig. 2) and the third one shows only two amino acid substitutions with respect to the second. Their closest relatives within the BPRPs clade are the lipocalins from *S. cynthia ricina*. Lip1 and Lip4 were identified as transcripts in the larval spicule and integument, respectively, being Lip1 postulated to play a role in the hemorrhagic syndrome. The third sequence has been reported as a prothrombin activator. Their function as well as their possible association with a bilin awaits further characterization, but it would be interesting to see if a set of lipocalins with a functional relationship with the hemostatic system of vertebrates is also present in Lepidoptera as is the case of the saliva lipocalins from hematophagous arthropods (see below).

Expansions of the Lipocalin Family in Blood-Feeding Arthropods

We have described so far many interesting functional explorations that lipocalins have undergone through arthropod evolution. Now, two independent expansions of the family have been found when exploring proteins or mRNA expressed in the salivary gland of two distantly related arthropods, insects and arachnids, that share a common way of life: feeding on vertebrate's blood.

Their assignment to the lipocalin family has been a difficult task, since they share little or almost undetectable protein sequence similarity with other lipocalins. However, evidence of their common ancestry keeps accumulating, and these data are helping to frame the following evolutionary hypothesis: A history of intra-lineage duplications accompanied by high divergence, while both groups of lipocalins (from insects and arachnids) converge to control the hemostatic system of their host to make the most out of a blood meal.

Both, blood-feeding insects and ticks, produce in their saliva a cocktail of proteins specialized in the control of their host hemostatic machinery (reviewed by refs. 51,52). In Table 2, we relate each antihemostatic activity with the protein that plays that role in the cocktail. In Hemiptera and Acari, most of these functions are carried out by an independently evolved set of lipocalins. In Diptera (mosquitoes) these roles have been taken by completely unrelated proteins.[53,54] This is certainly a striking case of convergent evolution, in which each parasite organism has adopted a different solution to fulfill the same task, probably evolving in parallel to the evolution of hematophagy (dating around 90 My[55]). It also represents a case of coevolution with the host hemostasis and inflammation systems, which have evolved very complex and redundant reactions.

Table 2. Different strategies for the control of the host's hemostatic machinery by blood feeding arthropods

Antihemostatic Activity (Mechanism)	Insects: Blood-Sucking Bugs (Hemiptera)	Insects: Mosquitoes (Diptera)	Arachnida: Ticks (Acari)
Enhancement of vasodilation:			
Release of NO (involving the soluble guanylate cyclase (sGC) signal cascade)	**Nitrophorins**		
Degradation of norepinephrine (by tiol-oxydase activity)	**Nitrophorins**		
Vasodilatory peptides		Sialokinins (not lipocalins)	
Reduction of immune and inflammation response:			
Sequestration of histamine	**Nitrophorins**		**HBPs (histamine), SHBP (serotonin + histamine)**
Interference with the wound healing and blood coagulation cascade:			
Inhibition of collagen-induced platelet aggregation	**RPAIs, Pallidipin**		**Moubatin**
Reduction of platelet aggregation through cGMP cascade (by release of NO)	**Nitrophorins**		
Inhibition of ADP-induced platelet aggregation			**Moubatin** Savignygrin (not a lipocalin)
Inhibition of ADP-induced platelet aggregation (by hydrolysis of ADP)	Apyrase (not a lipocalin)	Apyrase (not a lipocalin)	Apyrase (not a lipocalin)
Inhibition of ADP-induced platelet aggregation (by sequestration of ADP)	**RPAIs**		
Sequestration of serotonin and epinephrine (mediators of platelet aggregation, vasoconstriction, and inflammatory processes)	**BABP**		**SHBP (serotonin + histamine)**
Direct inhibition of platelet aggregation by protein-platelet membrane interaction	**Nitrophorin 7**		**Moubatin**
Inhibition of proteolytic factor Xase complex.	**Nitrophorin 2**	Anticoagulant-factor Xa activities (not lipocalins)	
Inhibition of thrombin (coagulation factor and platelet activator)	Triabin	Anti-thrombin activities (not lipocalins)	Savignin (not a lipocalin)
Other Activities			
Allergenicity	**Procalin**		
Toxicity			**TSGPs**

The Saliva Lipocalins in Hematophagous Insects

At least 28 proteins that share the lipocalin fold are expressed in the salivary glands of hematophagous insects. They comprise the Nitrophorins (NPs), the Biogenic Amine Binding Protein (BABP), the Rhodnius Platelet Aggregation Inhibitors (RPAIs), all of them found up to date only in one species (*Rhodnius prolixus*); and the Triabins, Pallidipins, and Procalins (found in three species of blood-feeders).

Their amino acid sequences have low similarity with the lipocalins. However, their eight-stranded β-barrel structure shows a striking similarity and constitutes the first hint to family relationship.[56-59] When secondary structure constraints are imposed into the alignment, a few conserved sites are then evident, and the SCR1 is recognized, although with some variations. An attempt to place them within the family phylogeny is shown in Figure 2, and proposes a biliprotein-like lipocalin as the common ancestor of the blood feeding-related lipocalins. However, the tree position of this clade must await confirmation from other sources (such as intron-exon structure[3,4]).

The Nitrophorins: Carriers of NO

Four Nitrophorins (NPs) were purified out of the salivary glands of *R. prolixus*[60-64] that constitute ~50% of the total protein content in the insect saliva. They were characterized as NO-transporters, histamine binding and anti-coagulation factors (see Table 2). Recent transcriptome and proteome analyses of the salivary glands have revealed up to 11 NPs (or NP-related proteins[65]).

The expression of NPs is developmentally regulated[66] through molting. New NPs are added in each molting step, been the fifth instar nymph the stage with a higher number of NPs. Some NPs, like NP3, are specific of this stage. Other NPs are more abundant in the first instar nymph and decrease through molting (like NP5 and NP6), been undetectable in adulthood. NP2, on the contrary, is the only one described to be present in all stages. As we will see below, NP2 has a functional specialization that would make it indispensable.

The crystal structure of several NPs have been resolved.[56-59,67] They show the typical lipocalin barrel fold with three short alpha-helices and two disulfide bonds (at conserved positions with respect to other insect lipocalins).

Functional Interactions through the Binding Pocket of Nitrophorins

NPs have a ferric heme that binds NO or histamine[62] (reviewed by ref. 51). The heme is deeply buried in their hydrophobic pocket, and one histidine residue constitutes the fifth iron ligand.[51] Binding and stabilization of NO is analogous to oxygen-binding myoglobins from vertebrate. However, NPs use the lipocalin fold, different from the α helical globin fold.

The complex NP-NO is stable at low pH (around 5), being the NO moiety buried by hydrophobic side chains and protected during storage at the entrance of the NP pocket.[67] However, when the NPs encounter a pH of ~7.5 in the host blood, a conformational change debilitates the NO binding.[60,62,68] NO is released and exchanged for histamine, that binds in the same location with a 100-fold higher affinity.[67] These properties make NPs excellent bifunctional shuttles bringing in NO and removing histamine from the wound, fulfilling two objectives at the same time: vasodilation and avoidance of the inflammatory response of the host.

The shuttle function of NPs depends on the properties of the hydrophobic pocket, but variations in the NPs binding pocket have been described.[56,69,70] Thus, NPs are not completely redundant transporters, and a potential for functional specialization exist.

In addition, the binding pocket of NPs could carry out other functions, like enzymatic activities[71,72] where NPs oxidize cysteine with H_2O_2 as reaction product. During this thiol-oxidase reaction NPs can also destroy norepinephrine, conferring additional vasodilation potential. NPs have also been shown to have a heme-peroxidase activity in vitro. However, the physiological significance of these enzymatic activities remains to be confirmed.

Molecular Interactions of Nitrophorins not Involving the Hydrophobic Pocket

NP2 is able to interfere in a heme-independent manner with the blood coagulation cascade at the factor X maturation step:[61] NP2 inhibits the conversion of factor X to factor Xa. Even though the NP2 mechanism of action has to be elucidated, its proteolysis regulation role is carried out through protein-protein interactions.

NP7 binds anionic phospholipids with high affinity in activated platelets.[73] Modeling of the NP7 three-dimensional structure shows a positively charged area on the surface of the protein that could bind negatively charged membrane surfaces containing phosphatidylserine. This binding competitively blocks the prothrombin activation, since the prothrombinase complex needs to bind to the same membrane surfaces.

These roles, not related to the binding pocket of lipocalins, seem to have evolved independently many times along the evolution of the family, showing once again the versatility of the lipocalin fold for the acquisition of novel functions.[74]

In summary, the cocktail of NPs released in blood-feeding insects saliva shows a high degree of molecular and functional diversification, from minor changes in the binding pocket leading to differences in affinity for their ligands, to completely novel interactions through their external protein surface.

The Biogenic Amine Binding Protein of *Rhodnius prolixus*

The biogenic amine binding protein (BABP) was identified in the salivary glands of fifth instar nymphs.[65,75] Its amino acid sequence and biochemical properties set this protein aside of the NPs within the set of insect saliva lipocalins. As NPs, BABP has four cysteines probably forming two disulfide bonds.

BABP does not contain a heme moiety in the hydrophobic pocket, and is unable to bind ADP (a ligand for other insect saliva lipocalins, see below). Instead, it is able to bind norepinephrine (Kd = 24 nM), serotonin (Kd = 102 nM), and epinephrine (Kd = 354 nM). The molecular modeling of BABP shows a narrower binding pocket that easily accommodates serotonin, but that would explain its inability to bind heme.

By sequestering serotonin and catecholamines from the wound site, BABP acts as a vasodilator and platelet aggregation inhibitor. Its function has been tested in tissue and cell culture assays. Addition of BABP inhibits the serotonin-mediated contraction of the rat uterine horn, the norepinephrine-mediated contraction of the rabbit aorta, and it eliminates the potentiation by serotonin and epinephrine of ADP- or collagen-induced platelet aggregation.[75] Therefore, BABP adds a different activity to the cocktail of saliva lipocalins controlling the host hemostatic system.

The Rhodnius Platelet Aggregation Inhibitors

With at least eleven NPs and one BABP, three important mechanisms that the host implements to avoid blood loss are counterattacked: vasoconstriction, immune response, and blood coagulation (including platelet aggregation). However, another set of lipocalins is put forward by *Rhodnius prolixus* in the feeding site: the rhodnius platelet aggregation inhibitors (RPAIs). A total of 7 RPAIs have been identified,[65,76,77] that show a set of 6 conserved cysteines (filled and open dots in Fig. 1) and share sequence similarity with other saliva lipocalins (Triabin and Pallidipin, see below). No crystal structure is available for RPAIs, but biochemical analyses suggest they can bind two ADP molecules.

RPAIs inhibit platelet aggregation by strongly binding ADP (in the nanomolar range[76]). ADP, which is released from activated platelets, injured endothelial cells, and erythrocytes, is able to decrease the threshold for the collagen-induced platelet aggregation even at very low concentrations. Apyrase, an enzyme also present in the insect saliva (not a lipocalin, see Table 2), would degrade ADP to AMP and Pi. However, Apyrase is not an efficient scavenger of ADP due to its high Km (>20 μM). When the concentration of ADP is too low for the Apyrase to work, RPAIs do their job, sequestering ADP and increasing the threshold for collagen-induced platelet aggregation.

Triabins, Pallidipins and Procalins

Triabin, first described in the saliva of *Triatoma pallidipennis*[78] has also close relatives in *Rhodnius prolixus*[65] and in *Triatoma brasiliensis.*[79] Its protein sequence is most similar to RPAIs and Pallidipins, with six conserved Cys residues that form three disulfide bonds (filled and open dots in Fig. 1).

The crystal structure of Triabin has been solved,[80] and it shows an important peculiarity: the β-barrel has strands B and C exchanged, altering the antiparallel organization common to all other lipocalin structures solved to date.

The *T. pallidipennis* Triabin functions as an anti-coagulation factor by directly binding to Thrombin through a protein-protein interaction. The details of this interaction are well known thanks to the crystallization of the complex Triabin-Thrombin.[80] Triabin represents another example (like NP2 and NP7) of the use of lipocalins for functional interactions not related to the hydrophobic binding pocket.

Pallidipin is another inhibitor of the collagen-induced platelet aggregation,[81,82] found in *T. pallidipennis, R. prolixus,*[65] and *T. brasiliensis.*[79] It has six conserved cysteines and a few other conserved amino acid residues and are also related to RPAIs and Triabins. The mechanism of action of Pallidipin is not known.

Finally, Procalin, has been described as an allergen found in the saliva of *T. pallidipennis,*[83] and related proteins have been found in *R. prolixus*[65] and *T. brasiliensis.*[79] The role of this lipocalin during blood feeding is not known. However, because of their known allergenicity, they are playing a role in the host response to the blood-feeder bite. In this case, we are looking at the coevolution process from the point of view of the host, since the immune response to salivary proteins is detrimental for the blood-feeder. Turning these parasite-host interactions into something useful in the control of diseases transmitted by blood-feeders arthropods is an important area of biomedical research.

The Saliva Lipocalins in Hematophagous Arachnids (the tick saliva lipocalins)

Lipocalin-related proteins have also been found in the saliva of a different group of hematophagous arthropods, the ticks (Acari). They are the only lipocalins described so far in the class Arachnida. The saliva of ticks has been of interest both, for the study of parasite-host interactions in terms of hemostasis control,[84] and for the toxicity of some ticks, provoking important livestock losses in certain areas.[85]

A total of 20 tick lipocalins have been described to date, being the hard tick Histamine Binding Proteins (HBPs) the ones whose crystal structure established the first hint for a possible relationship with lipocalins.[86,87] Other lipocalins found in the saliva of hard ticks (family Ixodidae) are the serotonin and histamine binding protein (SHBP), and 11 new proteins from *Ixodes scapularis.* Finally, Moubatin and a set of four tick salivary gland proteins (TSGPs) have been found in soft ticks (family Argasidae).

The tick saliva lipocalins share some sequence similarity among them, but when comparing their sequences with more conventional lipocalins, or even with the salivary proteins of blood-feeding insects, any traces of common ancestry seem to have been erased. However, one can use characters derived from the intron-exon arrangement of the genes to perform phylogenetic inferences.[4] This method has been useful to suggest a common origin for tick saliva proteins and lipocalins.[3] Therefore, we currently view this group of tick saliva proteins as another highly divergent extension of the lipocalin family, that originated during adaptation to blood feeding. In Figure 2, we show the branching point of tick saliva lipocalins (resulting from a gene structure based phylogeny[3]) superimposed to the amino acid sequence based phylogeny.

The Hard Tick Histamine Binding Proteins and Their Relatives

Three closely related histamine binding proteins (HBPs) have been isolated from the salivary glands of *Rhipicephalus appendiculatus.*[86] They sequester histamine released by the host in response to tissue damage. Therefore, they fulfill one of the roles described for NPs in hematophagous insects: reducing the immune and inflammatory host responses.

However, the histamine binding of HBPs uses a different mechanism.[86,87] No heme moiety is present and, most unusually within lipocalins, HBPs have two internal and separate binding sites for histamine: the H (high-affinity) and L (low-affinity) sites. These two sites are lined with acidic residues, quite useful for binding a basic ligand. The hydrophilicity of these pockets represents another striking difference with the binding pockets of most lipocalins, suited for housing hydrophobic ligands. The H site occupies the position expected for other lipocalins, but the entrance of histamine to this site is anomalous when comparing it with the open side of other lipocalin pockets (reviewed by ref. 87). On the other hand, the L site occupies the closed end of the barrel in other lipocalins, a region that bears the structurally conserved regions (SCRs), three amino acid sequence motifs clustered together in the barrel closed end region (see Chapter 3). This region is completely remodeled in HBPs, since it accommodates the L site and its entrance.

As it was the case for NPs (see above) the expression of HBPs is stage and sex specific: HBP1 and 2 are secreted only by adult females, and HBP3 by larvae, nymphs and adult males. They also differ in their glycosylation and macromolecular complexes: HBP1 and 2 are nonglycosylated monomers, while HBP3 forms a disulfide-linked dimer. The functional significance of their temporal and sex-dependent regulation or of their other molecular attributes are not yet understood.

An HBP related protein, the Serotonin and Histamine Binding Protein (SHBP) has been described in the hard tick *Dermacentor reticularis*.[88] SHBP also has two internal binding sites. Binding of histamine to the H site has been tested, while the ligand for the L-like site is unknown. However, the analysis of its structure makes serotonin the most likely candidate ligand.

Up to 11 new sequences with similarity to previously known tick saliva lipocalins have been retrieved from database searches[85] in another hard tick, *Ixodes scapularis*. The analysis of their primary sequences suggests that they are secreted proteins. However, more information needs to be collected on their structure and biochemical properties in order to understand their role in blood-feeding and their potential for functional specialization.

Moubatin and the Tick Salivary Gland Proteins Expressed in the Saliva of Soft Ticks

Moubatin was found in the saliva of the soft tick *Ornithodoros moubata* because of its platelet aggregation inhibitory activity.[89,90] Its amino acid sequence shows some similarity with HBPs, placing Moubatin as a member of this expansion of the lipocalin family. Moubatin fulfills a role that blood-sucking insects perform with a different set of lipocalins: inhibition of the collagen-stimulated aggregation of platelets.

In the closely related species *Ornithodoros savignyi*, four highly abundant tick salivary gland proteins (TSGPs) have been proposed to have a role in salivary gland granule biogenesis.[91] All proteins destined to be secreted have to be tightly packed into the secretory vesicles (or granules). Mans and Neitz[92] have shown that TSGPs take up most of the accessible volume in secretory granules, and propose macromolecular compaction of TSGPs as an important part of the packaging mechanism under the slight acidity and high calcium concentration that prevail in the granules. Since the cocktail of material secreted upon feeding contains not only proteins destined to control the hemostatic system of the host, but can also contain pathogens and toxins, a close understanding of the mechanisms that sort which proteins are secreted, could help to control the pathogens transmitted by ticks.

The role of TSGPs once they are secreted is unknown. They do not bind histamine nor any of the other mediators involved in the control of the host response to the tick bite. They have been tested for effects on the blood coagulation cascade and on ADP- and collagen-induced platelet aggregation assays[85] with negative results. As summarized in Table 2, these activities are carried out in this tick by nonlipocalin proteins like Savignygrin and Savignin.

However, two of the TSGPs (TSGP2 and TSGP4) are toxins that affect the cardiovascular system of the host[93] and are therefore involved in the pathogenesis of toxicosis caused by *O. savignyi* bites. Since the closely related TSGPs and Moubatin are not toxic, the properties that render a lipocalin toxic seem to be a recent acquisition in the evolution of ticks. The toxicity of these lipocalins might be considered as detrimental for the feeding parasite, as was the case for the allergenic Procalin, especially for ticks that have to spend longer periods of time on the host to complete a meal.

While the three dimensional structures of Moubatin or TSGPs have not been solved, TSGPs models have been constructed using the known structure of HBP2 with a reasonable fit. It is therefore expected that their tertiary structure will show commonalities with lipocalins. When the phylogenetic relationship among tick saliva lipocalins is assayed,[85] the major conclusion is that they are the result of recent gene duplications since they become grouped into genus-specific clades, reflecting gene duplication after tick speciation.

Therefore, in spite of the structural and biochemical differences of tick saliva lipocalins, the remaining resemblance to the β-barrel of standard lipocalins and the data on their gene structure[3] have shown more similarities than previously thought, and therefore had been assigned as lipocalins. Quoting Paesen et al,[87] 'tick lipocalins are very eccentric members of the lipocalin family', which highlights the versatility of the lipocalin fold to carry out many functions. Lipocalins seem to be perfect 'moonlightlers'[94,95] and the phylum Arthropoda has certainly been a great arena for lipocalin functional explorations.

More Expansions of the Lipocalin Family? The Case of Cockroach Milk

Recently, Williford et al[96] have described a new set of lipocalins that are secreted by the brood sack epithelium of the viviparous cockroach *Diploptera punctata* as part of a nutritive secretion or "milk" for the embryos. In species of ovoviviparous cockroaches, embryos develop within an infolding of the intersegmental membrane at the posterior end of the abdomen and are born as first instar larvae. This poach provides protection against desiccation and parasitism. The appearance of viviparity is linked to the production of "milk", now known to be full of lipocalins. Although only one of them has been included in our phylogenetic tree (Dpun.Milk in Fig. 2), a total of 22 distinct lipocalin protein sequences have been identified and the intron-exon structure of one of them has been described.[96]

The presence of four introns intervening the coding sequence of this cockroach milk lipocalin would relate their gene structure to the tick saliva lipocalins[3] and to NLaz[4] (see Fig. 4 in Chapter 3). However, their protein sequence becomes grouped with the insect saliva lipocalins and other highly divergent lipocalin sequences like *Drosophila yakuba* lipocalin, or the cockroach lipocalins of *Blatella germanica* (with allergenic activity[97]) and of *Leucophaea maderae*. Curiously, the later is part of an aphrodisiac secretion from the male tergal gland,[98] another infolding of the intersegmental membrane that could be considered as a morphological and physiological equivalent to the brood sac of *D. punctata*, but serving a different function at the organismal level.

Multiple gene copies of the milk lipocalins are detected in the genome of *D. punctata*, therefore representing one more case of gene duplication associated to a recent co-option event:[74] the acquisition of the new nutritive function which requires the production of these proteins in large amounts, and that embryos develop the ability to drink. The analysis of this new expansion of the lipocalin family should help to understand the transition of ovoviviparity to viviparity in this group of arthropods, in which morphological, physiological and molecular changes are linked and result in a completely new reproductive strategy.

Concluding Remarks

The main conclusion of this work is that lipocalins show an extraordinary ability to acquire novel functions, even though they are apparently simple, single modular proteins. The pattern of duplication and divergence of this protein family in the phylum Arthropoda is characterized by an initial low number of lipocalins maintained in most species sampled, in contrast with large intra-lineage multiplication of genes clearly linked to the appearance of new functions (e.g., feeding or reproductive strategies). This pattern is the result of an evolutionary pathway independent and quite distinct from the one followed by lipocalins in the other best sampled phylum, the chordates (see Chapter 3). However, the properties of lipocalins that grant evolutionary access to such an amazing panoply of functions are the same: their robust folding combined with a flexibility for change which allows lipocalins to interact with different ligands in their binding pocket and make contact with different molecules through their protein surface.

References

1. Kayser H. Lipocalins and structurally related ligand-binding proteins. In: Gilbert LI, Iatrou K, Gill S, eds. Comprehensive Molecular Insect Science. Oxford: Elsevier, 2005:4:267-306.
2. Philippe H, Laurent J. How good are deep phylogenetic trees? Curr Opin Genet Dev 1998; 8:616-623.
3. Mans BJ, Neitz AWH. Exon-intron structure of outlier tick lipocalins indicate a monophyletic origin within the larger lipocalin family. Insect Biochem Mol Biol 2004; 34(6):585-594.
4. Sanchez D, Ganfornina MD, Gutierrez G et al. Exon-intron structure and evolution of the lipocalin gene family. Mol Biol Evol 2003; 20(5):775-783.
5. Ganfornina MD, Gutierrez G, Bastiani M et al. A phylogenetic analysis of the lipocalin protein family. Mol Biol Evol 2000; 17(1):114-126.
6. Gutierrez G, Ganfornina MD, Sanchez D. Evolution of the lipocalin family as inferred from a protein sequence phylogeny. Biochim Biophys Acta 2000; 1482(1-2):35-45.
7. Ganfornina MD, Sanchez D, Bastiani MJ. Lazarillo, a new GPI-linked surface lipocalin, is restricted to a subset of neurons in the grasshopper embryo. Development 1995; 121:123-134.
8. Sanchez D, Ganfornina MD, Bastiani MJ. Developmental expression of the lipocalin Lazarillo and its role in axonal pathfinding in the grasshopper embryo. Development 1995; 121:135-147.
9. Sanchez D, Ganfornina MD, Bastiani MJ. Lazarillo, a neuronal lipocalin in grasshoppers with a role in axon guidance. Biochim Biophys Acta 2000; 1482(1-2 SU-):102-109.
10. Ganfornina MD, Sanchez D, Bastiani MJ. Embryonic development of the enteric nervous system of the grasshopper Schistocerca americana. J Comp Neurol 1996; 372:581-596.
11. Graf S, Ludwig P, Boyan G. Lazarillo expression reveals a subset of neurons contributing to the primary axon scaffold of the embryonic brain of the grasshopper Schistocerca gregaria. J Comp Neurol 2000; 419(3):394-405.
12. Boyan G, Reichert H, Hirth F. Commissure formation in the embryonic insect brain. Arthropod Structure & Development 2003; 32(1):61-77.
13. Boyan GS, Braunig P, Posser S et al. Embryonic development of the sensory innervation of the clypeo-labral complex: Further support for serially homologous appendages in the locust. Arthropod Structure & Development 2003; 32(4 SU-):289-302.
14. Sanchez D, Ganfornina MD, Torres-Schumann S et al. Characterization of two novel lipocalins expressed in the Drosophila embryonic nervous system. Int J Dev Biol 2000;44(4):349-359.
15. Bailey UM. The drosophila lipocalin, Karl, is specifically expressed in the blood cell compartment. Paper presented at: Benzon symposium No. 50: The lipocalin protein superfamily. Copenhagen, Denmark: 2003.
16. Chayen NE, Cianci M, Grossmann JG et al. Unravelling the structural chemistry of the colouration mechanism in lobster shell. Acta Crystallogr D Biol Crystallogr 2003; 59(Pt 12):2072-2082, (Epub 2003 Nov 2027).
17. Dellisanti CD, Spinelli S, Cambillau C et al. Quaternary structure of alpha-crustacyanin from lobster as seen by small-angle X-ray scattering. FEBS Lett 2003; 544(1-3):189-193.
18. van Wijk AA, Spaans A, Uzunbajakava N et al. Spectroscopy and quantum chemical modeling reveal a predominant contribution of excitonic interactions to the bathochromic shift in alpha-crustacyanin, the blue carotenoprotein in the carapace of the lobster homarus gammarus. J Am Chem Soc 2005; 127(5):1438-1445.
19. Habash J, Helliwell JR, Raftery J et al. The structure and refinement of apocrustacyanin C2 to 1.3 Å resolution and the search for differences between this protein and the homologous apoproteins A1 and C1. Acta Crystallogr D Biol Crystallogr 2004; 60(Pt 3):493-498, (Epub 2004 Feb 2025).
20. Gordon EJ, Leonard GA, McSweeney S et al. The C1 subunit of alpha-crustacyanin: The de novo phasing of the crystal structure of a 40 kDa homodimeric protein using the anomalous scattering from S atoms combined with direct methods. Acta Crystallogr D Biol Crystallogr 2001; 57(Part 9):1230-1237.
21. Cianci M, Rizkallah PJ, Olczak A et al. The molecular basis of the coloration mechanism in lobster shell: Beta-crustacyanin at 3.2-Å resolution. Proc Natl Acad Sci USA 2002; 99(15):9795-9800.
22. Cianci M, Rizkallah PJ, Olczak A et al. Structure of lobster apocrustacyanin A1 using softer X-rays. Acta Crystallogr D Biol Crystallogr 2001; 57(Pt 9):1219-1229.
23. Krissansen GW, Trotman CNA, Tate WP. Identification of the blue-green chromophore of an abundant biliprotein from the haemolymph of Artemia. Comp Biochem Physiol 1984; 77B:249-252.

24. Peeters K, Brendonck L, Moens L. The occurrence of artemocyanin in Branchiopoda (Crustacea). Comp Biochem Physiol A Physiol 1994; 109(3):773-779.
25. Huber R, Schneider M, Epp O et al. Crystallization, crystal structure analysis and preliminary molecular model of the bilin binding protein from the insect Pieris brassicae. J Mol Biol 1987; 195(2):423-434.
26. Huber R, Schneider M, Mayr I et al. Molecular structure of the bilin binding protein (BBP) from Pieris brassicae after refinement at 2.0 Å resolution. J Mol Biol 1987; 198(3):499-513.
27. Schirmer T, Bode W, Huber R. Refined three-dimensional structures of two cyanobacterial C-phycocyanins at 2.1 and 2.5 Å resolution. A common principle of phycobilin-protein interaction. J Mol Biol 1987; 196(3):677-695.
28. MacColl R. Cyanobacterial phycobilisomes. J Struct Biol 1998; 124(2-3):311-334.
29. Schmidt FS, Skerra A. The bilin-binding protein of Pieris brassicae. cDNA sequence and regulation of expression reveal distinct features of this insect pigment protein. Eur J Biochem 1994; 219(3):855-863.
30. Suter F, Kayser H, Zuber H. The complete amino-acid sequence of the bilin-binding protein from Pieris brassicae and its similarity to a family of serum transport proteins like the retinol-binding proteins. Biol Chem Hoppe Seyler 1988; 369(6):497-505.
31. Kayser H. Pigmens. In: Kerkut GA, Gilbert LI, eds. Comprehensive Insect Biochemistry, Physiology and Pharmacology. Pergamon Press, 1985:10:367-415.
32. Scheer H, Kayser H. Conformational studies of biliproteins from the insects Pieris brassicae and Cerura vinula. Z Naturforsch 1988; 43c:84-90.
33. Kayser H. De novo synthesis and levels of cytochrome c and a biliprotein during pupal-adult development in a butterfly, Pieris brassicae. Z Naturforsch 1984; 39c:938-947.
34. Kayser H, Krull-Savage U. Development-specific incorporation of [14C]5-aminolevulinate and [3H]leucine into cytochrome c and biliprotein in the butterfly, Pieris brassicae. Correlation with the ecdysteroid titer in the pupa. Z Naturf 1984; 39c(948-957).
35. Schneider S, Baumann F, Geiselhart P et al. Biliproteins from the butterfly Pieris brassicae studied by time-resolved fluorescence and coherent anti-stokes Raman spectroscopy. Photochem Photobiol 1988; 48:239-242.
36. Cherbas P. Biochemical studies of insecticyanin. Ph.D. thesis, Harvard University, Cambridge, Massachusets: 1973.
37. Riley CT, Barbeau BK, Keim PS et al. The covalent protein structure of insecticyanin, a blue biliprotein from the hemolymph of the tobacco hornworm, Manduca sexta L. J Biol Chem 1984; 259(21):13159-13165.
38. Holden HM, Rypniewski WR, Law JH et al. The molecular structure of insecticyanin from the tobacco hornworm Manduca sexta L. at 2.6 Å resolution. EMBO J 1987; 6(6):1565-1570.
39. Riddiford LM, Palli SR, Hiruma K et al. Developmental expression, synthesis, and secretion of insecticyanin by the epidermis of the tobacco hornworm, Manduca sexta. Arch Insect Biochem Physiol 1990; 14(3):171-190.
40. Li W, Riddiford LM. Two distinct genes encode two major isoelectric forms of insecticyanin in the tobacco hornworm, Manduca sexta. Eur J Biochem 1992; 205(2):491-499.
41. Li WC, Riddiford LM. The two duplicated insecticyanin genes, ins-a and ins-b are differentially expressed in the tobacco hornworm, Manduca sexta. Nucleic Acids Res 1994; 22(15):2945-2950.
42. Kiely ML, Riddiford LM. Temporal programming of epidermal cell protein synthesis during the larval-pupal transformation of Manduca sexta. Roux's Arch Devel Biol 1985; 194:325-335.
43. Li W, Riddiford LM. Differential expression of the two duplicated insecticyanin genes, ins-a and ins-b, in the black mutant of Manduca sexta. Arch Biochem Biophys 1996; 330:65-70.
44. Kang Y, Kulakosky PC, Van Antwerpen R et al. Sequestration of insecticyanin, a blue hemolymph protein, into the egg of the hawkmoth Manduca sexta. Evidence for receptor-mediated endocytosis. Insect Biochem Mol Biol 1995; 25(4):503-510.
45. Kang Y, Ziegler R, van Antwerpen R et al. Characterization of the solubilized oocyte membrane receptor for insecticyanin, a biliprotein of the hawkmoth, Manduca sexta. Biochim Biophys Acta 1997; 1324(2):285-295.
46. Saito H. Purification and characterization of two insecticyanin-type proteins from the larval hemolymph of the Eri-silkworm, Samia cynthia ricini. Biochim Biophys Acta 1998; 1380:141-150.
47. Filippov VA, Filippova MA, Kodrík D et al. Two lipocalin-like peptides of insect brain. In: Suzuki A, Kataoka H, Matsumoto S, eds. Molecular Mechanisms of Insect Metamorphosis and Diapause. Tokyo: Industrial Publishing and Consulting, Inc., 1995:35-43.
48. Sakai M, Wu C, Suzuki K. Nucleotide and deduced amino acid sequences of a cDNA encoding a lipocalin protein in the central nervous system of Bombyx mori. Nippon Sanshi-gaku Zasshi 2001; 70:105-111.

49. Veiga ABG, Francischetti IMB, Guimaraes JA et al. A catalog of Lonomia obliqua transcripts and proteins probably involved in the hemorrhagic syndrome [GenBank database].
50. Reis CV, Ho PL, Ramos CR et al. r-LOPAP: A member of lipocalin family which shows serine-protease activity [GenBank database].
51. Montfort WR, Weichsel A, Andersen JF. Nitrophorins and related antihemostatic lipocalins from Rhodnius prolixus and other blood-sucking arthropods. Biochim Biophys Acta 2000; 1482(1-2 SU-):110-118.
52. Urata J, Shojo H, Kaneko Y. Inhibition mechanisms of hematophagous invertebrate compounds acting on the host blood coagulation and platelet aggregation pathways. Biochimie 2003; 85(5):493-500.
53. Ribeiro JMC, Charlab R, Pham VM et al. An insight into the salivary transcriptome and proteome of the adult female mosquito Culex pipiens quinquefasciatus. Insect Biochem Mol Biol 2004; 34(6):543-563.
54. Valenzuela JG, Pham VM, Garfield MK et al. Toward a description of the sialome of the adult female mosquito Aedes aegypti. Insect Biochem Mol Biol 2002; 32(9):1101-1122.
55. Gaunt MW, Miles MA. An insect molecular clock dates the origin of the insects and accords with palaeontological and biogeographic landmarks. Mol Biol Evol 2002; 19(5):748-761.
56. Andersen JF, Champagne DE, Weichsel A et al. Nitric oxide binding and crystallization of recombinant nitrophorin I, a nitric oxide transport protein from the blood-sucking bug Rhodnius prolixus. Biochemistry 1997; 36(15):4423-4428.
57. Andersen JF, Weichsel A, Balfour CA et al. The crystal structure of nitrophorin 4 at 1.5 Å resolution: Transport of nitric oxide by a lipocalin-based heme protein. Structure 1998; 6(10):1315-1327.
58. Andersen JF, Montfort WR. The crystal structure of nitrophorin 2. A trifunctional antihemostatic protein from the saliva of Rhodnius prolixus. J Biol Chem 2000; 275(39):30496-30503.
59. Weichsel A, Andersen JF, Champagne DE et al. Crystal structures of a nitric oxide transport protein from a blood-sucking insect. Nat Struct Biol 1998; 5(4):304-309.
60. Ribeiro JM, Hazzard JM, Nussenzveig RH et al. Reversible binding of nitric oxide by a salivary heme protein from a bloodsucking insect. Science 1993; 260(5107):539-541.
61. Ribeiro JM, Schneider M, Guimaraes JA. Purification and characterization of prolixin S (nitrophorin 2), the salivary anticoagulant of the blood-sucking bug Rhodnius prolixus. Biochem J 1995; 308(Pt 1):243-249.
62. Ribeiro JM, Walker FA. High affinity histamine-binding and antihistaminic activity of the salivary nitric oxide-carrying heme protein (nitrophorin) of Rhodnius prolixus. J Exp Med 1994; 180(6):2251-2257.
63. Champagne DE, Nussenzveig RH, Ribeiro JM. Purification, partial characterization, and cloning of nitric oxide-carrying heme proteins (nitrophorins) from salivary glands of the blood-sucking insect Rhodnius prolixus. J Biol Chem 1995; 270(15):8691-8695.
64. Sun J, Yamaguchi M, Yuda M et al. Purification, characterization and cDNA cloning of a novel anticoagulant of the intrinsic pathway, (prolixin-S) from salivary glands of the blood sucking bug, Rhodnius prolixus. Thromb Haemost 1996; 75(4):573-577.
65. Ribeiro JMC, Andersen J, Silva-Neto MAC et al. Exploring the sialome of the blood-sucking bug Rhodnius prolixus. Insect Biochem Mol Biol 2004; 34(1):61-79.
66. Moreira MF, Coelho HSL, Zingali RB et al. Changes in salivary nitrophorin profile during the life cycle of the blood-sucking bug Rhodnius prolixus. Insect Biochem Mol Biol 2003; 33(1):23-28.
67. Weichsel A, Andersen JF, Roberts SA et al. Nitric oxide binding to nitrophorin 4 induces complete distal pocket burial. Nat Struct Biol 2000; 7(7):551-554.
68. Gerczei T, Fazekas A, Keseru GM. Theoretical study on the nitric oxide binding to nitrophorin 1, an NO-carrier protein from a blood-sucking insect. J Mol Struct: THEOCHEM 2000; 503(1-2):51-58.
69. Andersen JF, Ding XD, Balfour C et al. Kinetics and equilibria in ligand binding by nitrophorins 1-4: Evidence for stabilization of a nitric oxide-ferriheme complex through a ligand-induced conformational trap. Biochemistry 2000; 39(33):10118-10131.
70. Kaneko Y, Yuda M, Iio T et al. Kinetic analysis on nitric oxide binding of recombinant Prolixin-S, a nitric oxide transport protein from the bloodsucking bug, Rhodnius prolixus. Biochim Biophys Acta 1999; 1431(2):492-499.
71. Ribeiro JC. Salivary thiol oxidase activity of Rhodnius prolixus. Insect Biochem Mol Biol 1996; 26(8-9):899-905.
72. Ribeiro JMC. Rhodnius prolixus salivary nitrophorins display heme-peroxidase activity. Insect Biochem Mol Biol 1998; 28(12):1051-1057.
73. Andersen JF, Gudderra NP, Francischetti IMB et al. Recognition of anionic phospholipid membranes by an antihemostatic protein from a blood-feeding insect. Biochemistry 2004; 43(22):6987-6994.

74. Ganfornina MD, Sánchez D. Generation of evolutionary novelties by functional shift. Bioessays 1999; 21:432-439.
75. Andersen JF, Francischetti IMB, Valenzuela JG et al. Inhibition of hemostasis by a high affinity biogenic amine-binding protein from the saliva of a blood-feeding insect. J Biol Chem 2003; 278(7):4611-4617.
76. Francischetti IM, Ribeiro JM, Champagne D et al. Purification, cloning, expression, and mechanism of action of a novel platelet aggregation inhibitor from the salivary gland of the blood-sucking bug, Rhodnius prolixus. J Biol Chem 2000; 275(17):12639-12650.
77. Francischetti IM, Valenzuela JG, Pham VM et al. Toward a catalog for the transcripts and proteins (sialome) from the salivary gland of the malaria vector Anopheles gambiae. J Exp Biol 2002; 205(Pt 16):2429-2451.
78. Noeske-Jungblut C, Haendler B, Donner P et al. Triabin, a highly potent exosite inhibitor of thrombin. J Biol Chem 1995; 270(48):28629-28634.
79. Sant Anna MR, Araujo JG, Pereira MH et al. Molecular cloning and sequencing of salivary gland-specific cDNAs of the blood-sucking bug Triatoma brasiliensis (Hemiptera: Reduviidae). Insect Mol Biol 2002; 11(6):585-593.
80. Fuentes-Prior P, Noeske-Jungblut C, Donner P et al. Structure of the thrombin complex with triabin, a lipocalin-like exosite-binding inhibitor derived from a triatomine bug. Proc Natl Acad Sci USA 1997; 94(22):11845-11850.
81. Noeske-Jungblut C, Kratzschmar J, Haendler B et al. An inhibitor of collagen-induced platelet aggregation from the saliva of Triatoma pallidipennis. J Biol Chem 1994; 269(7):5050-5053.
82. Haendler B, Becker A, Noeske-Jungblut C et al. Expression, purification and characterisation of recombinant pallidipin, a novel platelet aggregation inhibitor from the haematophageous triatomine bug Triatoma pallidipennis. Blood Coagul Fibrinolysis 1996; 7(2):183-186.
83. Paddock CD, McKerrow JH, Hansell E et al. Identification, cloning, and recombinant expression of procalin, a major triatomine allergen. J Immunol 2001; 167(5):2694-2699.
84. Mans BJ, Neitz AWH. Adaptation of ticks to a blood-feeding environment: Evolution from a functional perspective. Insect Biochem Mol Biol 2004; 34(1):1-17.
85. Mans BJ, Louw AI, Neitz AWH. The major tick salivary gland proteins and toxins from the soft tick, Ornithodoros savignyi, are part of the tick lipocalin family: Implications for the origins of tick toxicoses. Mol Biol Evol 2003; 20(7):1158-1167.
86. Paesen GC, Adams PL, Harlos K et al. Tick histamine-binding proteins: Isolation, cloning, and three-dimensional structure. Mol Cell 1999; 3(5 SU-):661-671.
87. Paesen GC, Adams PL, Nuttall PA et al. Tick histamine-binding proteins: Lipocalins with a second binding cavity. Biochim Biophys Acta 2000; 1482(1-2 SU-):92-101.
88. Sangamnatdej S, Paesen GC, Slovak M et al. A high affinity serotonin- and histamine-binding lipocalin from tick saliva. Insect Mol Biol 2002; 11(1):79-86.
89. Keller PM, Waxman L, Arnold BA et al. Cloning of the cDNA and expression of moubatin, an inhibitor of platelet aggregation. J Biol Chem 1993; 268(8):5450-5456.
90. Waxman L, Connolly TM. Isolation of an inhibitor selective for collagen-stimulated platelet aggregation from the soft tick Ornithodoros moubata. J Biol Chem 1993; 268(8):5445-5449.
91. Mans BJ, Venter JD, Vrey PJ et al. Identification of putative proteins involved in granule biogenesis of tick salivary glands. Electrophoresis 2001; 22(9):1739-1746.
92. Mans BJ, Neitz AW. Molecular crowding as a mechanism for tick secretory granule biogenesis. Insect Biochem Mol Biol 2004; 34(11):1187-1193.
93. Mans BJ, Steinmann CM, Venter JD et al. Pathogenic mechanisms of sand tampan toxicoses induced by the tick, Ornithodoros savignyi. Toxicon 2002; 40(7):1007-1016.
94. Jeffery CJ. Moonlighting proteins. Trends Biochem Sci 1999; 24:8-11.
95. Jeffery CJ. Moonlighting proteins: Old proteins learning new tricks. Trends Genet 2003; 19(8):415-417.
96. Williford A, Stay B, Bhattacharya D. Evolution of a novel function: Nutritive milk in the viviparous cockroach, Diploptera punctata. Evol Dev 2004; 6(2):67-77.
97. Arruda LK, Vailes LD, Hayden ML et al. Cloning of cockroach allergen, Bla g 4, identifies ligand binding proteins (or calycins) as a cause of IgE antibody responses. J Biol Chem 1995; 270(52):31196-31201.
98. Korchi A, Brossut R, Bouhin H et al. cDNA cloning of an adult male putative lipocalin specific to tergal gland aphrodisiac secretion in an insect (Leucophaea maderae). FEBS Lett 1999; 449(2-3 SU-):125-128.
99. Thompson JD, Gibson TJ, Plewniak F et al. The ClustalX windows interface: Flexible strategies for multiple sequence alignment aided by quality analysis tool. Nucleic Acids Res 1997; 24:4876-4882.

CHAPTER 7

Retinol Binding Protein and Its Interaction with Transthyretin

Marcia E. Newcomer* and David E. Ong

Abstract

Transport of vitamin A to the target cells is mediated by the lipocalin retinol-binding protein. In plasma, RBP is found in a complex with its carrier protein transthyretin (TTR). The structures of RBP free and in complex with TTR provide the details of the protein-protein interaction.

Introduction

Vitamin A is unique among the vitamins in having two separable functions. The parent compound, retinol (also known as vitamin A alcohol) is itself inactive but serves as the precursor for the production of various active forms that include the aldehyde retinal and retinoic acid. The light-induced *cis-trans* isomerization of the former is the initial signal required for vision, while the latter regulates gene expression by serving as a ligand for specific nuclear receptors of the steroid hormone super-family.[1] Thus the vitamin is essential not only for vision, but also has a hormonal action as retinoic acid in processes such as morphogenesis and proper differentiation and maintenance of various tissues,[2] particularly epithelium. All of the physiologically active forms of the vitamin are hydrophobic molecules that readily partition into lipid-rich sites and consequently the specific transport of the vitamin is mediated by both intra and extra cellular binding proteins. Only small amounts of the vitamin are required (several mg of retinol per day are adequate to support good health for humans) and these specific transport proteins allow for efficient mobilization of the vitamin by minimizing the loss to nonspecific interactions with membranes. The extra cellular transport protein for retinol (retinol binding protein, RBP) is a member of the lipocalin super-family. RBP circulates in the plasma bound to its carrier protein transthyretin (TTR) and this protein-protein complex is thought to interact with specific cell receptors to deliver retinol to the target cells. Because of its role in the specific transport of an essential vitamin, this lipocalin has been the focus of a great deal of research. This review represents a very brief summary of the structure and function of the RBP:TTR complex.

The body maintains a homeostatic level of retinol in plasma to serve as a constant source of precursor for active retinoids. This retinol is able to circulate throughout without appreciable loss because it is tightly bound to the 21000 Da plasma retinol binding protein (RBP), first described by Goodman and colleagues in 1968.[3] RBP is synthesized primarily in the liver, where it requires the binding of retinol to trigger its secretion.[4] Other sites of synthesis are known[5] and include the kidney, peritubular and Sertoli cells of the testis,[6,7] the retinal pigment epithelium,[8,9] and the choroid plexus of the brain.[10] Synthesis in the kidney is for the purpose of returning

*Corresponding Author: Marcia E. Newcomer—Department of Biological Sciences, Louisiana State University, Baton Rouge, Louisiana 70803, U.S.A. Email: newcomer@lsa.edu

Lipocalins, edited by Bo Åkerström, Niels Borregaard, Darren R. Flower and Jean-Philippe Salier. ©2006 Landes Bioscience.

salvaged retinol to the circulation while the other sites are for the purpose of moving retinol across those cells that form the blood-organ barriers of the testis, retina, and brain.

Whether vitamin A is to be ultimately utilized as retinoic acid, 11-*cis*-retinal or another retinoid, RBP delivers only all-*trans*-retin*ol*, and only retin*ol* can trigger secretion of RBP.[4] In the plasma, RBP binds to the larger protein, transthyretin (TTR, previously referred to as thyroxine binding prealbumin). The binding of RBP to TTR was suggested to prevent extensive loss of the low molecular weight RBP through glomerular filtration.[3] This hypothesis was supported by the much later experiments of Blaner and colleagues with TTR "knockout" mice that demonstrate that RBP is rapidly cleared from the plasma in TTR deficient mice.[11] In vitro one tetramer of TTR can bind two molecules of retinol binding protein. However, the concentration of RBP in the plasma is limiting and as a consequence the complex isolated from serum is composed of TTR and RBP in a one to one stoichiometry with a resultant molecular mass of about 80000 Da.

The RBP Structure Provided the First View of the Lipocalin Structure

RBP was the first lipocalin for which an X-ray structure was described[12] and thus serves as a sort of 'prototype' or reference for the description of subsequently reported lipocalin structures. The identification of the then novel fold and the fact that so much is known about the biological function of RBP make it a paradigm for the study of lipocalins. Structures have been determined for the human,[12-14] bovine,[15] porcine[16] and chicken[17] proteins. The basic structural framework of RBP is an eight-stranded up-and-down β-barrel onto which a carboxy-terminal α-helix is packed. Because of the characteristic twist of β-sheets and the orthogonal stacking of the two layers of β-sheet, a cylindrical structure is formed. The amino terminus of the protein, which includes one of the three regions of highly conserved amino acids found in lipocalins, seals one end of the cylinder, while the opposite end is the entrance to the binding cavity. It is this calyx or cup-like structure that inspired the name 'lipocalin' for this superfamily of proteins of which RBP is a member. All three regions of highly conserved amino acid sequence, which are spaced throughout in the linear sequence and constitute the 'motif' characteristic of this superfamily,[18,19] converge in the three-dimensional structure at the base of the calyx. The loops that connect the β-strands of the barrel are generally short, except for the loop between the first two strands which is longer and folds over the open end of the barrel.

The Retinol Binding Site

All-*trans*-retinol binds with the isoprene tail fully extended in the β-barrel with the trimethyl cyclohexylene ring innermost in a hand-in-glove-like fit in a hydrophobic cavity. Aromatic amino acids dominate at the ring end of the ligand. The ligand hydroxyl is at the protein surface and is solvent accessible. The high degree of complementarity in the shape of the binding site for the ligand is consistent with the fact that RBP is less tolerant of changes to the ring end of the ligand structure (for review see ref. 20) where structural differences would be difficult to accommodate without the rearrangement of amino acids in the protein core. In contrast, RBP binds retinal and retinoic acid with affinities comparable to that for retinol. The aldehyde and carboxylate can be accommodated in the binding site with only minor changes at the entrance of the β-barrel where the polar/charged groups can remain solvent exposed. Crystal structures of bovine RBP with all-*trans*-retinoic acid, *N*-ethyl retinamide, and 4-hydroxyphenyl retinamide[21,22] all illustrate the relative ease with which a ligand modified at the alcohol end can fit in the binding cavity.

A comparison of the apo and holo structures of bovine RBP revealed a ligand-induced conformational change which is confined to the entrance loop that includes amino acid residues 33-36. The most striking difference (illustrated in Fig. 1) is in the orientation of amino acids Leu-35 and Phe-36. Phe-36 moves into the space occupied by the isoprene tail in the holo protein.[15]

The Retinol Binding Protein Transthyretin Complex

As mentioned above, retinol-RBP in plasma is found in a complex with transthyretin. The affinity of RBP for TTR is roughly micromolar[20] and the two proteins readily copurify. The proteins co-elute on gel filtration columns if not dialyzed into low ionic strength buffer prior to the application of sample.

The structure of TTR has also been determined , as well as the structures of the protein:protein complexes of chicken RBP and human TTR[23] and human RBP:TTR.[24] Transthyretin is a homotetramer which is best described as a dimer of dimers. Monomers associate into dimers via the formation of an eight-stranded antiparallel β-sheet to which each monomer contributes four β-strands. The dimers are associated such that the two β-sheets are back-to-back at the center of the tetramer, creating a large solvent channel passing between the two sheets.[25] In the channel is the binding pocket for two molecules of thyroxine.

The open end of the RBP barrel docks up against TTR at a twofold axis of symmetry perpendicular to the large solvent channel that provides access to the thyroxine binding sites. That small portion of the retinol which remains solvent accessible when bound to uncomplexed RBP is fully concealed with the formation of the ternary complex. This serves to protect the reactive end of the molecule and also substantially slows release of retinol from its RBP binding site (unpublished observation, Ong and Kakkad). In the human:human TTR:RBP complex, the accessible surface area of 42 amino acids is reduced by complex formation, and RBP and TTR contribute 21 amino acids each to the interface. As is typical for protein:protein interaction sites, hydrophobic amino acids are at the center of the site and charged amino acids at the periphery of the site. Leu and Ile are the predominant amino acids in the interface. Half of the amino acid side chains are hydrophobic or aromatic (16 and five, respectively). The hydrophobic nature of the interface is consistent with the observation that complex dissociates only at low ionic strength. The docking interface is compatible for TTR and RBP-retinol, but less so for the non-physiological complex that RBP can form in vitro with retinoic acid, or even for RBP when free of retinol. As mentioned above, ligand binding to RBP induces a conformational change which results in the movement of the loop which includes amino acids 35 and 36.[15] These amino acids are found at the protein:protein interface and in the apo protein are not positioned for favorable interaction with TTR, hence the observed diminished affinity of apo-RBP for TTR.

As described above, the face that TTR presents in the formation of the complex is at the twofold axis which relates the dimers of TTR in the tetramer. Consequently, the binding site for RBP is symmetric (despite the fact that RBP itself is not) and of a total of 21 amino acid residues of TTR that are buried in the complex, 12 are duplicates, i.e., six amino acids occur twice in the binding site. A variety of naturally occurring mutations in transthyretin have been described, and the mutation of Ile(84) to Asn or Ser abrogates RBP:TTR complex formation.[26] Individuals with this variant of TTR have substantially lowered plasma concentrations of RBP due to kidney filtration,[27] an observation which confirms the importance of complex formation in vitamin A homeostasis. Ile-84 is one of the amino acids that is contributed by two monomers to the RBP recognition site. The duplication of Ile-84 in the RBP binding site and its location at the heart of the hydrophobic interface is consistent with the severity of the consequences of the mutation of this surface amino acid. RBP affinity for I84S-TTR is negligible.[26] While the structures of piscine TTR and RBP are highly conserved with respect to their human counterparts, RBP and TTR do not form a complex in sea bream.[28] One key sequence difference is that the Ile(84) is a serine in the fish TTR. The complementarity of the TTR:RBP interface and the central location of ile-84 are illustrated in Figure 2.

In addition to the interface created by the docking of the barrel entrance of RBP at a TTR dimer twofold axis, an interaction of TTR with the carboxy terminus of RBP was revealed in the crystal structure of the human RBP:TTR complex. In the previously determined X-ray structures of human RBP alone the carboxy terminus was not visible and therefore disordered.[12-14] This interaction was not observed for the structure of the heterologous complex of chicken RBP and human TTR[23] as the carboxy-terminal eight amino acids are found only in

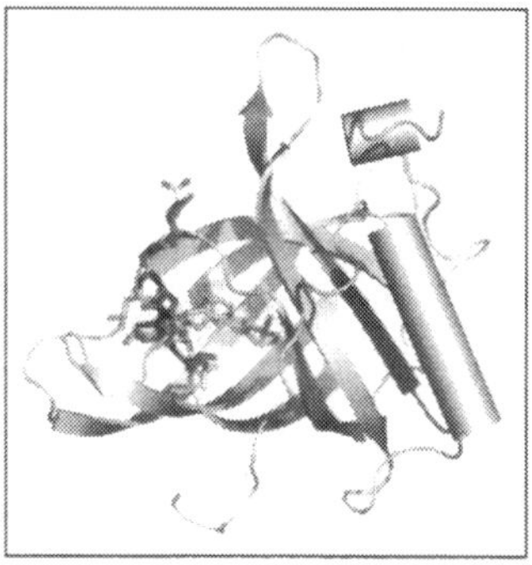

Figure 1. Please see legend on next page.

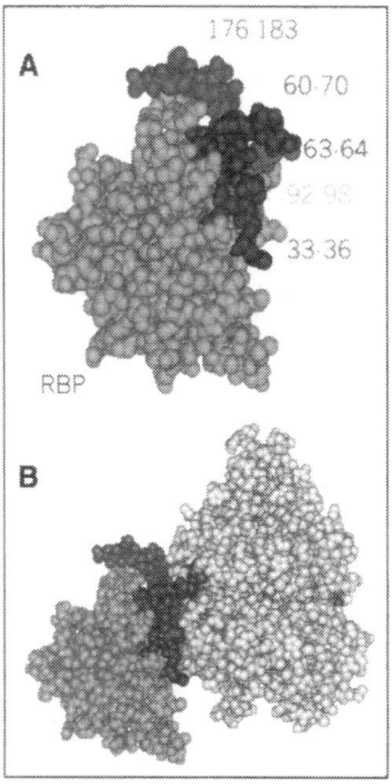

Figure 3. Please see legend on next page.

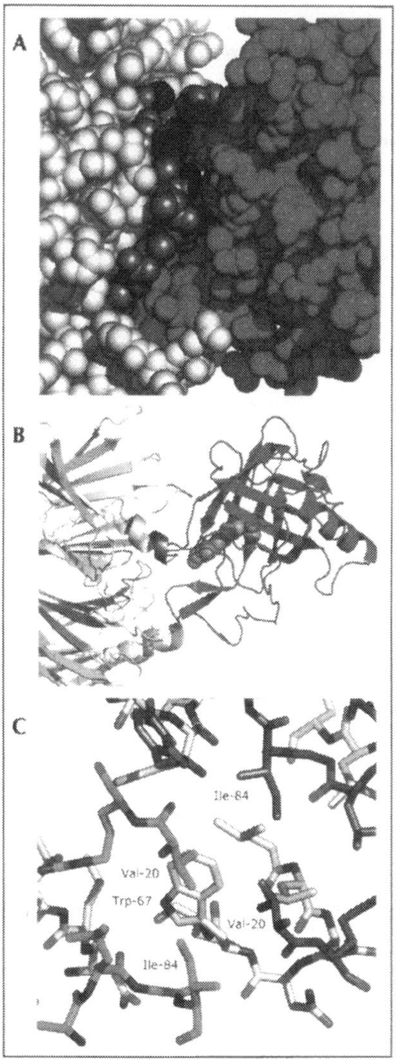

Figure 2. Please see legend on next page.

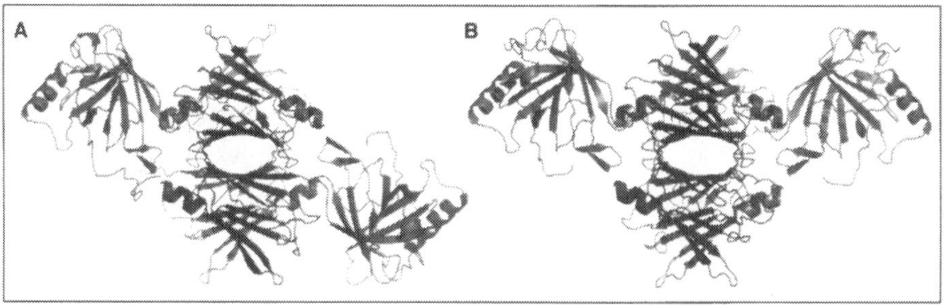

Figure 4. Please see legend on next page.

Figure 1. The structure of bovine retinol binding protein. A cartoon rendering of RBP (pdb code 1HBP) in which α-helices are shown as cylinders and β-strands as arrows. The conformational change that accompanies ligand binding is confined to amino acids 33-37;[15] this region is illustrated in stick rendering (holo-protein, carbon white, nitrogen blue, oxygen red; apo protein (1HBQ), carbon pink, nitrogen blue, oxygen red). Retinol is shown in orange.

Figure 2. The interaction of transthyretin and retinol-binding protein. A) A space-filled rendering of the transthyretin (white and blue and purple) and RBP (pink) interface in which shape complementarity is apparent. B) A ribbon drawing of the interaction. In this view the positions of the TTR residues that interact with RBP are indicated by shades of blue. Each independent chain is colored in a unique shade as residues from three monomers of TTR contribute to the interface. The amino acid side chains that RBP contributes to the interaction are from loops that flank the entrance to the binding site. C) Detail on the core of the interaction site with coloring by atom as above. The carbons of TTR are in blue and green, and those from RBP in white. Note that Ile-84 and Val-20 from two monomers of TTR are present.

Figure 3. Amino acids that have been shown to affect RBP:TTR complex formation experimentally. Regions of RBP that have been shown experimentally to be important for TTR binding are shown in purple, cyan, red, brown and magenta in free RBP (A) and the (B) RBP:TTR complex.

Figure 4. The quaternary structures of the RBP:TTR complexes. The human RBP:human TTR complex (top) and chicken RBP: human TTR complex (bottom). RBP is in green, and TTR is colored in shades of blue and purple.

mammalian RBPs. However, the interaction appears to be biologically relevant as it has been observed that naturally occurring truncated forms of RBP are more readily cleared from the plasma than full length RBP.[29,30] As indicated, complex formation prevents extensive loss of RBP through glomerular filtration and the loss of the two carboxy-terminal leucines (182-183) that are found nestled in a groove formed by the docking of RBP on to TTR may lead to a reduced affinity of RBP for TTR.

In Figure 3 space-filled renderings of the RBP and the RBP:TTR complex are presented in which regions of RBP have been highlighted to illustrate those stretches of the protein that various groups have demonstrated to be involved in TTR binding. As mentioned above, Zanotti et al[15] defined the conformational differences in the apo and holo protein that can account for the observation that apo-RBP has a reduced affinity for TTR. This region of the protein (shown in purple) is an integral part of the protein:protein recognition interface. Site directed mutagenesis experiments by Sivaprasadarao and Findlay[31] indicated that the loop which includes residues 92-98 (cyan in the drawing) is required for TTR binding. In contrast, point mutations at positions 63 and 64 (red) had no effect on TTR affinity. This is somewhat surprising in light of the fact these two residues, which in the native protein are leucines, are found at the interface, but entirely consistent with the results of Melhus et al[32] who observed that monoclonal antibodies to a synthetic peptide of RBP amino acids 60-70 (red and darker red), can immunoprecipitate TTR:RBP. In addition, Noy et al[33] observe that retinol dissociates from RBP prior to RBP:TTR complex dissociation. Such a sequence of events can be envisioned if the loop which includes amino acids 60-70 could peel away from the interface to allow ligand exit. Perhaps intermediate structures, which allow for interaction of RBP with a putative cell surface receptor, as well as TTR, exist. The carboxy terminus of RBP, which appears to modulate RBP:TTR affinity is shown in magenta. It is positioned in contact with the 60-70 loop, which is clearly part of the recognition surface.

Because TTR is a dimer of dimers, there are two equivalent binding sites for RBP. Furthermore, the twofold symmetry of the recognition interface itself results in two possible relative positionings of the two RBPs onto TTR. Both quaternary structures have been observed crystallographically. In the complex of chicken RBP with human TTR, the RBP molecules make the bulk of the contacts with the same dimer of TTR (where a dimer of TTR is the unit of monomers associated *via* the formation of the eight-stranded β-sheet) and in the human:human complex the RBPs make the bulk of the interactions with monomers of TTR from the two different dimers (Fig. 4). It is not possible to say whether the difference in quaternary structure reflects a true structural difference between the mixed species and human complexes, or whether they are simply the quaternary structure selected by the different crystallization conditions.

While the excellent surface complementarity of the interaction between RBP and TTR is illustrated in Figure 2A, from Figure 2B one can appreciate the fact that the TTR interaction sites on RBP are confined to loops at one end of the barrel. In an effort to confirm that TTR affinity is conferred by these loops, Sundaram et al[34] prepared a chimera of RBP and the rat epididymal retinoic acid binding protein (ERABP), also a lipocalin for which a structure has been determined.[35] The chimera was composed of an ERABP in which the two TTR-binding loops of RBP were substituted for their counterparts in ERABP. This hybrid protein has affinity for TTR.[36]

Naturally Occurring Mutant Forms of RBP

Vitamin A status is routinely evaluated by analysis of serum RBP levels. Only in extreme cases of malnourishment do the serum levels of RBP drop, but these can be readily restored to normal levels with ingestion of the vitamin. However, two instances of vitamin A deficiency which were not ameliorated by vitamin A intake have been reported.[37-39] Only one of these cases, in which no serum RBP could be detected, has been fully characterized, and in this case two point mutations in two separate alleles were identified.[39] One mutation results in the substitution of asparagine for isoleucine at position 41. This amino acid side chain forms part of the retinol binding site and is positioned roughly 5 Å from the ionone ring. The incorporation of a polar amino acid for a hydrophobic amino acid would surely result in a binding pocket which is less complementary in chemistry to its ligand and consequently a protein with diminished affinity for retinol. The second substitution is an aspartate for a glycine at position 75. It would appear that the protein fold would be much less forgiving of this latter change. This glycine is also found in the binding pocket and the substitution of the much larger aspartate which carries a negative charge at this position would undoubtedly have grave structural consequences.

The physiological consequence of the above mutations is that the sisters in whom it was identified suffer from night blindness. Although biological processes in addition to vision require vitamin A, these needs can apparently be met by less specific transport systems. Quadro et al[40] have generated $RBP^{-/-}$ mice and find that although these mice have markedly impaired retinal function during the first few months of life, they are otherwise healthy and capable of reproducing. If provided with a diet sufficient in vitamin A the mice have normal vision by five months of age. RBP thus appears to be required for efficient retinol mobilization in times of insufficient dietary intake. The alternative pathway is likely *via* hydrolysis of retinyl esters in the chylomicrons released to the circulation from the small intestine.[41] These esters are available from hydrolysis by lipoprotein lipase, releasing retinol for cellular uptake.[42] Interestingly, the retina takes up the retinol released in this way very poorly, compared to other tissues. Conversely, the retina takes up retinol from RBP avidly, much faster than most tissues except kidney.[11] This is consistent with a specific mechanism for uptake. As noted next, this is an important subject that is still unresolved.

Interaction with a Putative RBP Receptor

Retinol in the extracellular compartment must be taken up by the cell, but encapsulated in the RBP:TTR complex it is restricted from freely partitioning into the cell membrane. Undoubtedly some entry into the membrane fraction can occur by a non-specific process, and some investigators contend that such a process may account for all retinol uptake by cells. However, it is not clear that such an unregulated transport system can achieve the level of specificity dictated by the differing requirements of cell types for retinol. Experimental data support the existence of a cell surface receptor for RBP, and other data suggest that this interaction site may overlap with the TTR binding site. However, whether these cell surface receptors mediate internalization of free ligand or RBP-bound ligand remains an area of debate. In any event, the existence of a saturable cell surface receptor for RBP has been suggested in studies with placental brush border membranes[31] cultured Sertoli cells,[43] stellate cells,[44] peritubular cells,[7] retinal pigment epithelial cells,[45,46] embryonal carcinoma cells,[47] and the choroid plexus.[10]

However, such a protein remains to be unambiguously identified. Sundaram et al,[34] have put forth the hypothesis that the cell surface receptor mediates a transfer of retinol from RBP in the extracellular compartment to the intracellular retinol binding protein (cellular retinol binding protein, CRBP) a member of a distinct protein superfamily with a structural motif that is reminiscent of the lipocalin fold. The intra cellular binding proteins are composed of ten-stranded up-and-down -barrels with helical caps.[48]

Concluding Remarks

Although the lipocalins share a common structural motif, the biological functions of the proteins will no doubt prove to be as diverse as the sequences which adopt the characteristic fold. The serum retinol binding protein provides an excellent system in which to appreciate the complexity of the interactions a simple carrier molecule must participate in order to mediate the specific transport of a vital nutrient.

References

1. Steinmetz AC, Renaud JP, Moras D. Binding of ligands and activation of transcription by nuclear receptors. Annu Rev Biophys Biomol Struct 2001; 30:329-359.
2. Clagett-Dame M, DeLuca HF. The role of vitamin A in mammalian reproduction and embryonic development. Annu Rev Nutr 2002; 22:347-381.
3. Kanai M, Raz A, Goodman D. Retinol binding protein: The transport protein for vitamin A in human plasma. J Clin Invest 1968; 47:2025-2044.
4. Ronne H, Ocklind C, Wiman K et al. Ligand dependent regulation of intracellular protein transport: Effect of vitamin A on the secretion of retinol-binding protein. J Cell Biol 1983; 96(907-910).
5. Soprano DR, Blaner WS. Plasma retinol-binding protein. In: Sporn MB, Roberts AB, Goodman DS, eds. The Retinoids. 2nd ed. New York: Raven Press, 1994:257-282.
6. Davis JT, Ong DE. Synthesis and secretion of retinol-binding protein by cultured rat Sertoli cells. Biol Reprod 1992; 47(4):528-533.
7. Davis JT, Ong DE. Retinol processing by the peritubular cell from rat testis. Biol Reprod 1995; 52(2):356-364.
8. Ong DE, Davis JT, O'Day WT et al. Synthesis and secretion of retinol-binding protein and transthyretin by cultured retinal pigment epithelium. Biochemistry 1994; 33(7):1835-1842.
9. Jaworowski A, Fang Z, Khong TF et al. Protein synthesis and secretion by cultured retinal pigment epithelia. Biochim Biophys Acta 1995; 1245(1):121-129.
10. MacDonald PN, Bok D, Ong DE. Localization of cellular retinol-binding protein and retinol-binding protein in cells comprising the blood-brain barrier of rat and human. Proc Natl Acad Sci USA 1990; 87(11):4265-4269.
11. Vogel S, Piantedosi R, O'Byrne SM et al. Retinol-binding protein-deficient mice: biochemical basis for impaired vision. Biochemistry 2002; 41(51):15360-15368.
12. Newcomer ME, Jones TA, Aqvist J et al. The three-dimensional structure of retinol-binding protein. EMBO J 1984; 3(7):1451-1454.
13. Cowan SW, Newcomer ME, Jones TA. Crystallographic refinement of human serum retinol binding protein at 2A resolution. Proteins 1990; 8(1):44-61.
14. Zanotti G, Ottonello S, Berni R et al. Crystal structure of the trigonal form of human plasma retinol-binding protein at 2.5 A resolution. J Mol Biol 1993; 230(2):613-624.
15. Zanotti G, Berni R, Monaco HL. Crystal structure of liganded and unliganded forms of bovine plasma retinol-binding protein. J Biol Chem 1993; 268(15):10728-10738.
16. Zanotti G, Panzalorto M, Marcato A et al. Structure of pig plasma retinol-binding protein at 1.65 A resolution. Acta Crystallogr D Biol Crystallogr 1998; 54(Pt 5):1049-1052.
17. Zanotti G, Calderone V, Beda M et al. Structure of chicken plasma retinol-binding protein. Biochim Biophys Acta 2001; 1550(1):64-69.
18. Sawyer L. Protein structure. One fold among many [news]. Nature 1987; 327(6124):659.
19. Flower DR, North AC, Attwood TK. Structure and sequence relationships in the lipocalins and related proteins. Protein Sci 1993; 2(5):753-761.
20. Goodman DS. Plasma retinol-binding protein. Ann N Y Acad Sci 1980; 348:378-390.
21. Zanotti G, Marcello M, Malpeli G et al. Crystallographic studies on complexes between retinoids and plasma retinol-binding protein. J Biol Chem 1994; 269(47):29613-29620.
22. Zanotti G, Malpeli G, Berni R. The interaction of N-ethyl retinamide with plasma retinol-binding protein (RBP) and the crystal structure of the retinoid-RBP complex at 1.9-A resolution. J Biol Chem 1993; 268(33):24873-24879.

23. Monaco HL. Three-dimensional structure of the transthyretin-retinol-binding protein complex. Clin Chem Lab Med 2002; 40(12):1229-1236.
24. Naylor HM, Newcomer ME. The structure of human retinol-binding protein (RBP) with its carrier protein transthyretin reveals an interaction with the carboxy terminus of RBP. Biochemistry 1999; 38(9):2647-2653.
25. Blake CCF, Geisow MJ, Oatley SJ et al. Structure of prealbumin: secondary, tertiary and quaternary interactions determined by Fourier refinement at 1.8. J Mol Biol 1978; 121:339-356.
26. Berni R, Malpeli G, Folli C et al. The Ile-84→Ser amino acid substitution in transthyretin interferes with the interaction with plasma retinol-binding protein. J Biol Chem 1994; 269(38):23395-23398.
27. Waits RP, Yamada T, Uemichi T et al. Low plasma concentrations of retinol-binding protein in individuals with mutations affecting position 84 of the transthyretin molecule. Clin Chem 1995; 41(9):1288-1291.
28. Folli C, Pasquato N, Ramazzina I et al. Distinctive binding and structural properties of piscine transthyretin. FEBS Lett 2003; 555(2):279-284.
29. Jaconi S, Rose K, Hughes GJ et al. Characterization of two post-translationally processed forms of human serum retinol-binding protein: altered ratios in chronic renal failure. J Lipid Res 1995; 36(6):1247-1253.
30. Jaconi S, Saurat JH, Siegenthaler G. Analysis of normal and truncated holo- and apo-retinol-binding protein (RBP) in human serum: altered ratios in chronic renal failure. Eur J Endocrinol. 1996; 134(5):576-582.
31. Sivaprasadarao A, Findlay JB. The interaction of retinol-binding protein with its plasma-membrane receptor. Biochem J 1988; 255(2):561-569.
32. Melhus H, Bavik CO, Rask L et al. Epitope mapping of a monoclonal antibody that blocks the binding of retinol-binding protein to its receptor. Biochem Biophys Res Commun 1995; 210(1):105-112.
33. Noy N, Xu ZJ. Interactions of retinol with binding proteins: implications for the mechanism of uptake by cells. Biochemistry 1990; 29(16):3878-3883.
34. Sundaram M, Sivaprasadarao A, DeSousa MM et al. The transfer of retinol from serum retinol-binding protein to cellular retinol-binding protein is mediated by a membrane receptor. J Biol Chem 1998; 273(6):3336-3342.
35. Newcomer ME. Structure of the epididymal retinoic acid binding protein at 2.1 A resolution. Structure 1993; 1(1):7-18.
36. Sundaram M, van Aalten DM, Findlay JB et al. The transfer of transthyretin and receptor-binding properties from the plasma retinol-binding protein to the epididymal retinoic acid-binding protein. Biochem J 2002; 362(Pt 2):265-271.
37. Matsuo T, Matsuo N, Shiraga F et al. Keratomalacia in a child with familial hypo-retinol-binding proteinemia. Jpn J Ophthalmol 1988; 32(3):249-254.
38. Matsuo T, Matsuo N. Characterization of retinol-binding protein in familial hypo-retinol- binding proteinemia. Jpn J Ophthalmol 1988; 32(4):379-384.
39. Biesalski HK, Frank J, Beck SC et al. Biochemical but not clinical vitamin A deficiency results from mutations in the gene for retinol binding protein. Am J Clin Nutr 1999; 69(5):931-936.
40. Quadro L, Blaner WS, Salchow DJ, et al. Impaired retinal function and vitamin A availability in mice lacking retinol-binding protein. EMBO J 1999; 18(17):4633-4644.
41. Ross AC. Mutations in the gene encoding retinol binding protein and retinol deficiency: is there compensation by retinyl esters and retinoic acid? Am J Clin Nutr 1999; 69(5):829-830.
42. Blaner WS, Obunike JC, Kurlandsky SB et al. Lipoprotein lipase hydrolysis of retinyl ester. Possible implications for retinoid uptake by cells. J Biol Chem 1994; 269(24):16559-16565.
43. Shingleton JL, Skinner MK, Ong DE. Characteristics of retinol accumulation from serum retinol-binding protein by cultured Sertoli cells. Biochemistry 1989; 28(25):9641-9647.
44. Senoo H, Smeland S, Malaba L et al. Transfer of retinol-binding protein from HepG2 human hepatoma cells to cocultured rat stellate cells. Proc Natl Acad Sci USA 1993; 90(8):3616-3620.
45. Pfeffer BA, Clark VM, Flannery JG et al. Membrane receptors for retinol-binding protein in cultured human retinal pigment epithelium. Invest Ophthalmol Vis Sci 1986; 27(7):1031-1040.
46. Bavik CO, Eriksson U, Allen RA et al. Identification and partial characterization of a retinal pigment epithelial membrane receptor for plasma retinol-binding protein. J Biol Chem 1991; 266(23):14978-14985.
47. Eriksson U, Hansson E, Nilsson M et al. Increased levels of several retinoid binding proteins resulting from retinoic acid-induced differentiation of F9 cells. Cancer Res 1986; 46(2):717-722.
48. Banaszak L, Winter N, Xu Z et al. Lipid-binding proteins: a family of fatty acid and retinoid transport proteins. Adv Protein Chem 1994; 45:89-151.

Chapter 8

Siderocalins

Roland K. Strong*

Abstract

Siderocalin (lipocalin 2), first identified as a neutrophil granule component, is also found in uterine secretions, in serum and synovium during bacterial infection and secreted from epithelial cells in response to inflammation or tumorigenesis. Siderocalin is a potent bacteriostatic agent in vitro and, when knocked-out in mice, confers a remarkable susceptibility to bacterial infection in the absence of any other phenotype. However, siderocalin lacked any precise function until specific, high-affinity ligands were identified: bacterial ferric siderophores. Siderophores, small-molecule iron (III) chelators, are synthesized, secreted and reabsorbed by microorganisms in a competition to obtain iron, a scarce resource in the environment, and have been linked to virulence, though through previously undefined mechanisms. Siderocalin employs degenerate molecular recognition machinery to bind to two distinct families of siderophores: the catecholate siderophores of enteric bacteria and the mycobacterial carboxymycobactins. Siderocalin therefore functions as an anti-bacterial component of innate immune responses by sequestering iron away from invading pathogens; pathogens use siderophores that escape siderocalin capture to help establish virulence. However, the limited pattern of siderocalin siderophore specificity, the use of alternate or modified siderophores by bacteria and the possible existence of other siderophore-binding lipocalins ('siderocalins') clearly demonstrates that the battle for virulence is ongoing. Siderocalin may also have pleiotropic activities, having been implicated in diverse cellular processes such as apoptosis and differentiation.

Siderocalin

Lipocalin 2 (Lcn2), first identified as a lipocalin in neutrophil granules,[1-3] has been referred to by various names by various groups, reflecting the many contexts in which it has been found, including: neutrophil gelatinase-associated lipocalin (NGAL), human neutrophil lipocalin (HNL), 24p3, superinducible protein 24 kD (SIP24), uterocalin, *neu*-related lipocalin (NRL), α_2-microglobulin-related protein and, most recently, siderocalin.[4] The last appellation ties the protein to its established ligands and a specific function, rather than some subset of its expression pattern or the chromosomal location of its gene, and is thus our preferred label. Human siderocalin (NGAL, HNL), while also expressed by epithelial cells in response to inflammatory signals, is released from neutrophil granules as a 25 kiloDalton (kD) monomer, a 46 kD disulfide-linked homodimer and a disulfide-linked heterodimer with gelatinase-B (matrix metalloproteinase 9 (MMP-9)).[5,6] There is a single N-linked oligosaccharide site on siderocalin and a single, conserved internal disulfide bond shared by many lipocalins. The rat ortholog (NRL) was originally identified as a protein highly

*Roland K. Strong—Division of Basic Sciences, Fred Hutchinson Cancer Research Center, Mail Stop A3-025, 1100 Fairview Ave. North, Seattle, Washington 98109, U.S.A.
Email: rstrong@fhcrc.org

Lipocalins, edited by Bo Åkerström, Niels Borregaard, Darren R. Flower and Jean-Philippe Salier. ©2006 Landes Bioscience.

Table 1. Sequence identities between siderocalin and its nearest neighbor lipocalins

	Human Siderocalin	Rat Siderocalin	Murine Siderocalin	Human Lcn12	Human PDS	Murine Lcn12	Human HC	Chicken Ex-FABP	Quail Q83
Human Siderocalin	100	**62**	**60**	35	34	28	23	20	20
Rat Siderocalin		100	**81**	34	32	29	<20	20	20
Murine Siderocalin			100	31	31	28	<20	20	22
Human Lcn12				100	29	**56**	<20	28	<20
Human PDS					100	28	23	21	21
Murine Lcn12						100	<20	<20	<20
Human HC							100	<20	<20
Chicken Ex-FABP								100	**86**
Quail Q83									100

Identities between likely orthologs are in **bold**; pair-wise identities between human $C8\gamma$, human Lcn1 and all the remaining lipocalins above are all less than 20%. Only the closet PDS sequence is shown as a representative of the family. Candidate siderocalins are underlined.

overexpressed in *neu*-induced mammary cancers;[7] the murine ortholog (24p3, Sip24, uterocalin) was identified as a protein induced in response to various proliferative signals[8] and is highly expressed in uterine luminal fluids and by epithelial cells.[9] Human-mouse-rat Siderocalin pair-wise sequence identities are 60% or greater (Table 1) with marked conservation of calyx residues;[4] no other obvious orthologs have been identified so far. An equine lipocalin, also named uterocalin (on the basis of its presence in uterine secretions) or P19,[10] shows only weak sequence homology to siderocalin (16% to 18% pair-wise identity to the three orthologs).

Siderocalin has been implicated in processes as diverse as apoptosis[11,12] and kidney cell differentiation[13,14] and, like most lipocalins, is thought to modulate these cellular processes by binding to ligand(s) and interacting with specific cell-surface receptors. Evidence for mammalian cell-surface siderocalin receptors has been reported in two different systems,[11,13] though they remain uncharacterized, but the interaction with a third receptor, megalin (an extracellular matrix component), has been more fully characterized.[15] Megalin, a member of the low density lipoprotein family,[16] also interacts with lactoferrin[17,18] and other lipocalins: retinol binding protein, α_1-microglobulin/HC, mouse major urinary protein and odorant binding protein.[19,20] Megalin binds Siderocalin in a ligand-independent manner and can mediate its cellular uptake,[15] though the physiological role of this interaction remains speculative.

Lipocalins usually function through the ligands they bind and it is the shape of the lipocalin calyx, and the chemical character of the residues lining the calyx, that are the primary determinants of ligand specificity. [Equine uterocalin, for instance, displays a predicted calyx that is quite distinct in character from siderocalin; therefore, it is unlikely that equine uterocalin has an analogous ligand specificity or function.] However, some of these putative lipocalin functional assignments have been made on very indirect or circumstantial evidence, without determining the actual molecular mechanisms or ligands involved. For instance, the name 'lipocalin', referring to the lipophilic character of the stereotypical calyx, is reflected in the measurable affinity many family members have for simple fatty acids or retinoids, though many studies have failed to show that such interactions are specific and not due to trivial, non-specific, hydrophobic interactions. The term 'lipocalin' itself can be misleading, as a number of family members have now been shown to have distinctly polar calyces and/or bind preferentially to non-lipid or -lipophilic ligands.

Siderocalin was initially proposed to have immunomodulatory activity by binding and clearing lipophilic inflammatory mediators,[21] such as the neutrophil chemoattractant tripeptide N-formyl-Met-Leu-Phe.[22,23] However, the initial three-dimensional structures of human siderocalin, determined by X-ray crystallography[24] and NMR,[25] revealed that the calyx is shallower (only about 15Å deep) and broader (approximately 20Å wide at the brim) than is typical of most lipocalins and is also uncharacteristically lined with polar and positively-charged residues. The siderocalin calyx is highly sculpted, with three prominent pockets (#1, #2 & #3) outlined by the position of the side-chains of three positively-charged calyx residues (Arg81, Lys125 and Lys134).[24] The incompatible nature of the siderocalin calyx, together with binding data showing millimolar dissociation constants,[24,26] led to the conclusion that siderocalin does not specifically bind N-formylated tripeptides or other proposed hydrophobic ligands,[24] leaving its physiological function in question.

The initial crystallographic analysis of siderocalin, using baculovirus-expressed protein, did show the presence of what was likely a fatty acid in the calyx that could be, surprisingly, displaced by sulfate.[24] This weakly-bound bound fatty acid, modeled as n-capric acid (NCA), likely represented a serendipitous ligand co-purified with siderocalin, as has been seen in the structures of other lipocalins.[27] Distinct from the typical fatty acid-binding lipocalins, the carboxyl moiety of the fatty acid bound in the deepest pocket (#2) in the siderocalin calyx, with the aliphatic tail trailing outwards. However, paralleling the difficulty in identifying 'true' lipocalin ligands, the low affinity, small size relative to the calyx volume and poor overall shape/chemical compatibility of the fatty acid strongly argued that this compound was also not a physiologically-relevant ligand.

Siderocalin Ligands

Building on the unexpected observation that siderocalin co-purifies with a dark red chromophore when expressed in bacteria, our laboratory identified bacterial ferric siderophores as candidate siderocalin ligands. A 'siderophore' is defined as a low molecular weight, virtually ferric-specific chelator involved in iron acquisition. Microorganisms secrete siderophores in a competition to scavenge scarce environmental iron in a receptor-dependent manner.[28-31] Siderophores can be divided into three broad classes depending on the chemistry of chelation: hydroxamates, phenolates/catecholates or α-hydroxycarboxylates—though other liganding chemistries can be used (Tables 2-5). Siderophore affinities for iron (III) can exceed 10^{50} M^{-1}.

Subsequent biophysical studies show that siderocalin tightly binds (1) the catecholate-type ferric siderophore of enterobacteria, enterochelin (Ent; also known as enterobactin; K_D = 0.4 nM at 22°C),[4] (2) a panel of related catecholate-type siderophores[32] and (3) the soluble, mixed catecholate/hydroxamate carboxymycobactin siderophores of *Mycobacterium smegmatis* and *M. tuberculosis* (CMB-S and CMB-T; Table 2).[32] Complex structures[4] show that siderocalin binds ferric Ent (FeEnt) and related siderophores by intercalating the side-chains of the three positively-charged calyx residues (Arg81, Lys125 and Lys134; Fig. 1) between the three

Table 2. Siderophores demonstrated to bind to siderocalin[32,44]

Compound (references)	Organism(s)	Structure	Comments (K_a for Fe(III))
Carboxymyco-lipophilic bactins	mycobacteria	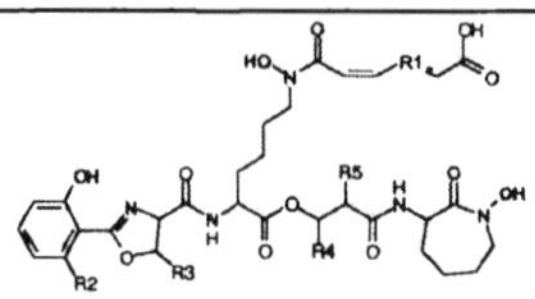	soluble form of the mycobactins (shorter acyl chain plus a carboxylate) R1 = $(\text{-CH}_2\text{-})_{1\text{-}9}$ R4 = -CH_3 or -H
2,3-Dihydroxy-benzoate (DHBA) and Serine-DHB (82)	*Brucella abortus* (DHBA (83))	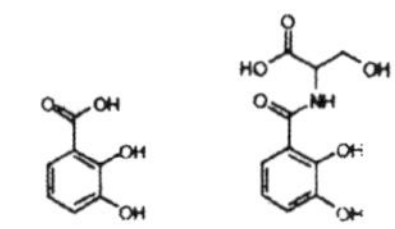	biosynthetic intermediates or breakdown products of Ent; three molecules together bind one iron
Enterochelin (Ent; Enterobactin) (84, 85)	enterobacteria	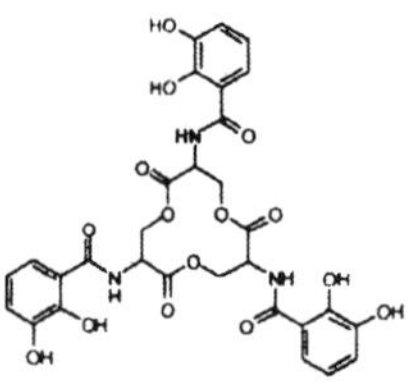	tris-catecholate; three DHBA groups coupled through amide linkages to a tri-serine, trilactone backbone; primary siderophore of many enterobacteria (10^{52})
MECAM (86)	synthetic Ent analog	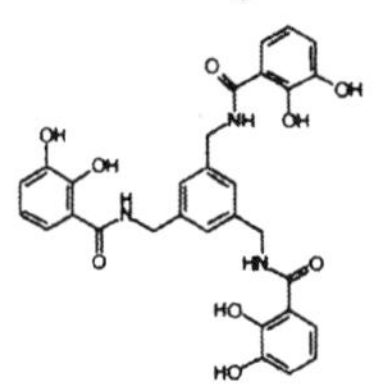	less-hydrolyzable Ent analog, differing only in the backbone
Parabactin (87)	*Paracoccus*	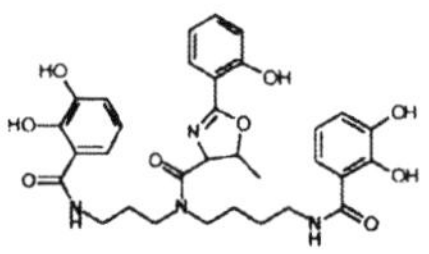	similar to Ent; oxazoline group replaces one phenolate; different backbone ($\sim 10^{50}$)
TRENCAM (86)	synthetic Ent analog	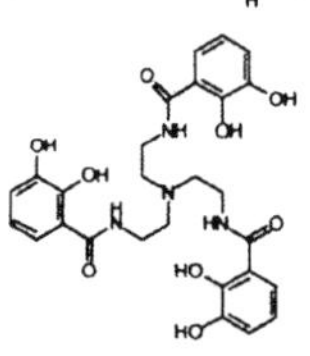	less-hydrolyzable Ent analog, differing only in the backbone
TREN-3,2-HOPO (86)	synthetic cepabactin analog	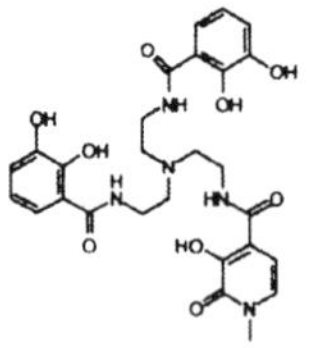	Related to cepabactin, with one hydroxypyridinone (HOPO) ring; the complex with iron is therefore charged –2

catecholate rings of FeEnt, generating a novel hybrid of ionic (Ent is uncharged, but FeEnt carries a net -3 charge[33] delocalized over the molecule[4]) and cation-π interactions, where the interacting groups are interlaced, cation-catecholate-cation-catecholate, in a cyclically-permuted manner around the iron atom.[4]

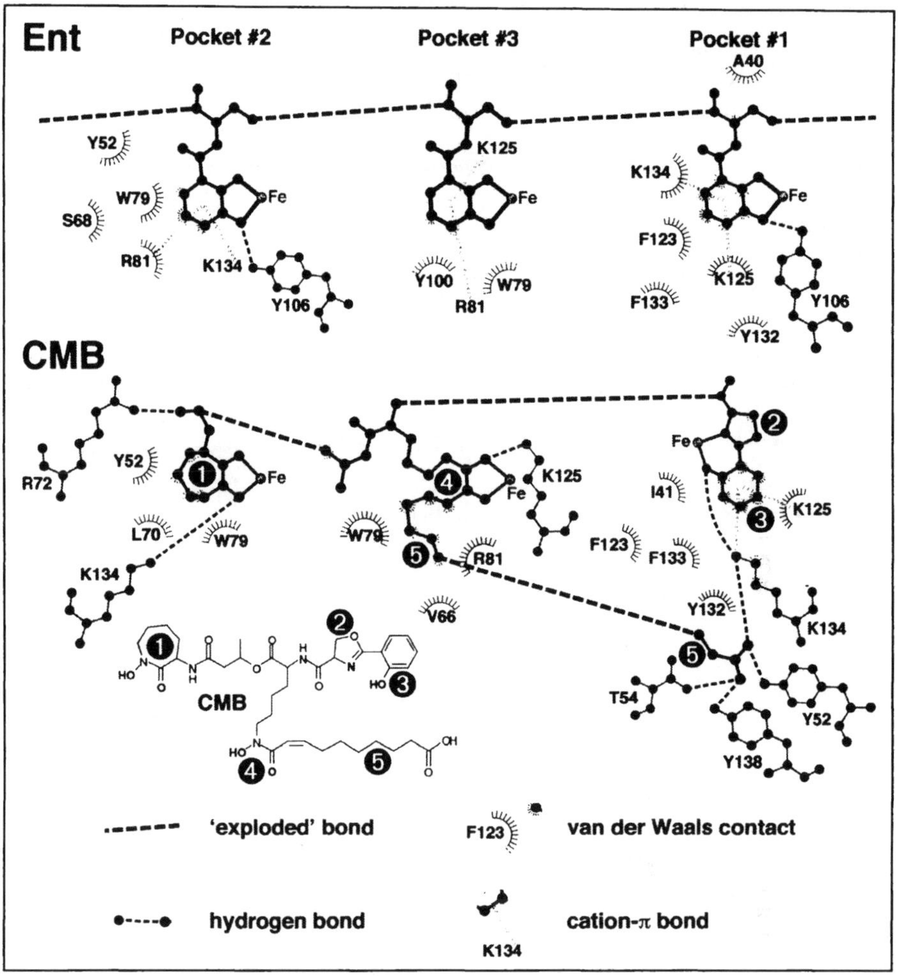

Figure 1. 'Exploded' LIGPLOT[113] representations of the interactions between siderocalin and Ent (*top*) or CMB-S (*bottom*), with bond types shown as indicated. Representations have been exploded to isolate the protein–ligand interactions in each pocket: #1, on the right; #2, on the left; and #3, in the middle. The chemical structure of the predominant CMB variant observed in the CMB-S crystal structures is also shown; CMB substituent groups (1, the cyclic hydroxamate; 2, the oxazoline; 3, the phenolate; 4, the linear hydroxamate; 5, the fatty acid tail) are numbered on the chemical structure and the corresponding parts of the LIGPLOT diagram for clarity.

Cation-π bonds in proteins are interactions between the positive charge of lysine or arginine side-chains and the quadrupole moment associated with the delocalized π-electrons of an aromatic functional group such as tryptophan, tyrosine or phenylalanine.[34] These interactions are seen in the binding of phosphotyrosine to an SH2 domain (for example, see ref. 35, where an arginine and a lysine interact with a phosphotyrosine-containing peptide). However, in SH2 complexes the actual aromatic ring remains uncharged. The nature of siderocalin/siderophore interactions, where the siderophore is centered in the siderocalin calyx making multiple, direct, polar interactions, clearly demonstrates specificity and cannot be the result of serendipity. Unusually, though, the FeEnt ligand makes few other stabilizing interactions with the protein, allowing it to wiggle around in the calyx, and fails to fill all of the calyx sub-pockets

Table 3. Siderophores predicted to bind to siderocalin, based on similarities to known siderocalin ligands and characterized elements of the recognition mechanism

Compound (references)	Organism(s)	Structure	Comments (K_a for Fe(III))
Agrobactin (88)	*Agrobacterium tumefaciens*	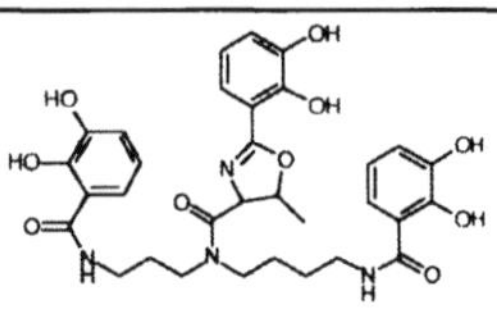	similar to parabactin (~10^{50})
Brucebactin (89)	*Brucella abortus*	(uncharacterized, but thought similar to enterochelin)	phenolate/catecholate-type
Cepabactin (90)	*Burkholderia cepacia*	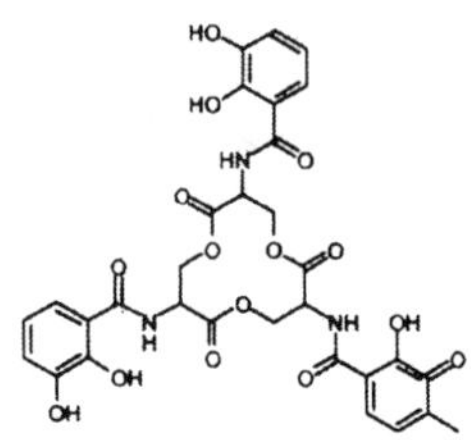	similar to Ent; an uncharged hydroxypyridinonate (HOPO) group replaces one phenolate, reducing overall charge of iron complex to –2
Corynebactin (91)	*Corynebacterium diphtheriae*	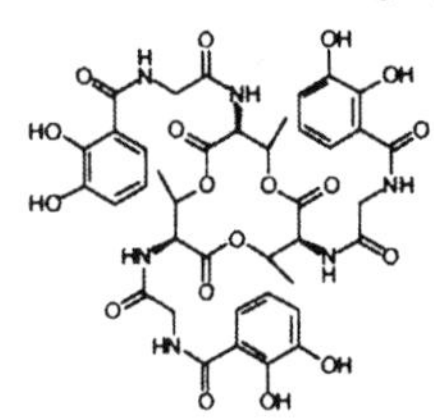	similar to Ent, with a tri-threonine, trilactone backbone; opposite chirality to Ent (91)
Fluvibactin (92)	*Vibrio fluvialis*	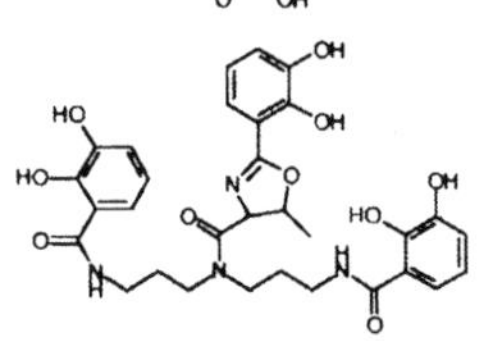	similar to parabactin
Vibriobactin (93)	*Vibrio cholerae*	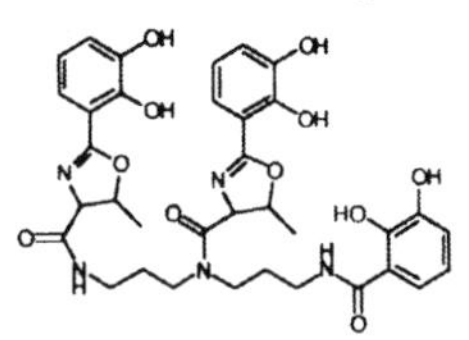	similar to parabactin; *vs.* Ent: oxazoline groups replace two phenolate groups; different backbone
Vulnibactin (94)	*Vibrio*	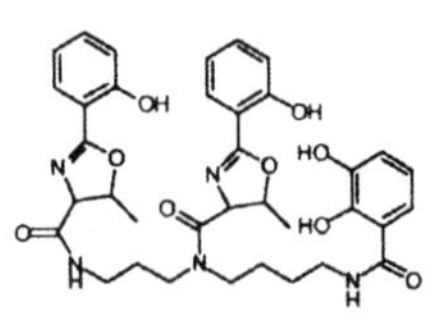	similar to vibriobactin, Ent

completely.[4,32] This behavior, suggesting that the calyx is optimized for some other ligand/s, lead to the search for other potential siderocalin siderophore ligands, culminating in the discovery that CMBs also bind while yet many other siderophores do not.[32]

Though considerably different in structure (Table 2), CMBs bind in the same position in the siderocalin calyx, though filling the calyx sub-pockets and crevices more completely and making more extensive interactions with the protein (Fig. 1). The CMB 2-hydroxyphenyloxazoline group is positioned in pocket #1, the cyclic hydroxamate group is positioned in the upper part of pocket #2 and the linear hydroxamate occupies pocket #3. The fatty acid tail curls under the rest of the siderophore, crossing from pocket #3 into pocket #1, positioning the carboxylate group into the bottom of pocket #2. Interestingly, the CMB carboxylate essentially superimposes on the carboxylate group of the NCA ligand in the original crystal structure of baculovirus-expressed siderocalin[24] and likely explains the retention of a fatty acid through purification. Lysines 125 and 134 participate in cation-π bonds to the CMB hydroxybenzoyl moiety completely analogous to those in Ent, with the hydroxybenzoyl of CMB superimposing almost identically onto the FeEnt catecholate in pocket #1. Therefore, the common element that is recognized by siderocalin in both complexes is the highly-polarized phenyl ring of one iron chelating group sitting in what is thus revealed as likely the key binding pocket, #1, between Lys125 and Lys134.

Siderocalin, Siderophores, Iron and Disease

What is the relevance of binding bacterial siderophores in a mammalian context? Iron is required by virtually all living things.[36] Within the body, the majority of iron is bound up in hemoglobin, though several proteins bind iron directly. Transferrin transports iron between cells and is normally 30 to 40% iron-saturated in the serum.[37] Iron is stored intracellularly in complex with ferritins.[38] Lactoferrin is a potent bacteriostatic agent, first discovered in milk, that is also released from neutrophil granules at sites of inflammation, directly inhibiting the growth of infecting pathogens by sequestering iron.[39,40] It has long been thought that the body generally lowers available iron in response to both infection and cancer in order to slow or stop the growth of pathogens and tumors.[40,41] The observation that giving iron supplements to patients with bacterial infections worsens their condition[40,41] demonstrates the scarcity of free iron in the body, with a serum concentration estimated to be as low as 10^{-24}M,[42] and the efficiency of iron sequestration as an antibiotic.[40] Pathogenic genera whose growth is stimulated by iron supplementation during infection in vertebrate hosts include: *Candida, Cryptococcus, Pneumocystis* and *Rhizopus* (Fungi); *Entamoeba, Leishmania, Naegleria, Plasmodium, Toxoplasma* and *Trypanosoma* (Protozoa); *Bacillus, Clostridium, Corynebacterium, Listeria, Mycobacterium, Staphylococcus* and *Streptococcus* (Gram-positive bacteria); and *Acinetobacter, Aeromonas, Alcaligenes, Campylobacter, Chlamydia, Ehrlichia, Enterobacter, Escherichia, Klebsiella, Legionella, Neisseria, Pseudomonas, Salmonella, Shigella* and *Yersinia* (Gram-negative bacteria).[41]

This ligand specificity therefore suggests a role for siderocalin in innate immune responses: a neutrophil granule protein, secreted in response to infection or inflammation, that sequesters iron, as ferric siderophore complexes, away from microbial pathogens, thus limiting their growth and virulence. Siderocalin complements the activity of lactoferrin by binding ferric siderophore complexes rather than iron directly.[4] Siderocalin is an acute phase protein, whose serum concentration can be used clinically to differentiate between bacterial and other types of infections.[43] During inflammation, concentrations of siderocalin can increase to levels, with concentrations approaching 20 to 30 nM in the serum, adequate to presumably bind all available iron as ferric siderophore complexes.[43] In direct support of this hypothesis, siderocalin is a potent bacteriostatic agent in vitro against *E. coli* cultured in iron-limiting conditions,[4] functioning specifically through its affinity for FeEnt, and, when the siderocalin gene is knocked-out, renders mice unable to fend off infections by virulent strains of *E. coli.*[44]

Siderocalin's binding properties may also contribute to the explanation of the association of certain siderophores with virulence. The pathogenesis islands of many bacteria (including species of *Yersinia, Shigella, Klebsiella, Salmonella* and *Neisseria*) encode proteins either associated with siderophore synthesis or uptake.[45-51] Biosynthesis of the siderophores aerobactin[51] (Table 4) or yersiniabactin[45] (Table 5), for instance, contributes to virulence for many bacteria. We have shown

Table 4. Siderophores demonstrated not to bind to siderocalin[32,44]

Compound (references)	Organism(s)	Structure	Comments (K_a for Fe(III))
Aerobactin (51)	enterobacteria	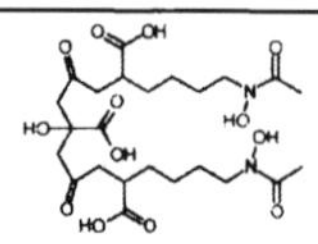	citrate-based hydroxamate type; virulence factor for pathogenic enteric bacteria (ColV encoded; ~10^{23})
Exochelins (95, 96)	mycobacteria	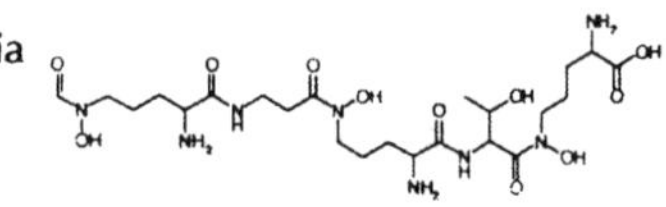	peptide-based; linear hydroxamate-type
Ferrichromes (97)	*Microsporum, Trichophyton* and *Aspergillus*	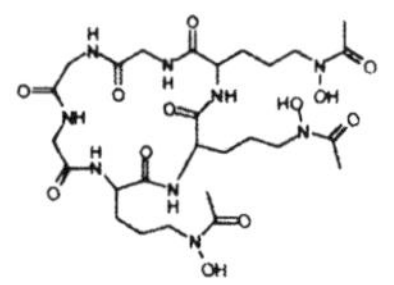	hydroxamate-type; cyclic peptides of ornithine derivatives and Gly, Ser or Ala (~10^{30})
Ferrioxamines (98, 99)	*Actinomyces*	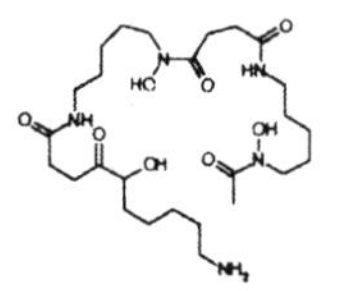	linear trihydroxamate-type yeast siderophores (~10^{32})
Fusarinines (Fusigens) (100, 101)	*Fusarium, Paecilomyces* and *Aspergillus*	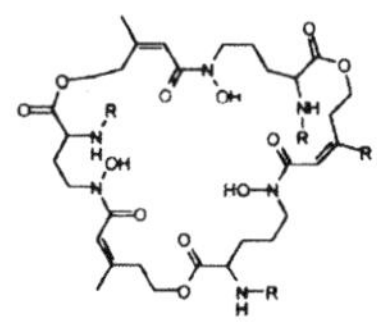	linear or cyclic trihydroxamates
Protocatechuic acid (3,4-dihydroxybenzoic acid)	*Bacillus anthracis*	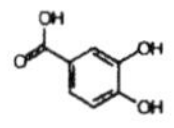	substituent of anthrachelin
Pyochelin (102)	*Pseudomonas aeruginosa, Burkholderia cepacia*	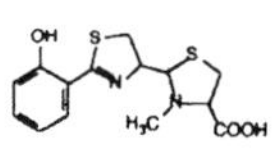	salicylate-based
Pyoverdin (also pyoverdine, pseudobactin) (103)	Pseudomonads	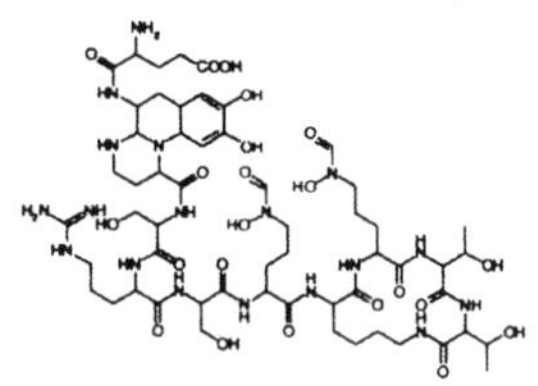	contains a dihydroxyquinoline chromophore, a variable peptide chain of 6 to 12 residues, and a dicarboxylic acid amide 'side-chain' (~10^{32})
Rhizoferrin (104)	Zygomycetes	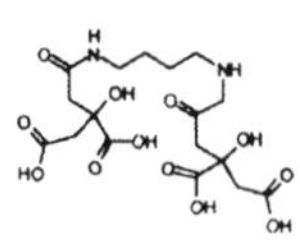	citric acid-based polycarboxylate

in qualitative binding assays that siderocalin has no appreciable affinity for aerobactin[4] or pyochelin[32] (Table 4)—therefore, because of its similarity to pyochelin, siderocalin very likely has limited affinity for yersiniabactin as well. These siderophores may thus confer virulence by allow-

Table 5. Siderophores predicted not to bind to siderocalin, based on similarities to demonstrated nonbinding siderophores and characterized elements of siderocalin recognition mechanism

Compound (references)	Organism(s)	Structure	Comments (K_a for Fe(III))
Alcaligin (105)	*Bordetella pertussis* and *bronchiseptica*	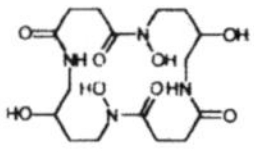	endomacrocyclic dihydroxamate; three molecules together bind two irons (M_2L_3) (~$10^{24}/10^{18}$)
Anthrachelin (106)	*Bacillus anthracis*	(uncharacterized, but based on protocatechuic acid)	phenolate/catecholate-type
Nannochelin (107)	myxobacteria	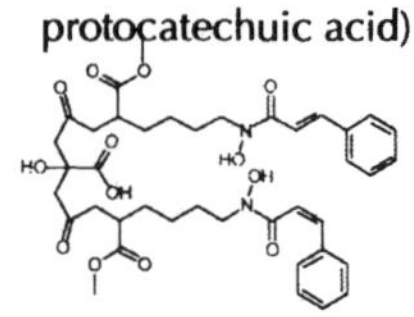	citrate-based, cinnamoyl; variant of aerobactin; dihydroxamate siderophore
Ornibactins (108)	*Burkholderia cepacia*	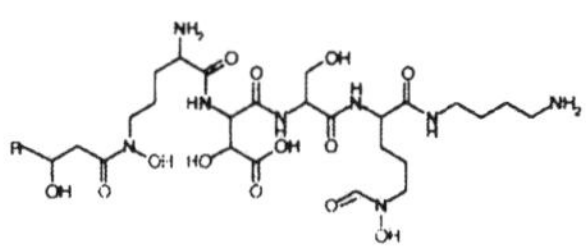	linear hydroxamate-hydroxycarboxylate (R = C4, C6, C8)
Salmochelins (53)	*Salmonella enterica; uropathogenic E. coli*	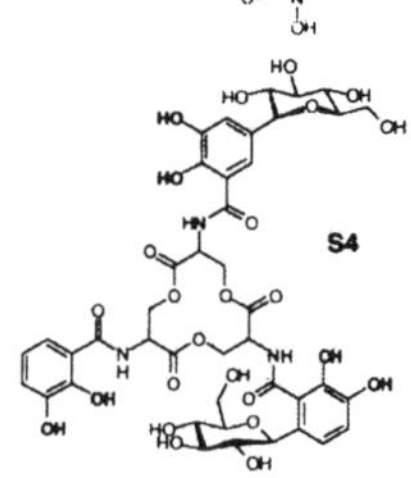	Ent with glucose adducts
Staphyloferrin (109, 110)	*Staphylococcus*	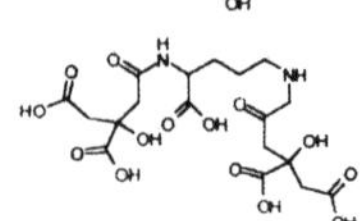	citric acid-containing polycarboxylate; lysine backbone
Yersiniabactin (111, 112)	*Yersinia*	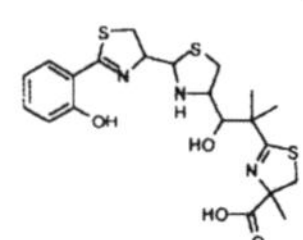	related to pyochelin

ing these bacteria to evade siderocalin-mediated iron sequestration. This hypothesis potentially explains the conundrum[30,52] of why production of a second, seemingly less efficient siderophore (aerobactin), one with a considerably lower affinity for iron (III) than Ent, would contribute to virulence. In support of this supposition, the relative susceptibility of siderocalin-KO versus wild-type mice to *E. coli* infection is only apparent with bacterial strains unable to synthesize aerobactin; strains that can secrete aerobactin do equally well in culture in the presence or absence of added siderocalin or in in vivo infections.[44] Infections with *Staphylococcus aureus* also show no difference in wild-type versus knock-out mice in terms of outcome, consistent with the utilization of siderophores (including staphyloferrin, Table 5) that are predicted to evade siderocalin binding on the basis of structural homology to siderophores (rhizoferrin, Table 4) that demonstrably do not bind.[32] *Salmonella* and uropathogenic strains of *E. coli* are also able to modify Ent by glucosylation[53] yielding the salmochelins (Table 5), siderophores that would also be predicted not to bind siderocalin because of significant steric clashes in the calyx.

However, siderocalin-mediated anti-mycobacterial responses may be limited by the high selectivity of siderocalin for particular CMB isoforms, in terms of fatty acid tail length, shown by the structural analysis of siderocalin/CMB complexes.[32] While siderocalin may be able to tolerate binding of CMB variants plus or minus one or perhaps two methylene groups in the fatty acid moieties from the optimum, though likely with significant concurrent reductions in affinity, it seems unlikely that siderocalin could accommodate the extremes of the reported CMB-T/S spectrum (Table 2), at least while retaining the overall ligand orientation seen in the co-crystal structures. Therefore, CMB variation may reflect mycobacterial responses to siderocalin-mediated defenses, evidenced by the obvious success of mycobacteria as human pathogens.

Siderocalin-mediated anti-bacterial responses through siderophore sequestration does not, a priori, require interactions with receptors. Therefore, the existence of siderocalin receptors suggests that the anti-bacterial iron depletion function of siderocalin may not be its sole physiological role.

Siderocalin and MMP-9

MMP-9 is a member of the matrix metalloproteinase family (multi-domain, zinc endopeptidases) that includes matrilysin (MMP-7), the collagenases (MMP-1, MMP-8 and MMP-13), the stromelysins (MMP 3, 10, 11 and 12), the gelatinases (-A, MMP-2 and -B, MMP-9) and a membrane associated MMP (MMP-14). MMP-9 efficiently cleaves gelatin, elastin and types V and X collagen, components of the extracellular matrix.[54] All members of the family share a homologous catalytic domain, containing the active-site zinc and are synthesized as inactive proenzymes. A cysteine in the propeptide coordinates the zinc, inactivating the proenzyme; the proenzyme can be activated in vitro through the addition of organomercurials and in vivo by plasma kallikrein.[55] MMP activity is regulated through control of the conversion of proenzyme to active enzyme and by the tight, non-covalent binding of 'tissue inhibitors of metalloproteinases' (TIMPs; ref. 56). Studies have shown that overexpression of MMPs (particularly MMP-9) can be correlated with tumor invasiveness and metastasis.[57-62] The only demonstrated consequence of a siderocalin/MMP-9 association is a slight acceleration of the direct activation of promatrix metalloproteinases through a non-physiological pathway.[63] There is no noncovalent component to siderocalin/MMP-9 association,[32] suggesting that the interaction may be serendipitous (murine siderocalin lacks the corresponding cysteine and is not known to associate with MMP-9). It is also hard to imagine a functional link between siderocalin's affinity for siderophores and the sequence-specific protease activity of MMP-9.

Siderocalin and an Alternate Iron Delivery Pathway in Mammals

Yang and coworkers[13] have shown that murine siderocalin, in a murine tissue culture system, also acts as an iron delivery protein, acting in concert with transferrin to convert mesenchymal progenitors into tubular epithelium, forming kidney nephrons. The effect is dependent upon iron bound by siderocalin as a complex with a chromophore—a mammalian 'siderophore'. Murine siderocalin also apparently recycles through sub-cellular compartments distinct from the intracellular trafficking of transferrin. Passage through low-pH intracellular compartments correlates with release of iron. The murine and rat orthologs are expected to have a similar ligand specificity to human siderocalin (experimentally confirmed for murine siderocalin[44]) as the majority of residues lining the calyx and approaching the ferric siderophore ligands are conserved or conservatively substituted between murine, rat and human siderocalin.[4] Therefore, the expectation is that the mammalian 'siderophore' will turn out to be something similar to one of the high-affinity bacterial siderophore ligands of siderocalin, probably more CMB-like, as these latter compounds are clearly more calyx-complementary—thus likely better representing the 'ideal' siderocalin ligand. If mammals do synthesize siderophores that are used to shuttle iron or act as growth factors, siderocalin may participate in a wide variety of cellular processes by playing a role in regulating their transport, thus potentially explaining siderocalin's association with tumorigenesis and apoptosis. However, these effects could alternately be directly mediated by siderocalin/receptor interactions, either ligand-dependent or -independent.

Other 'Siderocalins'?

While degenerately binding both Ent-like siderophores and CMBs, siderocalin fails to bind to many bacterial siderophores and essentially all types of fungal siderophores.[32] Therefore, the narrow range of siderocalin siderophore specificity leaves many holes in this potent innate immune defence, raising the question of whether there are other siderophore-binding proteins or peptides with complementary specificities. Siderocalin is the first non-bacterial siderophore-binding protein characterized and currently the only Ent-binding protein where the protein/ligand interactions have been clearly delineated. Therefore, initial attempts to identify other non-bacterial siderophore-binders have focused on lipocalins related to siderocalin.[32] Typical of the lipocalin family, sequence identities rapidly plummet as the alignments move from the siderocalins themselves (Table 1 and Fig. 2). A number of neighboring lipocalins are readily eliminated as having characterized functions or ligands that preclude siderophore binding: the prostaglandin D_2 synthases (PDS)[64] and human HC (also known as α_1-microglobulin),[65] a lipocalin associated with IgA that binds heme and heme-breakdown products.[66] However, candidate siderophore-binding lipocalins ('siderocalins') that display the hallmark features of the siderocalin calyx associated with siderophore binding, a triad of positively-charged side-chains, are identifiable (Fig. 2).[32]

Murine lipocalin 12 is found in seminal fluid,[67] but little else is known about its function; its human ortholog has only been identified through analysis of the human genome sequence.[68] Simplistic homology modelling of Lcn12 reveals a triad of positively charged side-chains in the murine Lcn12 calyx (two in human Lcn12) arranged analogously to siderocalin. The next most-related candidate siderocalins are the highly-homologous proteins chicken Ex-FABP[69] and quail Q83.[70] Ex-FABP is expressed during chicken embryo development in hypertrophic cartilage, muscle fibers and granulocytes and is also a component of egg whites. In chondrocyte and myoblast cultures, Ex-FABP expression is induced by inflammatory agents and inhibited by anti-inflammatory agents. Q83 is a protein strongly induced in v-*myc*-transformed avian fibroblasts, though no specific candidate ligands or functions have been proposed. The NMR structure of Q83[70] again shows a triad of positively-charged amino acids in the calyx (conserved in Ex-FABP) very reminiscent, in arrangement and character, of the key siderophore-binding residues of siderocalin. This arrangement is also echoed in the calyx of an even more distantly-related lipocalin, C8γ,[71] a well-studied member of the complement cascade and the subject of high-resolution crystallographic analyses.[72] However, preliminary binding and crystallographic analyses fail to convincingly demonstrate specific interactions between C8γ and any of the siderophores tested,[32] showing that calyx elements beyond positively-charged residues may be necessary for siderophore binding. Direct siderophore binding studies of the other candidate siderocalins, Lcn12, Ex-FABP and Q83 have yet to be reported.

Yet another lipocalin, Lcn1 (tear lipocalin, von Ebner's gland protein), a lipocalin even more distantly removed in sequence space, is also functionally a siderocalin, broadly inhibiting the growth of bacteria and fungi through ferric siderophore sequestration, though the nature of the recognition mechanism has yet to be elucidated and where the measured dissociation constants, in the millimolar range, are surprisingly weak.[73] The structure of Lcn1,[74] though containing several disordered sections that limit the conclusions, shows no immediately recognizable structural similarity to siderocalin in the calyx. Lcn1/siderophore complexes also have yet to be reported. It is even possible that there are siderophore-binding peptides; the structure of Hepcidin,[75] an anti-bacterial and -fungal peptide hormone,[76] involved in regulating iron homeostasis through interactions with ferroportin,[77] also displays a triad of positively-charged side-chains (Arg16, Lys18 and Lys24) on its concave surface almost superimposable on the positively-charged, ligand-interacting calyx residues of siderocalin.

```
Siderocalin
 human          --SDLIPAPPLSKVPLQQNFQDNQFQGKWYVVGLAGN-AILREDKDPQKMYATIYELKEDKSY
 rat            --QNLIPAPPLISVPLQPGFWTERFQGRWFVVGLAGN-AVQKERQSRFTMYSTIYELQEDNSY
 murine         --QNLIPAPSLLTVPLQPDFRSDQFRGRWYVVGLAGN-AVQKKTEGSFTMYSTIYELQENNSY

Lcn12
 human          KVLQAQTPTPLPLPPPMQSFQGNQFQGEWFVLGLAGN-SFRPEHRALLNAFTATFELSDDGRF
 murine         QILESQISAMSQGFPQMTSFQSDQFQGEWFVLGLADN-TFRREHRALLNFFTTLFELKEKSQF

Ex-FABP         ------HTEAAATVPDRS-----EVAGKWYIVALASNTDFFLREKGKMKMVMARISFLGEDEL
Q83             ------HAKAAATVPDRS-----EIAGKWYVVALASNTEFFLREKDKMKMAMARISFLGEDEL

C8γ             -------ASPISTIQPKANFDAQQFAGTWLLVAVGSA--------GRAEATTLHVAPQGTAMA

Lcn1
 human          AVSLGLIAALQAHHLLASDEEIQDVSGTWYLKAMTVDREFPEMNLESVTPMTLTTLEGGNLEA


Siderocalin                      •                                         •
 human          NVTSVLFRK--KKCDYWIRTFVPGCQPGEFTLGNIKSYPGLTSYLVRVVSTNYNQHAMVFFKK
 rat            NVTSILVRG--QGCRYWIRTFVPSSRPGQFTLGNIHSYPQIQSYDVQVADTDYDQFAMVFFQK
 murine         NVTSILVRDQDQGCRYWIRTFVPSSRAGQFTLGNMHRYPQVQSYNVQVATTDYNQFAMVFFRK

Lcn12                                         •                            •
 human          EVWNAMTRG--QHCDTWSYVLIPAAQPGQFTVDHG-VEPGADREETRVVDSDYTQFALMLSRR
 murine         QVTNSMTRG--KHCNTWSYTLIPATKPGQFTRDNRGSGPGADRENIQVIETDYITFALVLSLR
                                                           •               •
Ex-FABP         EVSYAAPSP--KGCRKWETTFKKTSDDGELY------YSEEAEKTVEVLDTDYKSYAVIFATR
Q83             KVSYAVPKP--NGCRKWETTFKKTSDDGEVY------YSEEAKKKVEVLDTDYKSYAVIYATR
                    •                                                      •
C8γ             VSTFRKLDG---ICWQVRQLYGDTGVLGRFLLQAR---GARGAVHVVVAETDYQSFAVLYLER

Lcn1
 human          KVTMLISGR----CQEVKAVLEKTDEPGKYT-------ADGGKHVAYIIRSHVKDHYIFYCEG


Siderocalin             •
 human          VSQNREYFKITLYGRTKELTSELKENFIRFSKYLGLPENHIVFPVPIDQCIDG---------
 rat            TSENKQYFKVTLYGRTKGLSDELKERFVSFAKSLGLKDNNIVFSVPTDQCIDNV--------
 murine         TSENKQYFKITLYGRTKELSPELKERFTRFAKSLGLKDDNIIFSVPTDQCIDN---------

Lcn12                   •
 human          HTSRLAVLRISLLGRSWLLPPGTLDQFICLGRAQGLSDDNIVFPDVTGNMVHLQACWAVGTG + 46 residues
 murine         QTSSQNITRVSLLGRNWRLSHKTIDKFICLTRTQNLTKDNFLFPDLSDWLPDPQVC------
                          •
Ex-FABP         VKDGRTLHMMRLYSRSREVSPTAMAIFRKLARERNYTDEMVAVLPSQEECSVDEV-------
Q83             VKDGRTLHMMRLYSRSPEVSPAATAIFRKLAGERNYTDEMVAMLPRQEECTVDEV-------
                          •
C8γ             AG----QLSVKLYARSLPVSDSVLSGFEQRVQEAHLTEDQIFYFPKYGFCEAADQFHVLDEV

Lcn1                      •
 human          ELHGKPVRGVKLVGRDPKNNLEALEDFEKAAGARGLSTESILIPRQSETCSPGSD-------
```

Figure 2. Unmodified CLUSTALW[114] alignments of the sequences of siderocalin with other potential siderocalins. Dots indicate positively-charged residues with side-chains extending into the calyx, based on either direct structure determinations, homology or homology modeling; dashes indicate gaps introduced to maximize homology.

Consideration of all these possibilities requires that proposals to use siderophores and siderophore analogs as therapeutics, either as antibiotics[78,79] or, through the bound iron, as oxygen radical scavengers in various clinical settings, such as the treatment of ischemia associated with congestive heart failure,[80] will need to be tempered by the possibility that endogenous siderophore-binding specificities will limit, defeat or confound such approaches.

References

1. Kjeldsen L, Johnsen AH, Sengelov H et al. Isolation and primary structure of NGAL, a novel protein associated with human neutrophil gelatinase. J Biol Chem 1993; 268(14):10425-32.
2. Kjeldsen L, Bainton DF, Sengelov H et al. Identification of neutrophil gelatinase-associated lipocalin as a novel matrix protein of specific granules in human neutrophils. Blood 1994; 83(3):799-807.
3. Kjeldsen L, Cowland JB, Borregaard N. Human neutrophil gelatinase-associated lipocalin and homologous proteins in rat and mouse. Biochim Biophys Acta 2000; 1482(1-2):272-83.
4. Goetz DH, Holmes MA, Borregaard N et al. The neutrophil lipocalin NGAL is a bacteriostatic agent that interferes with siderophore-mediated iron acquisition. Mol Cell 2002; 10(5):1033-43.
5. Triebel S, Bläser J, Reinke H et al. A 25 kDa α_2-microglobulin-related protein is a component of the 125 kDa form of human gelatinase. FEBS Lett 1992; 3:386-388.
6. Kjeldsen L, Johnsen AH, Sengeløv H et al. Isolation and primary structure of NGAL, a novel protein associated with human neutrophil gelatinase. J Biol Chem 1993; 268:10425-10432.
7. Stoesz SP, Gould MN. Overexpression of neu-related lipocalin (NRL) in neu-initiated but not ras or chemically initiated rat mammary carcinomas. Oncogene 1995; 11(11):2233-41.
8. Liu Q, Nilsen-Hamilton M. Identification of a new acute phase protein. J Biol Chem 1995; 270(38):22565-70.
9. Ryon J, Bendickson L, Nilsen-Hamilton M. High expression in involuting reproductive tissues of uterocalin/24p3, a lipocalin and acute phase protein. Biochem J 2002; 367(Pt 1):271-7.
10. Suire S, Stewart F, Beauchamp J et al. Uterocalin, a lipocalin provisioning the preattachment equine conceptus: fatty acid and retinol binding properties, and structural characterization. Biochem J 2001; 356(Pt 2):369-76.
11. Devireddy LR, Teodoro JG, Richard FA et al. Induction of apoptosis by a secreted lipocalin that is transcriptionally regulated by IL-3 deprivation. Science 2001; 293(5531):829-34.
12. Kamezaki K, Shimoda K, Numata A et al. The lipocalin 24p3, which is an essential molecule in IL-3 withdrawal-induced apoptosis, is not involved in the G-CSF withdrawal-induced apoptosis. Eur J Haematol 2003; 71(6):412-7.
13. Yang J, Goetz D, Li JY et al. An iron delivery pathway mediated by a lipocalin. Mol Cell 2002; 10(5):1045-56.
14. Yang J, Mori K, Li JY et al. Iron, lipocalin, and kidney epithelia. Am J Physiol Renal Physiol 2003; 285(1):F9-18.
15. Hvidberg V, Jacobsen C, Strong RK et al. The endocytic receptor megalin binds the iron transporting neutrophil-gelatinase-associated lipocalin (NGAL) with high affinity and mediates its cellular uptake. FEBS Lett, in press.
16. Saito A, Pietromonaco S, Loo AK et al. Complete cloning and sequencing of rat gp330/"megalin," a distinctive member of the low density lipoprotein receptor gene family. Proc Natl Acad Sci USA 1994; 91(21):9725-9.
17. Willnow TE, Goldstein JL, Orth K et al. Low density lipoprotein receptor-related protein and gp330 bind similar ligands, including plasminogen activator-inhibitor complexes and lactoferrin, an inhibitor of chylomicron remnant clearance. J Biol Chem 1992; 267(36):26172-80.
18. Meilinger M, Haumer M, Szakmary KA et al. Removal of lactoferrin from plasma is mediated by binding to low density lipoprotein receptor-related protein/alpha 2-macroglobulin receptor and transport to endosomes. FEBS Lett 1995; 360(1):70-4.
19. Flower DR. Beyond the superfamily: the lipocalin receptors. Biochim Biophys Acta 2000; 1482(1-2):327-36.
20. Leheste JR, Rolinski B, Vorum H et al. Megalin knockout mice as an animal model of low molecular weight proteinuria. Am J Pathol 1999; 155(4):1361-70.
21. Bundgaard JR, Sengelov H, Borregaard N et al. Molecular cloning and expression of a cDNA encoding NGAL: a lipocalin expressed in human neutrophils. Biochem Biophys Res Commun 1994; 202(3):1468-75.
22. Sengelov H, Boulay F, Kjeldsen L et al. Subcellular localization and translocation of the receptor for N-formylmethionyl-leucyl-phenylalanine in human neutrophils. Biochem J 1994; 299(Pt 2):473-9.
23. Chu ST, Lin HJ, Chen YH. Complex formation between a formyl peptide and 24p3 protein with a blocked N-terminus of pyroglutamate. J Pept Res 1997; 49(6):582-5.
24. Goetz DH, Willie ST, Armen RS et al. Ligand preference inferred from the structure of neutrophil gelatinase associated lipocalin. Biochemistry 2000; 39(8):1935-41.
25. Coles M, Diercks T, Muehlenweg B et al. The solution structure and dynamics of human neutrophil gelatinase-associated lipocalin. J Mol Biol 1999; 289(1):139-57.
26. Bratt T, Ohlson S, Borregaard N. Interactions between neutrophil gelatinase-associated lipocalin and natural lipophilic ligands. Biochim Biophys Acta 1999; 1472(1-2):262-9.

27. Wu SY, Perez MD, Puyol P et al. beta-lactoglobulin binds palmitate within its central cavity. J Biol Chem 1999; 274(1):170-4.
28. Neilands JB. Microbial iron compounds. Ann Rev Biochem 1981; 50:715-31.
29. Neilands JB. Iron absorption and transport in microorganisms. Ann Rev Nutrit 1981; 1:27-46.
30. Ratledge C, Dover LG. Iron metabolism in pathogenic bacteria. Ann Rev Microbiol 2000; 54:881-941.
31. Winkelmann G. Microbial siderophore-mediated transport. Biochem Soc Trans 2002; 30(4):691-6.
32. Holmes MA, Paulsene W, Jide X et al. Siderocalin (Lcn 2) Also binds carboxymycobactins, potentially defending against mycobacterial infections through iron sequestration. Structure (Camb) 2005; 13(1):29-41.
33. Raymond KN, Müller G, Matzanke BF. Complexation of iron by siderophores. A review of their solution and structural chemistry and biological function. In: Boschke FL, ed. Topics in Current Chemistry. Berlin, Heidelberg: Springer-Verlag; 1984:50-102.
34. Dougherty DA. Cation-pi interactions in chemistry and biology: a new view of benzene, Phe, Tyr, and Trp. Science 1996; 271(5246):163-8.
35. Waksman G. Crystal structure of the phosphotyrosine recognition domain SH2 of the Src oncogene product complexed with tyrosine-phosphorylated peptides. Cell Mol Biol 1994; 40(5):611-8.
36. Neilands JB. Siderophores: structure and function of microbial iron transport compounds. J Biol Chem 1995; 270(45):26723-6.
37. Ponka P, Beaumont C, Richardson DR. Function and regulation of transferrin and ferritin. Semin Hematol 1998; 35(1):35-54.
38. Harrison PM, Arosio P. The ferritins: molecular properties, iron storage function and cellular regulation. Biochim Biophys Acta 1996; 1275(3):161-203.
39. Ellison RT. The effects of lactoferrin on gram-negative bacteria. Adv Exp Med Biol 1994; 357:71-90.
40. Jurado RL. Iron, infections, and anemia of inflammation. Clin Infect Dis 1997; 25(4):888-95.
41. Weinberg ED. Iron withholding: a defense against infection and neoplasia. Physiol Rev 1984; 64(1):65-102.
42. Otto BR, Verweij-van Vught AM, MacLaren DM. Transferrins and heme-compounds as iron sources for pathogenic bacteria. Crit Rev Microbiol 1992; 18(3):217-33.
43. Xu S, Venge P. Lipocalins as biochemical markers of disease. Biochim Biophys Acta 2000; 1482(1-2):298-307.
44. Flo TH, Smith KD, Sato S et al. Lipocalin 2 mediates an innate immune response to bacterial infection by sequestrating iron. Nature 2004; 432(7019):917-21.
45. Carniel E. The Yersinia high-pathogenicity island: an iron-uptake island. Microbes Infect 2001; 3(7):561-9.
46. Carniel E. The Yersinia high-pathogenicity island. Int Microbiol 1999; 2(3):161-7.
47. Klee SR, Nassif X, Kusecek B et al. Molecular and biological analysis of eight genetic islands that distinguish Neisseria meningitidis from the closely related pathogen Neisseria gonorrhoeae. Infect Immun 2000; 68(4):2082-95.
48. Moss JE, Cardozo TJ, Zychlinsky A et al. The selC-associated SHI-2 pathogenicity island of Shigella flexneri. Mol Microbiol 1999; 33(1):74-83.
49. Vokes SA, Reeves SA, Torres AG et al. The aerobactin iron transport system genes in Shigella flexneri are present within a pathogenicity island. Mol Microbiol 1999; 33(1):63-73.
50. Zhou D, Hardt WD, Galan JE. Salmonella typhimurium encodes a putative iron transport system within the centisome 63 pathogenicity island. Infect Immun 1999; 67(4):1974-81.
51. Warner PJ, Williams PH, Bindereif A et al. ColV plasmid-specific aerobactin synthesis by invasive strains of Escherichia coli. Infect Immun 1981; 33(2):540-5.
52. Crosa JH. Genetics and molecular biology of siderophore-mediated iron transport in bacteria. Microbiol Rev 1989; 53(4):517-30.
53. Fischbach MA, Lin H, Liu DR et al. In vitro characterization of IroB, a pathogen-associated C-glycosyltransferase. Proc Natl Acad Sci USA 2004.
54. Davidson B, Reich R, Risberg B et al. The biological role and regulation of matrix metalloproteinases (MMP) in cancer. Arkh Patol 2002; 64(3):47-53.
55. Okada Y, Morodomi T, Enghild J et al. Matrix metalloproteinase 2 from human rheumatoid synovial fibroblasts. Purification and activation of the precursor and enzymatic properties. Eur J Biochem 1990; 194:721-730.
56. Murphy G, Willenbrock F. Tissue inhibitors of matrix metalloproteinases. Methods Enzymol 1995; 248:496-510.
57. Ballin M, Gomez DE, Sinha CC et al. Ras oncogene mediated induction of a 92 kDa metalloproteinase; strong correlation with the malignant phenotype. Biochem Biophys Res Commun 1988; 154:832-838.

58. Bernhard E, Muschel R, Hughes E. Mr 92,000 gelatinase release correlates with the metastatic phenotype in transformed rat embryo cells. Cancer Res 1990; 50:3872-3877.
59. Reich R, Thompson E, Iwamoto Y et al. Effects of inhibitors of plasminogen activator, serine proteinases, and collagenase IV on the invasion of basement membranes by metastatic cells. Cancer Res 1988; 48:3307-3312.
60. Nakajima M, Welch DR, Wynn DM et al. Serum and plasma M(r) 92,000 progelatinase levels correlate with spontaneous metastasis of rat 13762NF mammary adenocarcinoma. 1993.
61. Moscatelli D, Rifkin DB. Membrane and matrix localization of proteinases: A common theme in tumor cell invasion and angiogenesis. Biochem. Biophys. Acta 1988; 948:67-85.
62. Murphy G, Reynolds JJ, Hembry RM. Metalloproteinases and cancer invasion and metastasis. Int J Cancer 1989; 44:757-760.
63. Tschesche H, Zolzer V, Triebel S et al. The human neutrophil lipocalin supports the allosteric activation of matrix metalloproteinases. Eur J Biochem 2001; 268(7):1918-28.
64. Helliwell RJ, Adams LF, Mitchell MD. Prostaglandin synthases: recent developments and a novel hypothesis. Prostaglandins Leukot Essent Fatty Acids 2004; 70(2):101-13.
65. Åkerström B, Logdberg L, Berggard T et al. alpha(1)-Microglobulin: a yellow-brown lipocalin. Biochim Biophys Acta 2000; 1482(1-2):172-84.
66. Larsson J, Allhorn M, Kerstrom B. The lipocalin alpha(1)-microglobulin binds heme in different species. Arch Biochem Biophys 2004; 432(2):196-204.
67. Suzuki K, Lareyre J-J, Sánchez D et al. Molecular evolution of a multigene subfamily encoding epididymal lipocalins localized on the [A3] locus of mouse chromosome 2. Gene, in press.
68. Strausberg RL, Feingold EA, Grouse LH et al. Generation and initial analysis of more than 15,000 full-length human and mouse cDNA sequences. Proc Natl Acad Sci USA 2002; 99(26):16899-903.
69. Descalzi Cancedda F, Dozin B, Zerega B et al. Ex-FABP: a fatty acid binding lipocalin developmentally regulated in chicken endochondral bone formation and myogenesis. Biochim Biophys Acta 2000; 1482(1-2):127-35.
70. Hartl M, Matt T, Schuler W et al. Cell transformation by the v-myc oncogene abrogates c-Myc/Max-mediated suppression of a C/EBP beta-dependent lipocalin gene. J Mol Biol 2003; 333(1):33-46.
71. Schreck SF, Parker C, Plumb ME et al. Human complement protein C8 gamma. Biochim Biophys Acta 2000; 1482(1-2):199-208.
72. Ortlund E, Parker CL, Schreck SF et al. Crystal structure of human complement protein C8gamma at 1.2 A resolution reveals a lipocalin fold and a distinct ligand binding site. Biochemistry 2002; 41(22):7030-7.
73. Fluckinger M, Haas H, Merschak P et al. Human tear lipocalin exhibits antimicrobial activity by scavenging microbial siderophores. Antimicrob Agents Chemother 2004; 48(9):3367-72.
74. Breustedt DA, Korndorfer IP, Redl B et al. The 1.8-a crystal structure of human tear lipocalin reveals an extended branched cavity with capacity for multiple ligands. J Biol Chem 2005; 280(1):484-93.
75. Hunter HN, Fulton DB, Ganz T et al. The solution structure of human hepcidin, a peptide hormone with antimicrobial activity that is involved in iron uptake and hereditary hemochromatosis. J Biol Chem 2002; 277(40):37597-603.
76. McGrath H Jr, Rigby PG. Hepcidin: inflammation's iron curtain. Rheumatology (Oxford) 2004; 43(11):1323-5.
77. Nemeth E, Tuttle MS, Powelson J et al. Hepcidin regulates cellular iron efflux by binding to ferroportin and inducing its internalization. Science 2004; 306(5704):2090-3.
78. Roosenberg JM 2nd, Lin YM, Lu Y et al. Studies and syntheses of siderophores, microbial iron chelators, and analogs as potential drug delivery agents. Curr Med Chem 2000; 7(2):159-97.
79. Budzikiewicz H. Siderophore-antibiotic conjugates used as trojan horses against Pseudomonas aeruginosa. Curr Top Med Chem 2001; 1(1):73-82.
80. Horwitz LD, Sherman NA, Kong Y et al. Lipophilic siderophores of Mycobacterium tuberculosis prevent cardiac reperfusion injury. Proc Natl Acad Sci USA 1998; 95(9):5263-8.
81. Ratledge C, Ewing M. The occurrence of carboxymycobactin, the siderophore of pathogenic mycobacteria, as a second extracellular siderophore in Mycobacterium smegmatis. Microbiology 1996; 142(Pt 8):2207-12.
82. Cox GB, Gibson F. 2,3-Dihydroxybenzoic acid, a new growth factor for multiple aromatic auxotrophs. J Bacteriol 1967; 93(1):502-3.
83. Bellaire BH, Elzer PH, Baldwin CL et al. Production of the siderophore 2,3-dihydroxybenzoic acid is required for wild-type growth of Brucella abortus in the presence of erythritol under low-iron conditions in vitro. Infect Immun 2003; 71(5):2927-2832.
84. O'Brien IG, Gibson F. The structure of enterochelin and related 2,3-dihydroxy-N-benzoylserine conjugates from Escherichia coli. Biochim Biophys Acta 1970; 215(2):393-402.

85. Pollack JR, Neilands JB. Enterobactin, an iron transport compound from Salmonella typhimurium. Biochem Biophys Res Commun 1970; 38(5):989-92.
86. Thulasiraman P, Newton SM, Xu J et al. Selectivity of ferric enterobactin binding and cooperativity of transport in gram-negative bacteria. J Bacteriol 1998; 180(24):6689-96.
87. Bergeron RJ, Dionis JB, Elliott GT et al. Mechanism and stereospecificity of the parabactin-mediated iron-transport system in Paracoccus denitrificans. J Biol Chem 1985; 260(13):7936-44.
88. Ong SA, Peterson T, Neilands JB. Agrobactin, a siderophore from Agrobacterium tumefaciens. J Biol Chem 1979; 254(6):1860-5.
89. Gonzalez Carrero MI, Sangari FJ, Aguero J et al. Brucella abortus strain 2308 produces brucebactin, a highly efficient catecholic siderophore. Microbiology 2002; 148(Pt 2):353-60.
90. Meyer JM, Hohnadel D, Halle F. Cepabactin from Pseudomonas cepacia, a new type of siderophore. J Gen Microbiol 1989; 135(Pt 6):1479-87.
91. Bluhm ME, Kim SS, Dertz EA et al. Corynebactin and enterobactin: related siderophores of opposite chirality. J Am Chem Soc 2002; 124(11):2436-7.
92. Yamamoto S, Okujo N, Fujita Y et al. Structures of two polyamine-containing catecholate siderophores from Vibrio fluvialis. J Biochem (Tokyo) 1993; 113(5):538-44.
93. Griffiths GL, Sigel SP, Payne SM et al. Vibriobactin, a siderophore from Vibrio cholerae. J Biol Chem 1984; 259(1):383-5.
94. Okujo N, Saito M, Yamamoto S et al. Structure of vulnibactin, a new polyamine-containing siderophore from Vibrio vulnificus. Biometals 1994; 7(2):109-16.
95. Sharman GJ, Williams DH, Ewing DF et al. Determination of the structure of exochelin MN, the extracellular siderophore from Mycobacterium neoaurum. Chem Biol 1995; 2(8):553-61.
96. Sharman GJ, Williams DH, Ewing DF et al. Isolation, purification and structure of exochelin MS, the extracellular siderophore from Mycobacterium smegmatis. Biochem J 1995; 305(Pt 1):187-96.
97. Zalkin A, Forrester JD, Templeton DH. Ferrichrome-A tetrahydrate. Determination of crystal and molecular structure. J Am Chem Soc 1966; 88(8):1810-4.
98. van der Helm D, Poling M. The crystal structure of ferrioxamine E. J Am Chem Soc 1976; 98(1):82-6.
99. Dhungana S, White PS, Crumbliss AL. Crystal structure of ferrioxamine B: a comparative analysis and implications for molecular recognition. J Biol Inorg Chem 2001; 6(8):810-8.
100. Emery T. Isolation, characterization, and properties of fusarinine, a delta-hydroxamic acid derivative of ornithine. Biochemistry 1965; 4(7):1410-7.
101. Diekmann H. Metabolic products of microorganisms. 56. Fusigen—a new sideramine from fungi. Arch Mikrobiol 1967; 58(1):1-5.
102. Cox CD, Rinehart KL Jr, Moore ML et al. Pyochelin: novel structure of an iron-chelating growth promoter for Pseudomonas aeruginosa. Proc Natl Acad Sci USA 1981; 78(7):4256-60.
103. Abdallah MA, Pfestorf M, Doring G. Pseudomonas aeruginosa pyoverdin: structure and function. Antibiot Chemother 1989; 42:8-14.
104. Thieken A, Winkelmann G. Rhizoferrin: a complexone type siderophore of the Mucorales and entomophthorales (Zygomycetes). FEMS Microbiol Lett 1992; 73(1-2):37-41.
105. Hou Z, Raymond KN, O'Sullivan B et al. A preorganized siderophore: thermodynamic and structural characterization of alcaligin and bisucaberin, microbial macrocyclic dihydroxamate chelating agents(1). Inorg Chem 1998; 37(26):6630-6637.
106. Garner BL, Arceneaux JE, Byers BR. Temperature control of a 3,4-dihydroxybenzoate (protocatechuate)-based siderophore in Bacillus anthracis. Curr Microbiol 2004; 49(2):89-94.
107. Kunze B, Trowitzsch-Kienast W, Hofle G et al. Nannochelins A, B and C, new iron-chelating compounds from Nannocystis exedens (myxobacteria). Production, isolation, physico-chemical and biological properties. J Antibiot (Tokyo) 1992; 45(2):147-50.
108. Meyer JM, Van VT, Stintzi A et al. Ornibactin production and transport properties in strains of Burkholderia vietnamiensis and Burkholderia cepacia (formerly Pseudomonas cepacia). Biometals 1995; 8(4):309-17.
109. Meiwes J, Fiedler HP, Haag H et al. Isolation and characterization of staphyloferrin A, a compound with siderophore activity from Staphylococcus hyicus DSM 20459. FEMS Microbiol Lett 1990; 55(1-2):201-5.
110. Konetschny-Rapp S, Jung G, Meiwes J et al. Staphyloferrin A: a structurally new siderophore from staphylococci. Eur J Biochem 1990; 191(1):65-74.
111. Haag H, Hantke K, Drechsel H et al. Purification of yersiniabactin: a siderophore and possible virulence factor of Yersinia enterocolitica. J Gen Microbiol 1993; 139(Pt 9):2159-65.
112. Perry RD, Balbo PB, Jones HA et al. Yersiniabactin from Yersinia pestis: biochemical characterization of the siderophore and its role in iron transport and regulation. Microbiology 1999; 145(Pt 5):1181-90.
113. Wallace AC, Laskowski RA, Thornton JM. LIGPLOT: a program to generate schematic diagrams of protein-ligand interactions. Protein Engineering 1995; 8(2):127-34.
114. Higgins DG, Bleasby AJ, Fuchs R. CLUSTALW. Comput Applic Biosci 1992; 8(2):189-191.

CHAPTER 9

Lipocalin-Type Prostaglandin D Synthase as an Enzymic Lipocalin

Yoshihiro Urade,* Naomi Eguchi and Osamu Hayaishi

Abstract

Lipocalin-type prostaglandin (PG) D synthase (L-PGDS) is the first member of the lipocalin family to be recognized as an enzyme. L-PGDS catalyzes the isomerization of PGH_2, a common precursor of various prostanoids, to produce PGD_2, a potent endogenous somnogen and a nociceptive modulator as well as an allergic mediator. Mammalian L-PGDS is a highly glycosylated lipocalin of Mr = 26,000 with 2 N-glycosylation sites. L-PGDS is mainly localized in the central nervous system and male genital organs of various mammals and in the human heart. In the brain, L-PGDS is localized in the rough endoplasmic reticulum and nuclear membrane of oligodendrocytes and arachnoid trabecular cells, coupled with cyclooxygenase to produce PGD_2, and then secreted into the cerebrospinal fluid as β-trace, a major protein in human cerebrospinal fluid. L-PGDS/β-trace is also secreted from those production sites into the seminal plasma and plasma. L-PGDS binds various lipophilic ligands, such as PGD_2, biliverdin, bilirubin, retinoic acid, and retinal with high affinities of Kd = 20 to 80 nM, suggesting the multifunctionality of L-PGDS as a PGD_2-producing enzyme, an extracellular PGD_2-transporter, and a lipophilic ligand-binding protein. We generated L-PGDS gene knockout mice and human enzyme-overexpressing mice transgenic for L-PGDS and found them to be functionally abnormal in the regulation of sleep, pain, and cardiovascular responses. The X-ray crystallographic structure of L-PGDS was determined to possess the typical lipocalin fold, i.e., a β-barrel, with a hydrophobic interior in which Cys65 has been identified as the active thiol residue essential for the catalysis.

Introduction

In 1985, we isolated lipocalin-type prostaglandin (PG) D synthase (L-PGDS) from rat brain as a monomeric enzyme with a molecular weight of 26,000,[1] although the enzyme had previously been misidentified to be a protein with a molecular weight of 80,000 to 85,000.[2,3] L-PGDS catalyzes the isomerization of 9,11-endoperoxide of PGH_2 to produce PGD_2 with 9-hydroxy and 11-keto groups at the low turnover number of 170 min^{-1}in the presence of various sulfhydryl compounds, such as glutathione (GSH), dithiothreitol, β-mercaptoethanol, cysteine, and cysteamine (Fig. 1). L-PGDS was previously termed as brain-type PGD synthase or GSH-independent PGD synthase to distinguish it from GSH-requiring PGD synthase purified from rat spleen,[4,5] which is now named hematopoietic PGDS (H-PGDS).[6-8] L-PGDS and H-PGDS are quite different from each other in terms of amino acid sequence, tertiary structure, evolutional origin, cellular distribution etc., although both enzymes catalyze the

*Corresponding Author: Yoshihiro Urade—Department of Molecular Behavioral Biology, Osaka Bioscience Institute, 6-2-4 Furuedai, Suita City, Osaka 565-0874, Japan. Email: uradey@obi.or.jp

Lipocalins, edited by Bo Åkerström, Niels Borregaard, Darren R. Flower and Jean-Philippe Salier.

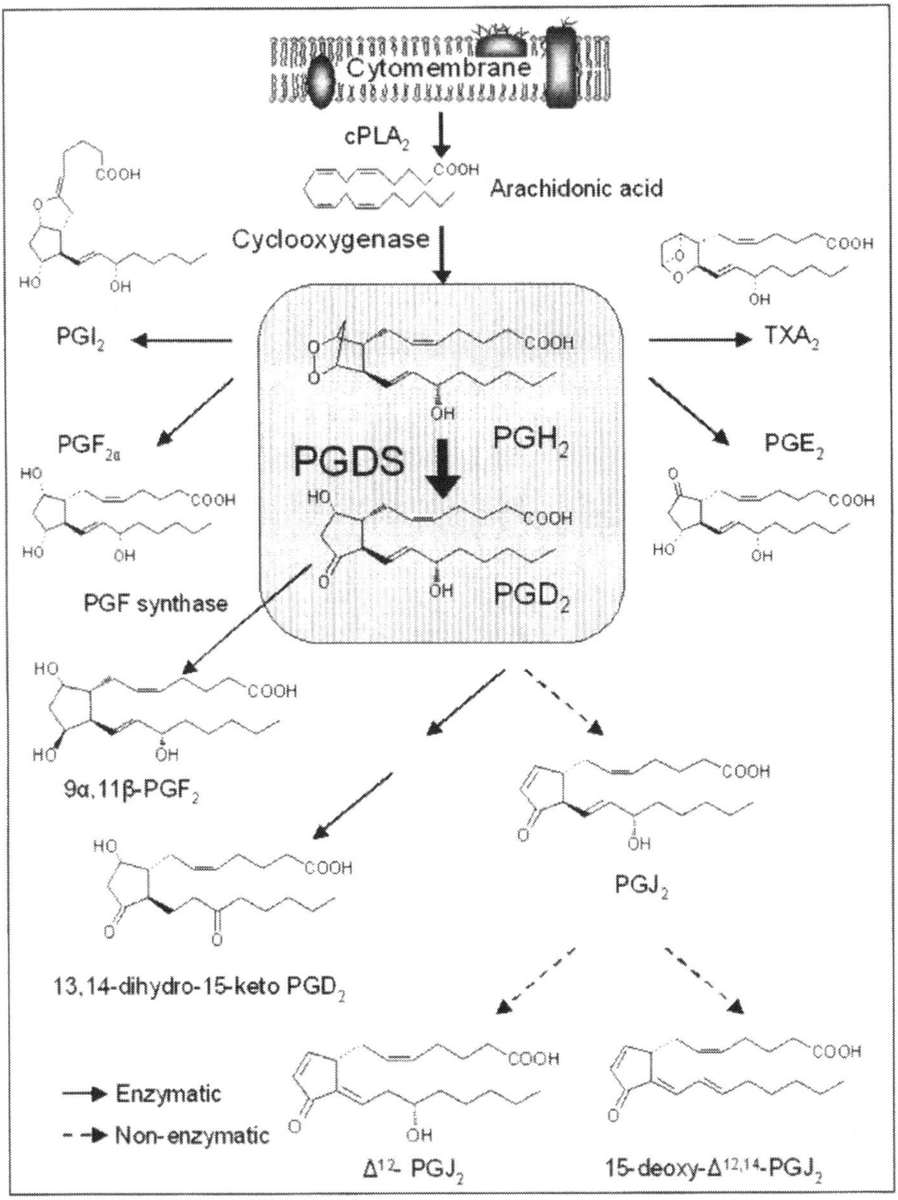

Figure 1. Biosynthetic and metabolic pathway of PGD_2. $cPLA_2$: cytosolic phospholipase A_2; TXA_2: thromboxane A_2.

same reaction. Thus, these 2 enzymes are a novel example of functional conversion.[6,7] During 20 years after the first report of the purification of L-PGDS,[1] we have extensively studied the chemical and functional properties of L-PGDS and reported cloning of the cDNA and the chromosomal gene of the human and mouse enzymes, its X-ray crystallographic structure and immunohistochemical localization, and functional abnormalities of L-PGDS gene knockout (KO) mice and human enzyme-overexpressing mice transgenic (TG) for L-PGDS. A part of

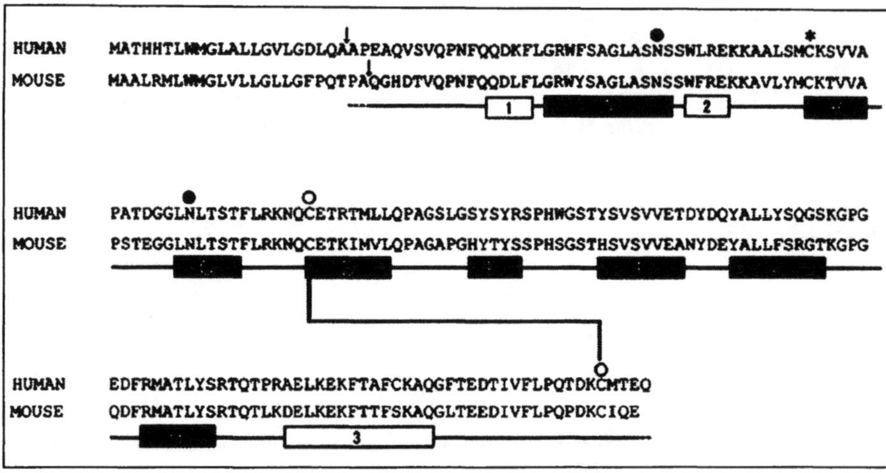

Figure 2. Sequence alignment of human and mouse L-PGDS's. Cleavage site of the signal sequence (arrows), the catalytic cysteine residue (star), a cystine (open circles) and 2 glycosylation sites (closed circles) of L-PGDS are shown on the top of sequence. Positions of α-helixes 1-3 and of β-strands A-H are indicated by open and closed boxes, respectively, under the L-PGDS sequence.

those findings have already been reviewed elsewhere.[6,7,9] In this chapter, we summarize the progress in the research of L-PGDS as a unique member of the lipocalin family.

Amino Acid Sequence and Secondary and Tertiary Structures

The cDNA for L-PGDS was first isolated from a rat brain cDNA library[10] and subsequently from many other mammalian species, including human[11] and mouse,[12] and also from 2 amphibians.[13,14] The cDNA for L-PGDS encodes a protein composed of 189 and 190 amino acid residues in mouse and human enzyme, respectively. Figure 2, shows sequence alignment of the human and mouse enzymes. L-PGDS is post-translationally modified by the cleavage of an N-terminal hydrophobic signal peptide comprising 24 and 22 amino acid residues in the mouse and human enzyme, respectively. Two N-glycosylation sites at positions of Asn51 and Asn78 of the mouse and human enzymes[10] are conserved in all mammalian enzymes thus far identified but were not found in the amphibian homologs. Mammalian L-PGDS is highly glycosylated with 2 N-glycosylated sugar chains, each with a molecular weight of 3,000. The carbohydrate structures of L-PGDS were determined in samples purified from human cerebrospinal fluid (CSF),[15] serum,[16] urine[16,17] and amniotic fluid.[17] However, the functional significance of sugar chains remains to be determined.

Protein chemical and structural properties of L-PGDS have been analyzed by using the recombinant protein heterologously expressed in E. coli or yeast. L-PGDS is a very stable enzyme and is highly resistant against heat treatment[1] and protease digestion.[18] For example, more than 50% of the activity was retained after heating the enzyme for 5 min at 100°C and alkaline pH.[1] L-PGDS was almost completely refolded by dilution and cooling of the enzyme denatured with 1% SDS at 100°C for 10 min or with 6 M guanidine hydrochloride.[19] Thus, L-PGDS is an interesting protein for the study of protein refolding.

Circular dichroism spectroscopy of the recombinant rat L-PGDS revealed that the enzyme is composed of mainly β-strands,[19] similar to other lipocalins. We have already crystallized the recombinant mouse L-PGDS.[6,9] However, the quality of the X-ray diffraction data obtained from the originally prepared crystals were insufficient to determine reliable coordinates of L-PGDS. Most recently, by modifying the crystallization conditions and using the selenomethionyl Cys65Ala mutant,[20] we successfully determined the X-ray crystallographic structure of L-PGDS with 2 different conformers of the open and closed calyxes at 2.1Å resolution (Kumasaka T, Irikura D, Ago H, Aritake K, Yamamoto M, Miyano M, Y.U. and O.H., unpublished results). X-ray crystallographic analysis revealed that L-PGDS possesses a typical lipocalin-fold, β-barrel structure. However, L-PGDS contains 2 hydrophobic pockets; one is the catalytic site corresponding to the ligand-binding pocket of other lipocalins and the other, the lipophilic ligand-binding site, located on the side of the L-PGDS molecule opposite to that containing the retinoid-binding site.

Ligand-Binding Properties

Similar to other lipocalins, L-PGDS binds retinoids,[21] bilirubin, and biliverdin[22] with high affinities (Kd = 30-80 nM). By monitoring the quenching of the intrinsic tryptophan fluorescence of L-PGDS and by circular dichroism spectroscopy of the bound ligands, we showed that L-PGDS binds all-trans- or 9-cis-retinoic acid and all-trans- or 13-cis-retinaldehyde, but not all-trans-retinol, at a molar ratio of 1:1 with a Kd of 70 to 80 nM.[21] The affinities of L-PGDS for retinoids are comparable to or slightly higher than those of other lipocalins acting as extracellular retinoid transporters, such as β-lactoglobulin, plasma retinol-binding protein, and plasma retinoic acid-binding protein (reviewed in other chapters). L-PGDS also binds thyroid hormone with a Kd of 0.7-2 μM and biliverdin and bilirubin with high affinities, i.e., with a Kd of 30-40 nM.[22] We recently found that L-PGDS binds its product PGD_2 with high affinity, (Kd of 20 nM; Aritake K, Y.U.; unpublished results), suggesting that L-PGDS acts not only as a PGD_2-producing enzyme but also as an extracellular transporter of PGD_2 to prevent its metabolism and nonenzymatic dehydration.

Among those nonsubstrate hydrophobic ligands, retinoids,[21] bilirubin, and biliverdin[22] inhibited the PGDS activity in a noncompetitive manner. However, their inhibitory potencies were remarkably weak (IC50 = 4~10 μM) in contrast to their high affinities (Kd = 30-80 nM) of binding to L-PGDS as assessed by quenching of the intrinsic tryptophan fluorescence. Therefore, the catalytic site and the binding site for those hydrophobic ligands are considered to be different from each other. On the other hand, we predicted that PGD_2 is bound to the catalytic pocket. Two distinct binding-pockets were finally identified by the crystallographic study of mouse L-PGDS, as described above.

Therefore, we propose that L-PGDS is a multifunctional protein; it acts as a PGD_2-producing enzyme by coupling with cyclooxygenase-1 or -2, the upstream enzymes in the PG cascade, within the cells and also functions as a lipophilic ligand-binding protein after having been secreted into the extracellular spaces and various body fluids. As described in the next paragraph, PGD_2 is less chemically stable than PGE_2 or $PGF_{2\alpha}$, being dehydrated to the J series of PGs. Once bound to L-PGDS, PGD_2 is remarkably stabilized to slow down the conversion to the J series PGs. Moreover, the binding affinity of L-PGDS for PGD_2 is slightly weaker than those affinities of 2 distinct types of PGD_2 receptor, i.e., the D type of prostanoid (DP, DP_1) receptors and CRTH2 (DP_2). Therefore, we predict that L-PGDS binds the chemically unstable PGD_2, transports PGD_2 within the extracellular space to the action sites, and transfer PGD_2 to its receptors.

Enzymatic Properties as PGD Synthase

L-PGDS is the first lipocalin recognized as an enzyme. L-PGDS has originally been purified from rat brain[1] as PGD synthase (PGH_2 D-isomerase, EC 5.3.99.2), which catalyzes the isomerization of the 9,11-endoperoxide group of PGH_2, a common precursor of various prostanoids, to produce PGD_2 with 9-hydroxy and 11-keto groups in the presence of sulfhydryl compounds. PGD_2 is the major PG produced in the central nervous system of various mammals and is involved in the regulation of sleep[23,24] and nociception[12] through DP (DP_1) receptors.[25,26] PGD_2 is also actively produced and secreted by mast cells,[18] basophils, and Th2 cells[27] by evolutionally different H-PGDS's,[6-8] acting as an allergic and inflammatory mediator through the DP (DP_1)[25,28] and CRTH2 (DP_2)[29] receptors in an autocrine and/or paracrine manner.

The intracellular localization of L-PGDS was most intensely investigated in the brain.[30-32] By immunoelectron microscopy, the immunoreactive deposits of L-PGDS were seen in rough-surfaced endoplasmic reticulum, outer nuclear membrane, Golgi apparatus, and secretory vesicles of oligodendroglial cells[30] and arachnoid trabecular cells in the adult rats[31] and of human arachnoid and meningioma cells.[32] The colocalization of L-PGDS and cyclooxygenase, which produce PGH_2, was demonstrated in virtually all cells of the leptomeninges, choroid plexus epithelial cells, and perivascular microglial cells, suggesting that these cells synthesize PGD_2 actively.[31]

As mentioned above, PGD_2 is chemically unstable and nonenzymically dehydrated to produce the J series of PGs with a cyclopentenone structure, such as PGJ_2, Δ^{12}-PGJ_2, and 15-deoxy-$\Delta^{12,14}$-PGJ_2, all of which show pharmacological activities quite different from those of PGD_2. Since 15-deoxy-$\Delta^{12,14}$-PGJ_2 has been demonstrated to act as a ligand for PPARγ, a nuclear receptor involved in differentiation of adipocytes, macrophages, and monocytes,[33,34] many researchers have studied the possible involvement of 15-deoxy-$\Delta^{12,14}$-PGJ_2 in the various types of physiological function. However, those J-series PGs have not been detected in fresh physiological samples. The physiological relevance of the J series of PGs is thus unlikely.[35] PGD_2 is also metabolized by 11-keto PGD_2 reductase (PGF_2 synthase), belonging to the aldo-keto reductase family, to 9α,11β-PGF_2,[36-38] a stereo isomer of $PGF_{2\alpha}$, having pharmacological activities different from those of $PGF_{2\alpha}$.

L-PGDS requires free sulfhydryl compounds, such as β-mercaptoethanol, DTT or GSH, for its reaction. The enzyme activity is inhibited by $SeCl_4$[39] and SH modifiers, such as N-ethyl maleimide and iodo acetoamide, indicating that Cys residue is involved in the catalytic reaction of L-PGDS.[1] Three Cys residues, Cys65, Cys89, and Cys186, in human L-PGDS, are conserved among all species. Two of these Cys residues, Cys89 and Cys186, form a disulfide bridge, which is highly conserved among most, but not all, lipocalins. On the other hand, Cys65 is unique to L-PGDS, as it has never been found in other lipocalins. Chemical modification and site-directed mutagenesis revealed that the Cys65 residue is the key residue for the catalytic reaction of L-PGDS.[20,40] Thus, L-PGDS is considered to have evolved from a common ancestral nonenzymic lipocalin to the enzyme by acquiring an active residue, Cys65.

Inhibition of sleep by central or systemic administration of Se^{4+} was demonstrated in freely moving rats[41,42] and in unanesthetized fetal sheep,[43] suggesting that PGD_2 produced by L-PGDS plays an important role in the induction and maintenance of sleep.

Gene Structure and Regulation

The gene for L-PGDS was cloned from rat,[44] human,[45] and mouse[12] sources and shown to span about 3 kb and to contain 7 exons split by 6 introns. The gene organization is remarkably analogous to that of other lipocalins in terms of number and size of exons and phase of splicing of introns.[44,45] The human and mouse genes were mapped to chromosome 9q34.2-34.3[45] and 2B-C1,[46] respectively, both of which were localized within the lipocalin gene cluster.

The transcriptional regulation of the L-PGDS gene has been studied after stimulation with various hormones. For example, the thyroid hormone activates L-PGDS expression through a thyroid hormone response element in human brain-derived TE671 cells.[47] Dexamethasone, a synthetic glucocorticoid, induces L-PGDS expression via glucocorticoid receptors in mouse neuronal GT1-7 cells.[48] 17β-Estradiol regulates L-PGDS gene expression in a tissue and region-specific manner. It activates the expression via estrogen β receptors in the mouse heart[49] and increases the L-PGDS expression in the arcuate and ventromedial nuclei of the rat hypothalamus but decreases it in the ventrolateral preoptic area of the hypothalamus, which area is a sleep center.[50,51]

L-PGDS gene expression is down-regulated by the binding of Hes-1, a mammalian homolog of *Drosophila Hairy and enhancer of split*, to the N-box of the promoter in rat primary cultured leptomeningeal cells[52] and human TE671 cells.[53] We recently demonstrated that human L-PGDS gene expression is activated by protein kinase C signaling through de-repression of Notch-HES signaling and enhancement of AP-2β function in TE671 cells.[53] Fluid shear stress induces L-PGDS gene expression in human vascular endothelial cells[54] by binding of c-Fos and c-Jun to the AP-1 binding site of the promoter.[55]

L-PGDS (β-Trace) as a Clinical Marker

L-PGDS is localized in the central nervous system and male genital organs of various mammals,[53] as well as in the human heart[54] and the cellular localization of L-PGDS has been extensively studied in these tissues of various mammals. For example, L-PGDS is dominantly localized in the leptomeninges, choroids plexus, and oligodendrocytes of the rat, mouse, and human brain[12,30-32,58] and is secreted into the CSF. In the testis and epididymis of humans[59] and other mammals,[60-68] L-PGDS is localized in Leydig cells, Serotoli cells, and ductal epithelial cells and is secreted from them into the seminal plasma. Among various human tissues, the heart expresses L-PGDS mRNA the most intensely.[57] In the human heart, L-PGDS is localized in myocardial cells and atrial endocardial cells, and most interestingly has been detected in smooth muscle cells (having the synthesis phenotype) in the arteriosclerotic intima and in the atherosclerotic plaque of coronary arteries with severe stenosis, being secreted by these cells into the plasma.[57]

L-PGDS is the same protein as β-trace,[69,70] which was originally discovered in 1961 as a major protein of human CSF[71] and later identified in the seminal plasma, serum, and urine. Therefore, the L-PGDS/β-trace concentration in body fluids may be a useful clinical marker for various diseases.[7,9] The L-PGDS/β-trace concentrations in seminal plasma, serum, and urine have been extensively evaluated in recent years as a biomarker for diagnosis of several neurological disorders,[72-78] dysfunction of sperm formation,[59] and cardiovascular[57,79-81] and renal[82-87] diseases. The serum L-PGDS/β-trace concentration shows a circadian change with a nocturnal increase, which is suppressed during total sleep deprivation but not affected by rapid eye movement (REM) sleep deprivation.[88] The L-PGDS concentration in cervicovaginal secretions of pregnant women with ruptured membranes was reported to be significantly higher than that of normal pregnant women.[89]

Moreover, the upregulation of L-PGDS gene expression was reported in a genetic demyelinating model mice, twitcher[90] and in patients with multiple sclerosis,[91,92] Tay-Sachs or Sandohof disease.[93] A single nucleotide polymorphism found in the 3'-untranslated region of the human L-PGDS gene (4111 A>C) was shown to be associated with carotid atherosclerosis in Japanese hypertensive patients.[94] In them, the serum levels of high-density lipoprotein cholesterol were higher in subjects with the A/A genotype than in those with the A/C or C/C genotype and the maximum intima-media thickness in the common carotid artery was smaller in the A/A group than in the A/C and C/C groups. RT-PCR analysis in microdissected rat nephron segments revealed that L-PGDS mRNA is widely expressed in the cortex and outer medulla, and mainly in the thick ascending limb and the collecting duct.[95] In a mouse model of adriamycin-induced nephropathy, urinary L-PGDS excretion was shown to precede overt albuminuria.[96]

Functional Abnormalities of L-PGDS KO Mice and Human L-PGDS-Overexpressing TG-Mice

We generated L-PGDS KO mice with the null mutation by homologous recombination[12] and demonstrated that the KO mice grow normally but show several functional abnormalities in the regulation of nociception,[12] sleep,[24,97] and energy metabolism.[98] L-PGDS KO mice do not exhibit allodynia (touch-evoked pain), which is a typical phenomenon of neuropathic pain, after an intrathecal administration of PGE_2 or bicuculline, a γ-aminobutyric acid $(GABA)_A$ receptor antagonist.[12] The KO mice do not accumulate PGD_2 in their brain during sleep deprivation nor show the nonREM sleep rebound after sleep deprivation; whereas the wild-type mice show an increase in the PGD_2 content in their brain during sleep deprivation, which induces the nonREM sleep rebound.[24,97] L-PGDS KO mice become glucose intolerant and insulin resistant at an accelerated rate as compared with the wild-type mice.[98] The KO mice possess adipocytes of larger size than do wild-type mice and develop nephropathy and an aortic thickening when fed a high-fat diet.[98]

We also generated TG mice[99] that over-expressed the human L-PGDS under the control of the β-actin promoter. We serendipitously discovered that these TG mice showed a transient increase in nonREM sleep after their tails had been clipped for DNA sampling used for genetic analysis.[97,99] We showed that the noxious stimulation of tail clipping induced a remarkable increase in the PGD_2 content in the brain of the TG mice but not in that of the wild-type ones,[97,99] although we do not yet understand in detail the mechanism responsible for this increase. Alternatively, in an ovalbumin-induced asthma model, the TG-mice showed a remarkably increased PGD_2 production in the lung after the antigen challenge and developed pronounced eosinophilic lung inflammation and Th2 cytokine release as compared with their wild-type littermates.[100] These TG mice also exhibited accelerated adipogenesis (Fujitani Y, Aritake K, and Y.U., unpublished results). Therefore, L-PGDS TG mice are useful as a unique animal model to study the functional abnormalities caused by the overproduction of PGD_2.

Closing Remarks

Recently, exogenously administered L-PGDS was demonstrated to inhibit the growth of vascular smooth muscle cells obtained from spontaneously hypertensive rats, but not from normotensive control animals[80] and L-PGDS was also identified as a cellular target of the immediate-early protein, BICP0, of bovine herpesvirus 1.[101] However, the action mechanisms of L-PGDS operating in those processes remain to be elucidated. Most recently we determined the three-dimensional coordinates of L-PGDS complexed with retinoic acid and also found an orally active inhibitor of L-PGDS. These results are useful for designing inhibitors of L-PGDS, which will promote further pharmacological evaluation of L-PGDS as an enzyme and a retinoid-transporter. Several groups are now trying to identify endogenous ligands of L-PGDS in various body fluids. The screening for functional abnormalities of L-PGDS gene-manipulated mice is still on going by collaborative researches with many groups.

References

1. Urade Y, Fujimoto N, Hayaishi O. Purification and characterization of rat brain prostaglandin D synthetase. J Biol Chem 1985; 260:12410-12415.
2. Shimizu T, Yamamoto S, Hayaishi O. Purification and properties of prostaglandin D synthetase from rat brain. J Biol Chem 1979; 254:5222-5228.
3. Shimizu T, Yamamoto S, Hayaishi O. Purification of PGH-PGD isomerase from rat brain. Methods Enzymol 1982; 86:73-77.
4. Christ-Hazelhof E, Nugteren DH. Purification and characterization of prostaglandin endoperoxide D-isomerase, a cytoplasmic, glutathione-requiring enzyme. Biochim Biophys Acta 1979; 572:43-51.
5. Urade Y, Fujimoto N, Ujihara M et al. Biochemical and immunological characterization of rat spleen prostaglandin D synthetase. J Biol Chem 1987; 262:3820-3825.
6. Urade Y, Hayaishi O. Prostaglandin D synthase: Structure and function. Vitam Horm 2000; 58:89-120.
7. Urade Y, Eguchi N. Lipocalin-type and hematopoietic prostaglandin D synthases as a novel example of functional convergence. Prostaglandins Other Lipid Mediat 2002; 68-69:375-382.
8. Kanaoka Y, Urade Y. Hematopoietic prostaglandin D synthase. Prostaglandins Leukot Essent Fatty Acids 2003; 69:163-167.
9. Urade Y, Hayaishi O. Biochemical, structural, genetic, physiological, and pathophysiological features of lipocalin-type prostaglandin D synthase. Biochim Biophys Acta 2000; 1482:259-271.
10. Urade Y, Nagata A, Suzuki Y et al. Primary structure of rat brain prostaglandin D synthetase deduced from cDNA sequence. J Biol Chem 1989; 264:1041-1045.
11. Nagata A, Suzuki Y, Igarashi M et al. Human brain prostaglandin D synthase has been evolutionarily differentiated from lipophilic-ligand carrier proteins. Proc Natl Acad Sci USA 1991; 88:4020-4024.
12. Eguchi N, Minami T, Shirafuji N et al. Lack of tactile pain (allodynia) in lipocalin-type prostaglandin D synthase-deficient mice. Proc Natl Acad Sci USA 1999; 96:726-730.
13. Achen MG, Harms PJ, Thomas T et al. Protein synthesis at the blood-brain barrier. The major protein secreted by amphibian choroid plexus is a lipocalin. J Biol Chem 1992; 267:23170-23174.

14. Lepperdinger G, Strobl B, Jilek A et al. The lipocalin Xlcpl1 expressed in the neural plate of Xenopus laevis embryos is a secreted retinaldehyde binding protein. Protein Sci 1996; 5:1250-1260.
15. Hoffmann A, Nimtz M, Wurster U et al. Carbohydrate structures of beta-trace protein from human cerebrospinal fluid: Evidence for "brain-type" N-glycosylation. J Neurochem 1994; 63:2185-2196.
16. Hoffmann A, Nimtz M, Conradt HS. Molecular characterization of beta-trace protein in human serum and urine: A potential diagnostic marker for renal diseases. Glycobiology 1997; 7:499-506.
17. Manya H, Sato Y, Eguchi N et al. Comparative study of the asparagine-linked sugar chains of human lipocalin-type prostaglandin D synthase purified from urine and amniotic fluid, and recombinantly expressed in Chinese hamster ovary cells. J Biochem (Tokyo) 2000; 127:1001-1011.
18. Urade Y, Ujihara M, Horiguchi M et al. Mast cells contain spleen-type prostaglandin D synthase. J Biol Chem 1990; 265:371-375.
19. Inui T, Ohkubo T, Emi M et al. Characterization of the unfolding process of lipocalin-type prostaglandin D synthase. J Biol Chem 2003; 278:2845-2852.
20. Irikura D, Kumasaka T, Yamamoto M et al. Cloning, expression, crystallization, and preliminary X-ray analysis of recombinant mouse lipocalin-type prostaglandin D synthase, a somnogen-producing enzyme. J Biochem (Tokyo) 2003; 133:29-32.
21. Tanaka T, Urade Y, Kimura H et al. Lipocalin-type prostaglandin D synthase (β-trace) is a newly recognized type of retinoid transporter. J Biol Chem 1997; 272:15789-15795.
22. Beuckmann CT, Aoyagi M, Okazaki I et al. Binding of biliverdin, bilirubin, and thyroid hormones to lipocalin-type prostaglandin D synthase. Biochemistry 1999; 38:8006-8013.
23. Hayaishi O, Urade Y. Prostaglandin D_2 in sleep-wake regulation: Recent progress and perspectives. Neuroscientist 2002; 8:12-15.
24. Hayaishi O, Urade Y, Eguchi N et al. Genes for prostaglandin D synthase and receptor as well as adenosine A2A receptor are involved in the homeostatic regulation of NREM sleep. Arch Ital Biol 2004; 142:533-539.
25. Hirata M, Kakizuka A, Aizawa M et al. Molecular characterization of a mouse prostaglandin D receptor and functional expression of the cloned gene. Proc Natl Acad Sci USA 1994; 91:11192-11196.
26. Mizoguchi A, Eguchi N, Kimura K et al. Dominant localization of prostaglandin D receptors on arachnoid trabecular cells in mouse basal forebrain and their involvement in the regulation of nonrapid eye movement sleep. Proc Natl Acad Sci USA 2001; 98:11674-11679.
27. Tanaka K, Ogawa K, Sugamura K et al. Cutting edge: Differential production of prostaglandin D_2 by human helper T cell subsets. J Immunol 2000; 164:2277-2280.
28. Matsuoka T, Hirata M, Tanaka H et al. Prostaglandin D_2 as a mediator of allergic asthma. Science 2000; 287:2013-2017.
29. Hirai H, Tanaka K, Yoshie O et al. Prostaglandin D_2 selectively induces chemotaxis in T helper type 2 cells, eosinophils, and basophils via seven-transmembrane receptor CRTH2. J Exp Med 2001; 193:255-261.
30. Urade Y, Fujimoto N, Kaneko T et al. Postnatal changes in the localization of prostaglandin D synthetase from neurons to oligodendrocytes in the rat brain. J Biol Chem 1987; 262:15132-15136.
31. Beuckmann CT, Lazarus M, Gerashchenko D et al. Cellular localization of lipocalin-type prostaglandin D synthase (β-trace) in the central nervous system of the adult rat. J Comp Neurol 2000; 428:62-78.
32. Yamashima T, Sakuda K, Tohma Y et al. Prostaglandin D synthase (beta-trace) in human arachnoid and meningioma cells: Roles as a cell marker or in cerebrospinal fluid absorption, tumorigenesis, and calcification process. J Neurosci 1997; 17:2376-2382.
33. Forman BM, Tontonoz P, Chen J et al. 15-Deoxy-delta 12, 14-prostaglandin J_2 is a ligand for the adipocyte determination factor PPAR gamma. Cell 1995; 83:803-812.
34. Kliewer SA, Lenhard JM, Willson TM et al. A prostaglandin J_2 metabolite binds peroxisome proliferator-activated receptor gamma and promotes adipocyte differentiation. Cell 1995; 83:813-819.
35. Bell-Parikh LC, Ide T, Lawson JA et al. Biosynthesis of 15-deoxy-delta12,14-PGJ_2 and the ligation of PPARgamma. J Clin Invest 2003; 112:945-955.
36. Liston TE, Roberts IInd LJ. Transformation of prostaglandin D_2 to 9 alpha, 11 beta-(15S)-trihydroxyprosta-(5Z,13E)-dien-1-oic acid (9 alpha, 11 beta-prostaglandin F_2): A unique biologically active prostaglandin produced enzymatically in vivo in humans. Proc Natl Acad Sci USA 1985; 82:6030-6034.
37. Watanabe K, Iguchi Y, Iguchi S et al. Stereospecific conversion of prostaglandin D_2 to (5Z,13E)-(15S)-9 alpha-11 beta,15-trihydroxyprosta-5,13-dien-1-oic acid (9 alpha,11 beta-prostaglandin F_2) and of prostaglandin H_2 to prostaglandin F_2 alpha by bovine lung prostaglandin F synthase. Proc Natl Acad Sci USA 1986; 83:1583-1587.

38. Urade Y, Watanabe K, Eguchi N et al. 9 Alpha,11 beta-prostaglandin F_2 formation in various bovine tissues. Different isozymes of prostaglandin D_2 11-ketoreductase, contribution of prostaglandin F synthetase and its cellular localization. J Biol Chem 1990; 265:12029-12035.
39. Islam F, Watanabe Y, Morii H et al. Inhibition of rat brain prostaglandin D synthase by inorganic selenocompounds. Arch Biochem Biophys 1991; 289:161-166.
40. Urade Y, Tanaka T, Eguchi N et al. Structural and functional significance of cysteine residues of glutathione-independent prostaglandin D synthase. Identification of Cys65 as an essential thiol. J Biol Chem 1995; 270:1422-1428.
41. Matsumura H, Takahata R, Hayaishi O. Inhibition of sleep in rats by inorganic selenium compounds, inhibitors of prostaglandin D synthase. Proc Natl Acad Sci USA 1991; 88:9046-9050.
42. Takahata R, Matsumura H, Kantha SS et al. Intravenous administration of inorganic selenium compounds, inhibitors of prostaglandin D synthase, inhibits sleep in freely moving rats. Brain Res 1993; 623:65-71.
43. Lee B, Hirst J, Walker DW. Prostaglandin D synthase in the prenatal ovine brain and effects of its inhibition with selenium chloride on fetal sleep/wake activity in utero. J Neurosci 2002; 22:5679-5686.
44. Igarashi M, Nagata A, Toh H et al. Structural organization of the gene for prostaglandin D synthase in the rat brain. Proc Natl Acad Sci USA 1992; 89:5376-5380.
45. White DM, Mikol DD, Espinosa R et al. Structure and chromosomal localization of the human gene for a brain form of prostaglandin D_2 synthase. J Biol Chem 1992; 267:23202-23208.
46. Chan P, Simon-Chazottes D, Mattei MG et al. Comparative mapping of lipocalin genes in human and mouse: The four genes for complement C8γ chain, prostaglandin-D-synthase, oncogene-24p3 and progestagen-associated endometrial protein map to HSA9 and MMU2. Genomics 1994; 23:145-150.
47. White DM, Takeda T, DeGroot LJ et al. β-Trace gene expression is regulated by a core promoter and a distal thyroid hormone response element. J Biol Chem 1997; 272:14387-14393.
48. Garcia-Fernandez LF, Iniguez MA, Eguchi N et al. Dexamethasone induces lipocalin-type prostaglandin D synthase gene expression in mouse neuronal cells. J Neurochem 2000; 75:460-470.
49. Otsuki M, Gao H, Dahlman-Wright K et al. Specific regulation of lipocalin-type prostaglandin D synthase in mouse heart by estrogen receptor beta. Mol Endorinol 2003; 17:1844-1855.
50. Mong JA, Devidze N, Frail DE et al. Estradiol differentially regulates lipocalin-type prostaglandin D synthase transcript levels in the rodent brain: Evidence from high-density oligonucleotide arrays and in situ hybridization. Proc Natl Acad Sci USA 2003; 100:318-323.
51. Mong JA, Devidze N, Goodwillie A et al. Reduction of lipocalin-type prostaglandin D synthase in the preoptic area of female mice mimics estradiol effects on arousal and sex behavior. Proc Natl Acad Sci USA 2003; 100:15206-15211.
52. Fujimori K, Fujitani Y, Kadoyama K et al. Regulation of lipocalin-type prostaglandin D synthase gene expression by Hes-1 through E-box and interleukin-1 beta via two NF-kappa B elements in rat leptomeningeal cells. J Biol Chem 2003; 278:6018-6026.
53. Fujimori K, Kadoyama K, Urade Y. Protein kinase C activates human lipocalin-type prostaglandin D synthase gene expression through De-repression of notch-Hes signaling and enhancement of AP-2beta function in brain-derived TE671 cells. J Biol Chem 2005; 280:18452-18461.
54. Taba Y, Sasaguri T, Miyagi M et al. Fluid shear stress induces lipocalin-type prostaglandin D_2 synthase expression in vascular endothelial cells. Circ Res 2000; 86:967-973.
55. Miyagi M, Miwa Y, Takahashi-Yanaga F et al. Activator protein-1 mediates shear stress-induced prostaglandin D synthase gene expression in vascular endothelial cells. Arterioscler Thromb Vasc Biol 2005; 25:970-975.
56. Ujihara M, Urade Y, Eguchi N et al. Prostaglandin D_2 formation and characterization of its synthetases in various tissues of adult rats. Arch Biochem Biophys 1988; 260:521-531.
57. Eguchi Y, Eguchi N, Oda H et al. Expression of lipocalin-type prostaglandin D synthase (β-trace) in human heart and its accumulation in the coronary circulation of angina patients. Proc Natl Acad Sci USA 1997; 94:14689-14694.
58. Urade Y, Kitahama K, Ohishi H et al. Dominant expression of mRNA for prostaglandin D synthase in leptomeninges, choroid plexus, and oligodendrocytes of the adult rat brain. Proc Natl Acad Sci USA 1993; 90:9070-9074.
59. Tokugawa Y, Kunishige I, Kubota Y et al. Lipocalin-type prostaglandin D synthase in human male reproductive organs and seminal plasma. Biol Reprod 1998; 58:600-607.
60. Gerena RL, Irikura D, Urade Y et al. Identification of a fertility-associated protein in bull seminal plasma as lipocalin-type prostaglandin D synthase. Biol Reprod 1998; 58:826-833.
61. Gerena RL, Eguchi N, Irikura Y et al. Immunocytochemical localization of lipocalin-type prostaglandin D synthase in the bull testis, epididymis and on ejaculated sperm. Biol Reprod 2000; 62:547-556.

62. Rodriguez CM, Day JR, Killian GJ. Expression of the lipocalin-type prostaglandin D synthase gene in the reproductive tracts of Holstein bulls. J Reprod Fertil 2000; 120:303-309.
63. Gerena RL, Eguchi N, Urade Y et al. Stage and region-specific localization of lipocalin-type prostaglandin D synthase in the adult murine testis and epididymis. J Androl 2000; 21:848-854.
64. Samy ET, Li JC, Grima J et al. Serotoli cell prostaglandin D_2 synthase is a multifunctional molecule: Its expression and regulation. Endocrinology 2000; 141:710-721.
65. Fouchecourt S, Charpigny G, Reinaud P et al. Mammalian lipocalin-type prostaglandin D_2 synthase in the fluids of the male genital tract: putative biochemical and physiological functions. Biol Reprod 2002; 66:458-467.
66. Fouchecourt S, Chaurand P, DaGue BB et al. Epididymal lipocalin-type prostaglandin D_2 synthase: Identification using mass spectrometry, messenger RNA localization, and immunodetection in mouse, rat, hamster, and monkey. Biol Reprod 2002; 66:524-533.
67. Zhu H, Ma H, Ni H et al. Expression and regulation of lipocalin-type prostaglandin D synthase in rat testis and epididymis. Biol Reprod 2004; 70:1088-1095.
68. Malki S, Nef S, Cécile N et al. Prostaglandin D_2 induces nuclear import of the sex-determining factor SOX9 via its cAMP-PKA phosphorylation. EMBO J 2005; 24:1798-1809.
69. Hoffmann A, Conradt HS, Gross G et al. Purification and chemical characterization of β-trace protein from human cerebrospinal fluid: Its identification as prostaglandin D synthase. J Neurochem 1993; 61:451-456.
70. Watanabe K, Urade Y, Mäder M et al. Identification of β-trace as prostaglandin D synthase. Biochem Biophys Res Commun 1994; 203:1110-1116.
71. Clausen J. Proteins in normal cerebrospinal fluid not found in serum. Proc Soc Exp Biol Med 1961; 107:170-172.
72. Melegos DN, Freedmann MS, Diamandis EP. Prostaglandin D synthase concentration in cerebrospinal fluid and serum of patients with neurological disorders. Prostaglandins 1997; 54:463-474.
73. Hiraoka A, Arato T, Tominaga I et al. Sodium dodecyl sulfate-capillary gel electrophoretic analysis of molecular mass microheterogeneity of β-trace protein in cerebrospinal fluid from patients with central nervous system diseases. J Chromatogr 1998; 802:143-148.
74. Tumani H, Nau R, Felgenhauer K. β-Trace protein in cerebrospinal fluid: A blood-CSF barrier-related evaluation in neurological diseases. Ann Neurol 1998; 44:882-889.
75. Mase M, Yamada K, Iwata A et al. Acute and transient increase of lipocalin-type prostaglandin D synthase (beta-trace) level in cerebrospinal fluid of patients with aneurismal subarachnoid hemorrhage. Neurosci Lett 1999; 270:188-190.
76. Hiraoka A, Seiki K, Oda H et al. Charge microheterogeneity of the beta-trace proteins (lipocalin-type prostaglandin D synthase) in the cerebrospinal fluid of patients with neurological disorders analyzed by capillary isoelectrofocusing. Electrophoresis 2001; 22:3433-3437.
77. Mase M, Yamada K, Shimazu N et al. Lipocalin-type prostaglandin D synthase (beta-trace) in cerebrospinal fluid: A useful marker for the diagnosis of normal pressure hydrocephalus. Neurosci Res 2003; 47:455-459.
78. Brettschneider J, Riepe MW, Petereit HF et al. Meningeal derived cerebrospinal fluid proteins in different forms of dementia: Is a meningopathy involved in normal pressure hydrocephalus? J Neurol Neurosurg Psychiatry 2004; 75:1614-1616.
79. Hirawa N, Uehara Y, Yamakado M et al. Lipocalin-type prostaglandin D synthase in essential hypertension. Hypertension 2002; 39:449-454.
80. Ragolia L, Palaia T, Paric E et al. Prostaglandin D_2 synthase inhibits the exaggerated growth phenotype of spontaneously hypertensive rat vascular smooth muscle cells. J Biol Chem 2003; 278:22175-22181.
81. Cipollone F, Fazia M, Iezzi A et al. Balance between PGD synthase and PGE synthase is a major determinant of atherosclerotic plaque instability in humans. Arterioscler Thromb Vasc Biol 2004; 24:1259-1265.
82. Melegos DN, Grass L, Pierratos A et al. Highly elevated levels of prostaglandin D synthase in the serum of patients with renal failure. Urology 1999; 53:32-37.
83. Priem F, Althaus H, Birnbaum M et al. Beta-trace protein in serum: A new marker of glomerular filtration rate in the creatinine-blind range. Clin Chem 1999; 45:567-568.
84. Giessing M. Beta-trace protein as indicator of glomerular filtration rate. Urology 1999; 54:940-941.
85. Hirawa N, Uehara Y, Ikeda T et al. Urinary prostaglandin D synthase (β-trace) excretion increases in the early-stage of diabetes mellitus. Nephron 2001; 87:321-327.
86. Oda H, Shiina Y, Seiki K et al. Development and evaluation of a practical ELISA for human urinary lipocalin-type prostaglandin D synthase. Clin Chem 2002; 48:1445-1453.
87. Hamano K, Totsuka Y, Ajima M et al. Blood sugar control reverses the urinary excretion of prostaglandin D synthase in diabetic patients. Nephron 2002; 92:77-85.

88. Jordan W, Tumani H, Cohrs S et al. Prostaglandin D synthase (bata-trace) in healthy human sleep. Sleep 2004; 27:867-874.
89. Shiki Y, Shimoya K, Tokugawa Y et al. Changes of lipocalin-type prostaglandin D synthase level during pregnancy. J Obstet Gynaecol Res 2004; 30:65-70.
90. Taniike M, Mohri I, Eguchi N et al. Perineuronal oligodendrocytes protect against neuronal apoptosis through the production of lipocalin-type prostaglandin D synthase in a genetic demyelinating model. J Neurosci 2002; 22:4885-4896.
91. Chabas D, Baranzini SE, Mitchell D et al. The influence of the proinflammatory cytokine, osteopontin, on autoimmune demyelinating disease. Science 2000; 48:1731-1735.
92. Kagitani Shimono K, Mohri I, Oda H et al. Lipocalin-type prostaglandin D synthase (β-trace) is upregulated in the αB-crystallin-positive oligodendrocytes and astrocytes in the chronic multiple sclerosis. Neuropathol Appl Neurobiol 2005; in press.
93. Myerowitz R, Lawson D, Mizukami H et al. Molecular pathophysiology in Tay-Sachs and Sandhoff diseases as revealed by gene expression profiling. Hum Mol Genet 2002; 15:1343-1350.
94. Miwa Y, Takiuchi S, Kamide K et al. Identification of gene polymorphism in lipocalin-type prostaglandin D synthase and its association with carotid atherosclerosis in Japanese hypertensive patients. Biochem Biophys Res Commun 2004; 322:428-433.
95. Vitzthum H, Abt I, Einhellig S et al. Gene expression of prostanoid-forming enzymes along the rat nephron. Kidney Int 2002; 62:1570-1581.
96. Tsuchida T, Eguchi N, Eguchi Y et al. Lipocalin-type prostaglandin D synthase in urine in adriamycin-induced nephropathy of mice. Nephron Physiol 2004; 96:42-51.
97. Eguchi N, Pinzar E, Kuwahata Y et al. Sleep of transgenic and gene-knockout mice for lipocalin-type prostaglandin D synthase. In: Ishimura Y, Nozaki M, Yamamoto S et al, eds. Oxygen and Life - Oxygenases, Oxidases, and Lipid Mediators. International Congress Series 1233. Excerpta Medica, 2002:429-433.
98. Ragolia L, Palaia T, Hall CE et al. Accelerated glucose intolerance, nephropathy, and atherosclerosis in prostaglandin D_2 synthase knockout mice. J Biol Chem 2005; in press.
99. Pinzar E, Kanaoka Y, Inui T et al. Prostaglandin D synthase gene is involved in the regulation of nonrapid eye movement sleep. Proc Natl Acad Sci USA 2000; 97:4903-4907.
100. Fujitani Y, Kanaoka Y, Aritake K et al. Pronounced eosinophilic lung inflammation and Th2 cytokine release in human lipocalin-type prostaglandin D synthase transgenic mice. J Immunol 2002; 168:443-449.
101. Saydam O, Abril C, Vogt B et al. Transactivator protein BICP0 of bovine herpesvirus 1 (BHV-1) is blocked by prostaglandin D_2 (PGD_2), which points to a mechanism for PGD_2-mediated inhibition of BHV-1 replication. J Virol 2004; 78:3805-3810.

Chapter 10

α_1-Microglobulin

Bo Åkerström* and Lennart Lögdberg

Abstract

α_1-Microglobulin is one of the three original members of the lipocalin superfamily. It has been found in mammals, birds, amphibians and fish and is distributed in plasma and extravascular compartments of all organs. α_1-Microglobulin has a free cysteine side-chain located in a flexible loop, giving the protein reductase and dehydrogenase properties with a broad biological substrate specificity. Three lysyl residues located around the opening of the lipocalin pocket carry yellow-brown modifications originating from the binding and degradation of heme and kynurenin, the latter a tryptophan metabolite. We have suggested that α_1-microglobulin is involved in defending tissues against oxidation by heme, kynurenin and reactive oxygen species.

Introduction

α_1-Microglobulin (α_1m) was discovered in human urine 40 years ago and was named in the tradition of plasma proteins, reflecting its small size (26 kDa) and its electrophoretic migration slightly behind albumin.[1] The protein was characterized by several research groups and given the alternative names protein HC, "human complex-forming protein, heterogeneous in charge"[2] and α_1-microglycoprotein.[3] α_1M, retinol-binding protein (RBP) and β-lactoglobulin were the three original members when the lipocalin family was defined in 1985.[4]

The physiological role of α_1m has only recently been clarified. Immunoregulatory (mainly suppressive) properties of α_1m were identified (see below), but did not seem distinct or strong enough to constitute the major function of the molecule in vivo. However, several very recent reports have suggested that α_1m may play a biological role as an anti-oxidant with oxidant-scavenging and enzymatic reductase properties. In this review, we will describe the structural features of α_1m, its distribution among tissues and species and the anti-oxidation and immunoregulatory properties of this lipocalin.

Structure

Peptide Chain

The full sequence of human α_1m was first reported by Kaumeyer et al.[5] The protein was found to consist of 183 amino acid residues. Since then, ten additional α_1m cDNAs and/or proteins have been detected, isolated and/or sequenced, from other mammals,[6-17] birds,[18] amphibians,[19] and fish[20-22] (Table 1). The length of the peptide chain of α_1m differs slightly among species, due mainly to variations in the C-terminus. Alignment comparisons of the different deduced amino acid sequences show that the percentage of identity varies from approximately 75-80% between rodents or ferungulates and man, down to approximately

*Corresponding Author: Bo Åkerström—Department of Clinical Sciences, Lund University, BMC, B14, 221 84 Lund, Sweden. Email: Bo.Akerstrom@med.lu.se

Lipocalins, edited by Bo Åkerström, Niels Borregaard, Darren R. Flower and Jean-Philippe Salier. ©2006 Landes Bioscience.

Table 1. α_1M and bikunin are encoded by a single gene and co-expressed as a precursor-protein. This genetic construction has been found in many species.

Species	Protein		mRNA
	$\alpha_1 m$	bikunin	
Man	α_1m[1-3]	bikunin	α_1m-bikunin[5]
Pig			α_1m-bikunin[6]
Cow	α_1m[7]		α_1m-bikunin[7]
Guinea pig	α_1m[8]		α_1m-bikunin[9]
Rat	α_1m[10]	bikunin[11]	α_1m-bikunin[12,13]
Rabbit	α_1m[14]		
Mouse	α_1m[15]	bikunin[15]	α_1m-bikunin[16]
Hamster			α_1m-bikunin[17]
Gerbil			α_1m-bikunin[17]
Chicken	α_1m[18]		
Frog			α_1m-bikunin[19]
Plaice	α_1m[20]		α_1m-bikunin[21]
Salmon			α_1m-bikunin[22]

45% between fish and mammals. A free cysteine side-chain at position 34 is conserved. This group has been shown to be involved in redox reactions, in complex formation with other plasma proteins and in binding to a yellow-brown chromophore (see below). Computerised 3D models[23] based on the known X-ray crystallographic structures of other lipocalins suggest that Cys34 is solvent exposed and located near the opening of the lipocalin pocket (Fig. 1). Complement factor C8γ, another lipocalin, also carries an unpaired Cys in position 34 that is involved in the formation of the active C8 complex.[24] In yet another lipocalin, prostaglandin D synthase, a free Cys65 has been shown to be of importance for the catalytic activity of the enzyme.[25]

Carbohydrates

Human α_1m is substituted with oligosaccharides in three positions, two sialylated complex-type, probably diantennary carbohydrates linked to Asn17 and Asn96 and one more simple oligosaccharide linked to Thr5.[26-28] The carbohydrate content of α_1m proteins from different species varies greatly, though, ranging from no glycosylation at all in *Xenopus leavis*[19] over a spectrum of different glycosylation patterns. However, one glycosylation site, corresponding to Asn96 in man, is conserved in mammals, suggesting that this specific carbohydrate may be functionally important.

Chromophore

α_1M is charge- and size-heterogeneous. Tightly bound, heterogeneous, brown-coloured prosthetic groups have been proposed as responsible for the heterogeneity.[29] The heterogeneity and brown colour are universal properties of α_1m from all species and at least parts of the brown materials are attached intracellularly to the protein.[20,30] Several structurally different chromophores attached to multiple residues may explain the observed charge and size heterogeneity of the protein. Covalently linked coloured substances have been localized to Cys34,[29] and Lys92, Lys118 and Lys130, the latter with molecular masses between 100 and 300 Da.[31] Molecular modelling suggests that all four residues are located at the entrance of the lipocalin cavity (Fig. 1). Recently, the tryptophan metabolite kynurenine was found covalently attached to lysyl residues in α_1m from urine of hemodialysis patients and appears to be the source of the brown colour of the protein in this case.[32]

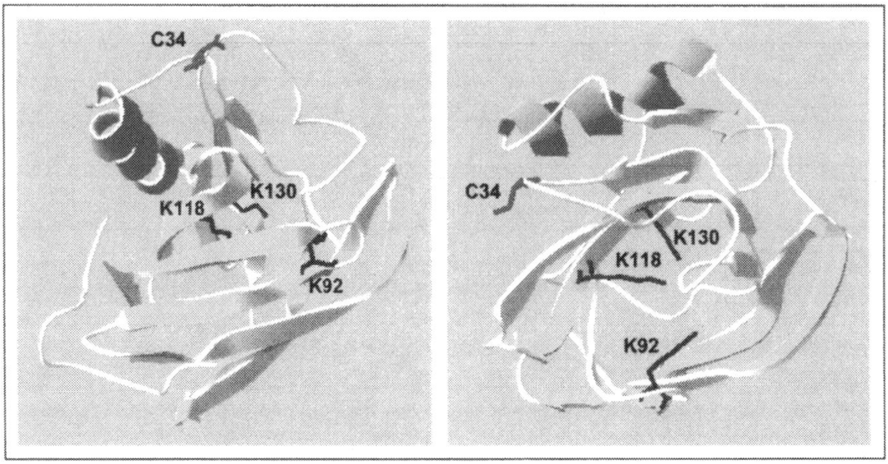

Figure 1. Three-dimensional model of α_1m. The model was generated using Swiss-Pdb Viewer and Swiss Model by alignment of the human α_1m amino acid sequence with rat epididymal retinoic acid-binding protein (E-RABP). β-strands are shown in yellow, α-helix in orange, Cys-34 in red and Lys-92, Lys-118 and Lys-130 in blue. Viewed from the side of the β-barrel (left) and looking down into the pocket (right).

Lipocalin Ligand

Available data suggest that many different ligands can be fitted into the pocket of α_1m. Several hydrophobic substances, including retinol, have been extracted from α_1m, but in molar quantities never more than approximately 1/1000 of the protein.[33] A 282 Da lipophilic substance was copurified with the peptides containing Lys92, Lys118 and Lys130.[31] Recently, several papers demonstrated that heme binds specifically to α_1m in plasma, urine and other tissue fluids.[34-36] The affinity constant (K_a) was approximately 1-2 x 10^6 M^{-1}, and the binding evolutionarily conserved.[36] A physiological role as a heme-scavenger was therefore proposed for α_1m (see below).

Synthesis

Cosynthesis with Bikunin

The gene for α_1m is called *AMBP* (Alpha-1-Microglobulin-Bikunin Precursor gene) because it also encodes bikunin, another plasma protein[5] (Fig. 2). Bikunin[37] is a common subunit of a group of protein/carbohydrate complexes that constitute the inter-α-inhibitor family. Its members are plasma and tissue proteins that have proteinase inhibitor activity[38] and serve as structural components of extracellular matrix.[39] Curiously, transcription of the *AMBP* gene, which takes place mainly in the liver, produces an mRNA that is translated into a precursor protein consisting of a 19 amino acid residue signal peptide and the α_1m and bikunin proteins connected by a linker tripeptide. The linker tripeptide and the last amino acid in α_1m, an arginine, constitute a basic cleavage site, R-V-R-R, that is recognized by furin and other subtilisin-like proprotein convertases (SPCs).[40] Before secretion, the α_1m- and bikunin-components are separated by proteolytic cleavage in the late phase of post-translational processing in the Golgi-system.[41] The bikunin-part is modified by attachment of a glycosaminoglycan (chondroitin sulfate chain) to Ser10[42,43] and the bikunin molecules are linked to larger subunits, so-called heavy chains, via N-acetylgalactosamine residues in the glycosaminoglycan.[44,45] Both α_1m and bikunin then leave the hepatocyte and enter the blood. No connection has been found between α_1m and bikunin after they leave the hepatocyte and the reason for the cosynthesis of α_1m and bikunin is not understood. Moreover, it has been shown

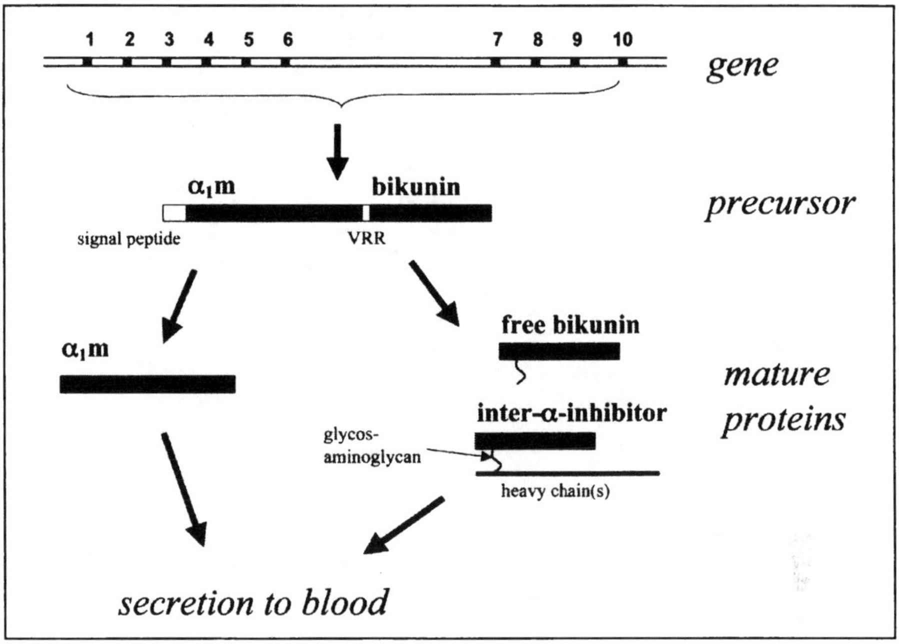

Figure 2. Synthesis of α_1m and bikunin in human liver. The α_1-microglobulin/bikunin precursor gene (*AMBP*), codes for the precursor-protein α_1m/bikunin. The signal peptide and the α_1m-part are encoded by exons 1-6 and the linker peptide VRR and bikunin-part by exons 7-10. Post-translational modification of the precursor includes glycosylation and cross-linking of bikunin to heavy chains via the glykosaminoglycan, forming the various members of the inter-α-inhibitor family. The α_1m- and bikunin-parts are separated in the Golgi apparatus via proteolytic cleavage and then secreted separately to the blood.

in different expression systems that both α_1m and bikunin can be expressed alone.[20,46-49] In spite of this, the α_1m/bikunin genetic construction is conserved in all species where α_1m has been found.

Gene

The *AMBP* gene from man[50] and mouse[51] has been cloned. Of its ten exons, the first six code for α_1m (Fig. 2). The *AMBP* gene has been mapped to the 9q32-33 region in man[52] and to chromosome 4 in mouse,[53] sites in both species where other lipocalin genes are clustered (reviewed in ref. 54). Intron F, which separates the α_1m-exons from the bikunin-exons, contains retroposons and other repeated structures, suggesting that it is a recombinatorial hot-spot.[51] This could have provided the basis for a fusion between an ancestral lipocalin gene (α_1m) with an ancestral Kunitz inhibitor gene (bikunin).

Expression and Distribution

Liver, Blood and Kidney

Early quantitative tissue distribution studies revealed liver, blood plasma, and kidney as major sites of α_1m localization.[55] This pattern reflects the major phases of the metabolism of the protein. (1) The liver is a major site of synthesis of α_1m in adult tissues in all species studied.[56-59] (2) α_1m is then secreted to the blood, where the protein exists in free form as well as in a variety of high molecular weight complexes (see below). Both free α_1m and the complexed forms are very rapidly equilibrated between the intra- and extra-vascular compartments and their half-lifes in blood are 2-3 min.[60,61] (3) Free monomeric α_1m passes relatively freely through the glomerular membranes out into the primary urine, where it is reabsorbed by the proximal tubular cells and catabolized.[62] The binding of α_1m to the tubular cells is mediated by the multiligand receptor megalin, a member of the LDL-receptor gene family.[63] Megalin also binds albumin and the lipocalins retinol-bindning protein (RBP) and major urinary protein (MUP).[63]

Expression

The liver is the predominant site of synthesis of α_1m. *AMBP* gene expression in liver is tightly regulated by a unique set of cis-elements and transcription factors known as hepatocyte nuclear factors (HNF 1-4).[64-66] Several secondary and minor locations of α_1m expression have been reported. Thus, α_1m mRNA has been detected in adult kidney,[12,21] pancreas,[67,68] stomach[6] and blood cells.[21] Traditionally, the expression of α_1m has been considered as constitutive, but lately this view been challenged. Thus, induction of α_1m-bikunin expression in kidneys by oxalate was recently demonstrated,[69] suggesting a regulatory role of α_1m in oxalate-containing kidney stone formation. Moreover, hemoglobin and pro-oxidants up-regulated the α_1m-expression in hepatoma and blood cell lines,[70] supporting the view that α_1m is involved in anti-oxidation and heme-protection (see below and Fig. 3).

Plasma

In human plasma, approximately 50% of α_1m forms a one-to-one complex with monomeric IgA by a reduction-resistant bond between the penultimate cysteine in the α-chain and Cys34 of α_1m.[71-73] Approximately 7% is linked to albumin[2,74] and 1% to prothrombin by a disulfide bond.[74] High molecular weight (HMW) forms of α_1m have been found in all species examined. In rat serum, α_1m is found covalently linked by a disulfide link to fibronectin[75] and by a reduction-resistant bond to the proteinase inhibitor α_1-inhibitor-3, a homologue of human α_2-macroglobulin.[76] α_1M complexes have also been detected in plaice serum.[20] Thus, the complex-forming ability of α_1m is conserved from fish to man, although the identity of the complex partners is not conserved. It is not known in what compartment any of the α_1m-complexes are formed.

Unusual forms of HMW α_1m have been found in plasma from patients with various pathologies; α_1m has a tendency to bind to mutated forms of coagulation factors that include a free Cys residue. The conserved unpaired Cys34 of α_1m is probably involved in the formation of these complexes. Thus, circulating complexes have been described between α_1m and factor IX Zutphen,[77] factor XII Tenri[78] and several protein C mutants.[79] In all mutants the unusual free, unpaired cysteine residue is located in an N-terminal Gla-domain (γ-carboxy glutamic acid-domain). The Gla-domains are believed to mediate the binding of the coagulation factors to membrane surfaces. It was shown by molecular modelling of α_1m and the protein C Gla-domain that electrostatic and hydrophobic interactions may attract the two proteins to each other and orient them in such a way that formation of a disulfide bond between the two free cysteines is favoured.[22]

Determination of the α_1m concentration in human plasma or serum is complicated by the presence of complex forms of the protein (see above). Consequently, reports on normal α_1m-concentrations in human plasma/serum have varied. Several investigators have measured free α_1m and IgA-α_1m separately in normal serum.[64,80-82] For example, DeMars et al[82] found a mean concentration of 33 mg/L for free α_1m and 248 mg/L for IgA-α_1m, corresponding to a molar ratio of approximately 1:1 (~1 µM of each). Without distinguishing between free and complexed α_1m, the values for "total" α_1m in rat, guinea pig and plaice serum were determined to be 9-16 mg/L, 26 mg/L and 20 mg/L, respectively.[8,12,20]

Other Tissues

Besides its predominant localization in plasma and liver and kidney cells, α_1m has been identified in the perivascular connective tissue of most organs[67,83,84] and is especially abundant in epidermis of skin[35,85] and epithelium of the gut.[61,86,87] It is often colocalized with elastin

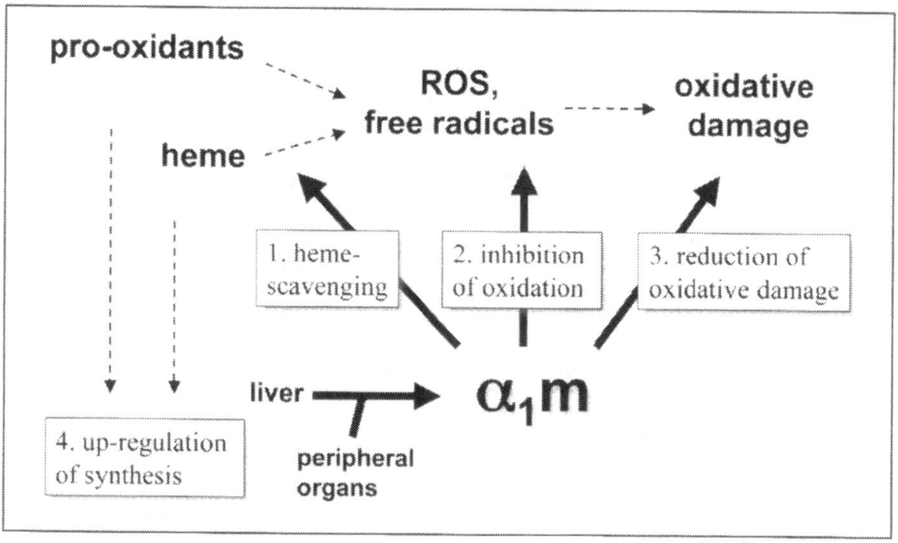

Figure 3. Anti-oxidant properties of α_1m. Pro-oxidants, exemplified by heme, the iron-containing prosthetic group of hemoglobin, myoglobin, cytochrome c and other heme-proteins, induce formation of free radicals and reactive oxygen species (ROS). These cause oxidative damage by undergoing harmful oxidation reactions with tissue components. α_1M interferes with this process as summarized in the figure.

and collagen[83,85] and was reported to bind to collagen in vitro.[88] A distribution of matrix-α_1m at various interfaces between the cells of the body and the external environment (blood/tissue, air/tissue, intestinal lumen/villi), as well as at the interface between maternal blood and fetal tissues in placenta,[89] is suggestive of a protective role of α_1m in vivo (see below). A similar distribution was seen in human[90] and mouse fetuses.[15] In addition, fetal α_1m was seen in the cytoplasm of many epithelial cells that do not express the protein post-term.[15,90]

Anti-Oxidant Properties

Recent reports suggest that α_1m is involved in the defense against oxidative tissue damage (oxidative stress). The proposed anti-oxidant mechanisms are summarized in Figure 3. Pro-oxidants are constantly introduced to the human body via the environment (air, food, etc) but are also produced endogenously as metabolites in the normal homeostasis.[91] Increased amounts of pro-oxidants are seen during inflammation, and oxidative stress is considered to be a major factor in the development of many conditions such as atherosclerosis, rheumatoid arthritis, ischemia/reperfusion injury, and diabetes. The pro-oxidants undergo reactions forming reactive oxygen species (ROS) and free radicals. These react with proteins, DNA and other molecules of human tissues by oxidation. These oxidation reactions are often harmful and may destroy the function of the target molecules (oxidative damage). Heme, the prosthetic group of heme-proteins, is a prominent example of an endogenous pro-oxidant and is harmful when released into the extracellular environment.

Thus, α_1m uses three major mechanisms to achieve anti-oxidation: (1) scavenging heme and other pro-oxidants, (2) inhibiting oxidation reactions, and (3) enzymatic reduction of harmful oxidation products. Additionally, a fourth mechanism enhance its anti-oxidant action: heme- and pro-oxidant-induced up-regulation of the synthesis of α_1m[70] (see above).

Heme-Scavenging

Both α_1m and its IgA-complex bind to the heme group.[34-36] The exposure of α_1m to erythrocyte membranes or purified hemoglobin leads to the binding of heme and the formation of a truncated form of the protein (t-α_1m) that lacks the C-terminal tetrapeptide LIPR and has heme-degrading properties. A pronounced yellow-brown colour was formed by incubating t-α_1m with heme, suggesting that at least some of the α_1m-chromophores may be heme degradation products.[34] The t-α_1m form is present in urine and thus is formed in vivo.[34,69,92] In chronic venous ulcers, an inflammatory condition where free heme and iron released after hemolysis are considered to be pathogenic factors, α_1m was colocalized with heme and t-α_1m was continuously formed.[35] Based on these findings, a role as an extracellular heme-scavenger was proposed for α_1m. As mentioned above, the tryptophan metabolite kynurenine is attached to lysyl residues in α_1m from the urine of hemodialysis patients.[32] Kynurenine metabolites are formed in tryptophan metabolism and are pro-oxidants, i.e., may induce oxidative stress.[93-95] The binding of heme and kynurenin and the transformation of these substances to chromophores may be two examples of a pro-oxidant scavenging mechanism of α_1m.

Inhibition of Oxidation and Enzymatic Reductase Activity

α_1M inhibited the heme- and ROS-induced oxidation of collagen, low-density lipoproteins (LDL), membrane lipids and whole cells.[96] α_1M also removed preformed oxidation products present on collagen and LDL. This suggests that α_1m may act as an oxidation repair factor. A possible mechanism for this may be the enzymatic reductase/dehydrogenase properties recently described for α_1m.[97] Thus, the protein was capable of reducing heme proteins, free iron and the synthetic compound nitroblue tetrazolium (NBT) using the electron donors ascorbate and NADH/NADPH as cofactors. The thiol group of Cys34 and the three lysyl residues of K92, K118 and K130 were found in the active site, suggesting that the chromophore or chromophore formation is linked to the reductase activities.

Immunoregulatory Properties

Due to its immunoregulatory properties, α_1m is categorized as an immunocalin.[98] It inhibits central events of the immune response in vitro. Thus, the antigen-induced cell division of peripheral blood lymphocytes was inhibited by α_1m.[99,100] The effects were species independent, i.e., similar effects on human cells were obtained with human, rat, rabbit or guinea pig α_1m.[14] It was also shown that the antigen-induced interleukin-2 (IL-2) production by mouse T helper cell hybridomas was inhibited by human α_1m.[101] Furthermore, inflammatory responses of blood cells were inhibited by α_1m; these included migration[98] and chemotaxis[102] of neutrophil granulocytes and the production of free radicals and IL-1β by peripheral lymphocytes/monocytes.[88] Finally, at low serum concentrations, a strong direct mitogenic effect by α_1m on resting guinea pig and human lymphocytes was seen.[100,103,104] At the high serum concentrations (10-20%) used to measure the inhibition of antigen-stimulated lymphocytes, no direct stimulation of the cell division was seen. It is possible that the immunoregulatory effects of α_1m are related to its anti-oxidant and reductase activities, as ROS have been shown to be involved as (positive) factors in cell signalling during lymphocyte activation (reviewed in refs. 105,106).

Cell Receptor

α_1M has been shown to bind to the surface of various white blood cells, including human peripheral B and T lymphocytes, human NK cells,[60] the human histiocytic cell-line U937,[107] mouse peripheral B and T lymphocytes,[104] and mouse T helper cell hybridomas.[101] The binding is species independent, specific for α_1m, saturable and trypsin-sensitive, suggesting that a protein that recognizes a conserved part of α_1m is present on the surface of white blood cells. The α_1m-receptor on blood cells has not yet been identified, however.

Concluding Remarks

Above, we have reviewed the current knowledge about the lipocalin α_1-microglobulin, its structural features, its unusual synthesis as a diprotein, its tissue expression and distribution, and we have particularly highlighted recent elucidation of its anti-oxidant properties and how these may be employed in tissue protection.

Acknowledgements

The text has been revised by Linda Lögdberg, PhD.

References

1. Ekström B, Peterson PA, Berggård I. A urinary and plasma α_1-glycoprotein of low molecular weight: Isolation and some properties. Biochem Biophys Res Commun 1975; 65(4):1427-33.
2. Tejler L, Grubb AO. A complex-forming glycoprotein heterogeneous in charge and present in human plasma, urine and cerebrospinal fluid. Biochim Biphys Acta 1976; 439(1):82-94.
3. Seon BK, Pressman D. Unique human glycoprotein, α_1-microglycoprotein: Isolation from the urine of a cancer patient and its characterization. Biochemistry 1978; 17(14):2815-21.
4. Pervaiz S, Brew K. Homology of β-lactoglobulin, serum retinol-binding protein, and protein HC. Science 1985; 228(4697):335-7.
5. Kaumeyer JF, Polazzi JO, Kotick MP. The mRNA for a proteinase inhibitor related to the HI-30 domain of inter-α-trypsin inhibitor also encodes α_1-microglobulin (protein HC). Nucleic Acids Res 1986; 14(20):7839-50.
6. Tavakkol A. Molecular cloning of porcine α_1-microglobulin/HI-30 reveals developmental and tissue-specific expression of two variant messenger ribonucleic acids. Biochim Biophys Acta 1991; 1088(1):47-56.
7. Lindqvist A, Åkerström B. Bovine α_1-microglobulin/bikunin. Isolation and characterization of liver cDNA and urinary α_1-microglobulin. Biochim Biophys Acta 1996; 1306(1):98-106.
8. Åkerström B, Berggård I. Guinea-pig α_1-microglobulin. Isolation and properties in comparison with human α_1-microglobulin. Eur J Biochem 1979; 101(1):215-23.
9. Yoshida K, Suzuki Y, Yamamoto K et al. Guinea pig α_1-microglobulin/bikunin: cDNA sequencing, tissue expression and expression during acute phase. Comp Biochem Physiol B Biochem Mol Biol 1999; 122(2):165-72.
10. Vincent C, Bouic P, Revillard JP. Characterization of rat α_1-microglobulin. Biochem Biophys Res Commun 1983; 116(1):180-8.
11. Slota A, Sjöquist M, Wolgast M et al. Bikunin in rat plasma, lymph and bile. Biol Chem Hoppe Seyler 1994; 375(2):127-33.
12. Kastern W, Björck L, Åkerström B. Developmental and tissue-specific expression of α_1-microglobulin mRNA in the rat. J Biol Chem 1998; 261(32):15070-4.
13. Lindqvist A, Bratt T, Altieri M et al. Rat α_1-microglobulin: Coexpression in liver with the light chain of inter-α-trypsin inhibitor. Biochim Biophys Acta 1992; 1130(1):63-7.
14. Åkerström B, Lögdberg L, Babiker-Mohamed H et al. Structural relationship between α_1-microglobulin from man, guinea-pig, rat and rabbit. Eur J Biochem 1987; 170(1-2):143-8.
15. Sanchez D, Martinez S, Lindqvist A et al. Expression of the AMBP gene transcript and its two protein products, α_1-microglobulin and bikunin, in mouse embryogenesis. Mech Dev 2002; 117(1-2):293-8.
16. Chan P, Salier J-P. Mouse α_1-microglobulin/bikunin precursor: cDNA analysis, gene evolution and physical assignment of the gene next to the orosomucoid locus. Biochim Biophys Acta 1993; 1174(2):195-200.
17. Ide H, Itoh H, Nawa Y. Sequencing of cDNAs encoding α_1-microglobulin/bikunin of Mongolian gerbil and Syrian golden hamster in comparison with man and other species. Biochim Biophys Acta 1994; 1209(2):286-92.
18. Åkerström B. Immunological analysis of α_1-microglobulin in different mammalian and chicken serum. α_1-Microglobulin is 5-8 kilodaltons larger in primates. J Biol Chem 1985; 260(8):4839-44.
19. Kawahara A, Hikosaka A, Sasado T et al. Thyroid hormone-dependent repression of α_1-microglobulin/bikunin precursor (AMBP) gene expression during amphibian metamorphosis. Dev Genes Evol 1997; 206(6):355-62.
20. Lindqvist A, Åkerström B. Isolation of plaice (Pleuronectes platessa) α_1-microglobulin: Conservation of structure and chromophore. Biochim Biophys Acta 1999; 1430(2):222-33.
21. Leaver MJ, Wright J, George SG. Conservation of the tandem arrangement of α_1-microglobulin/bikunin mRNA: Cloning of a cDNA from plaice (Pleuronectes platessa). Comp Biochem Physiol Biochem Mol Biol 1994; 108(3):275-81.
22. Hanley S, Powell R. Sequence of a cDNA clone encoding the Atlantic salmon α_1-microglobulin/bikunin protein. Gene 1994; 147(2):297-8.
23. Villoutreix B, Åkerström B, Lindqvist A. Structural model of human α_1-microglobulin: proposed scheme for the interaction with the Gla domain of anticoagulant protein C. Blood Coagul Fibrinolysis 2000; 11(3):261-75.
24. Schreck SF, Parker C, Plumb ME et al. Human complement protein C8γ. Biochim Biophys Acta 2000; 1482(1-2):199-208.

25. Urade Y, Tanaka T, Eguchi N et al. Structural and functional significance of cysteine residues of glutathione-independent prostaglandin D synthase. Identification of Cys65 as an essential thiol. J Biol Chem 1995; 270(3):1422-8.
26. Ekström B, Lundblad A, Svensson S. Structural studies on the carbohydrate portion of human α_1-microglobulin. Eur J Biochem 1981; 114(3):663-6.
27. Escribano J, Lopez-Otin C, Hjerpe A et al. Location and characterization of the three carbohydrate prosthetic groups of human protein HC. FEBS Lett 1990; 266(1-2):167-70.
28. Amoresano A, Minchiotti L, Cosulich ME et al. Structural characterization of the oligosaccharide chains of human α_1-microglobulin from urine and amniotic fluid. Eur J Biochem 2000; 267(7):2105-12.
29. Escribano J, Grubb A, Calero M et al. The protein HC chromophore is linked to the cysteine residue at position 34 of the polypeptide chain by a reduction-resistant bond and causes the charge heterogeneity of protein HC. J Biol Chem 1991; 266(24):15758-63.
30. Åkerström B, Bratt T, Enghild JJ. Formation of the α_1-microglobulin chromophore in mammalian and insect cells: A novel post-translational mechanism? FEBS Lett 1995; 362(1):50-4.
31. Berggård T, Cohen A, Persson P et al. α_1-microglobulin chromophores are located to three lysine residues semiburied in the lipocalin pocket and associated with a novel lipophilic compound. Protein Sci 1999; 8(12):2611-20.
32. Sala A, Campagnoli M, Perani E et al. Human α_1-microglobulin is covalently bound to kynurenine-derived chromophores. J Biol Chem 2004; 279(49):51033-41.
33. Escribano J, Grubb A, Mendez E. Identification of retinol as one of the protein HC chromophores. Biochem Biophys Res Commun 1988; 155(3):1424-9.
34. Allhorn M, Berggård T, Nordberg J et al. Processing of the lipocalin α_1-microglobulin by hemoglobin induces heme-binding and heme-degradation properties. Blood 2002; 99(6):1894-901.
35. Allhorn M, Lundqvist K, Schmidtchen A et al. Heme-scavenging role of α_1-microglobulin in chronic ulcers. J Invest Dermatol 2003; 121(3):640-6.
36. Larsson J, Allhorn M, Åkerström B. The lipocalin α_1-microglobulin binds heme in different species. Arch Biochem Biophys 2004; 432(2):196-204.
37. Blom A, Fries E. Bikunin-not just a plasma proteinase inhibitor. Int J Biochem Cell Biol 2000; 32(2):125-37.
38. Salier J-P, Rouet P, Raguenez G et al. The inter-α-inhibitor family: From structure to regulation. Biochem J 1996; 315(Pt 1):1-9.
39. Chen L, Mao SJT, Larsen WJ. Identification of a factor in fetal bovine serum that stabilizes the cumulus extracellular matrix. A role for a member of the inter-α-trypsin inhibitor family. J Biol Chem 1992; 267(17):12380-6.
40. Molloy SS, Bresnahan PA, Leppla S et al. Human furin is a calcium-dependent serine endoprotease that recognizes the sequence Arg-X-X-Arg and efficiently cleaves anthrax toxin protective antigen. J Biol Chem 1992; 267(23):16396-402.
41. Bratt T, Olsson H, Sjöberg EM et al. Cleavage of the α_1-microglobulin-bikunin precursor is localized to the Golgi apparatus of rat liver cells. Biochim Biophys Acta 1993; 1157(2):147-54.
42. Sjöberg EM, Fries E. Biosynthesis of bikunin (urinary trypsin inhibitor) in rat hepatocytes. Biochem J 1990; 272(2):113-8.
43. Chirat F, Balduyck M, Mizon C et al. A chondroitin-sulfate chain is located on serine-10 of the urinary trypsin inhibitor. Int J Biochem 1991; 23(11):1201-3.
44. Enghild JJ, Salvesen G, Hefta SA et al. Chondroitin 4-sulfate covalently cross-links the chains of the human blood protein preα-inhibitor. J Biol Chem 1989; 266(2):747-51.
45. Morelle W, Capon C, Balduyck M et al. Chondroitin sulphate covalently cross-links the three polypeptide chains of inter-α-trypsin inhibitor. Eur J Biochem 1994; 221(2):881-8.
46. Bratt T, Cedervall T, Åkerström B. Processing and secretion of rat α_1-microglobulin-bikunin expressed in eukaryotic cell lines. FEBS Lett 1994; 354(1):57-61.
47. Thuveson M, Fries E. Intracellular proteolytic processing of the heavy chain of rat preα-inhibitor. The COOH-terminal propeptide is required for coupling to bikunin. J Biol Chem 1999; 274(10):6741-6.
48. Wester L, Johansson MU, Åkerström B. Physicochemical and biochemical characterization of human α_1-microglobulin expressed in baculovirus-infected insect cells. Protein Expr Purif 1997; 11(1):95-103.
49. Falkenberg C, Wester L, Belting M et al. Expression of a functional proteinase inhibitor capable of accepting xylose: Bikunin. Arch Biochem Biophys 2001; 387(1):99-106.
50. Diarra-Mehrpour M, Bourguignon J, Sesboüé R et al. Structural analysis of the human inter-α-trypsin inhibitor light-chain gene. Eur J Biochem 1990; 191(1):131-9.
51. Lindqvist A, Rouet P, Salier J-P et al. The α_1-microglobulin/bikunin gene: Characterization in mouse and evolution. Gene 1999; 234(2):329-36.
52. Diarra-Mehrpour M, Bourguignon J, Sesboüé R et al. Human plasma inter-α-trypsin inhibitor is encoded by four genes on three chromosomes. Eur J Biochem 1989; 179(1):147-54.

53. Salier J-P, Verga V, Doly J et al. The genes for the inter-α -inhibitor family share a homologous organization in human and mouse. Mamm Genome 1992; 2(4):233-9.
54. Salier J-P. Chromosomal location, exon/intron organization and evolution of lipocalin genes. Biochim Biophys Acta 2000; 1482(1-2):35-45.
55. Åkerström B. Tissue distribution of guinea pig α_1-microglobulin. Cell Mol Biol 1983; 29(6):489-95.
56. Tejler L, Eriksson S, Grubb A et al. Production of protein HC by human fetal liver explants. Biochim Biophys Acta 1978; 542(3):506-14.
57. Åkerström B. Synthesis of α_1-microglobulin by guinea-pig liver. Eur J Biochem 1983; 133(1):235-9.
58. Åkerström B, Landin B. Rat α_1-microglobulin. Purification from urine and synthesis by hepatocyte monolayers. Eur J Biochem 1985; 146(2):353-8.
59. Vincent C, Marceau M, Blangarin P et al. Purification of α_1-microglobulin produced by human hepatoma cell lines. Biochemical characterization and comparison with alpha 1-microglobulin synthesized by human hepatocytes. Eur J Biochem 1987; 165(3):699-704.
60. Wester L, Fast J, Labuda T et al. Carbohydrate groups of α_1-microglobulin are important for secretion and tissue localization but not for immunological properties. Glycobiology 2000; 10(9):891-900.
61. Larsson J, Wingårdh K, Berggård T et al. Distribution of iodine 125-labeled α_1-microglobulin in rats after intravenous injection. J Lab Clin Med 2001; 137(3):165-75.
62. Strober W, Waldmann TA. The role of the kidney in the metabolism of plasma proteins. Nephron 1974; 13(1):35-66.
63. Leheste JR, Rolinski B, Vorum H et al. Megailin knockout mice as an animal model of low molecular weight proteinuria. Am J Pathol 1999; 155(4):1361-70.
64. Rouet P, Raguenez G, Tronche F et al. A potent enhancer made of clustered liver-specific elements in the transcription of control sequences of human α_1-microglobulin/bikunin gene. J Biol Chem 1992; 267(29):20765-73.
65. Rouet P, Raguenez G, Tronche F et al. Hierarchy and positive/negative interplays of the hepatocyte nuclear factors HNF1, -3 and –4 in the liver-specific enhancer for the human α_1-microglobulin/ bikunin precursor. Nucleic Acids Res 1995; 23(3):395-404.
66. Rouet P, Raguenez G, Ruminy P et al. An array of binding sites for hepatocyte nuclear factor 4 of high and low affinities modulates the liver-specific enhancer for the human α_1-microglobulin/bikunin precursor. Biochem J 1998; 334(Pt 3):577-84.
67. Berggård T, Oury TD, Thøgersen IB et al. α_1-microglobulin is found both in blood and in most tissues. J Histochem Cytochem 1998; 46(8):887-93.
68. Itoh H, Tomita M, Kobayashi T et al. Expression of inter-α-trypsin inhibitor light chain (bikunin) in human pancreas. J Biochem 1996; 120(2):271-5.
69. Grewal JS, Tsai JY, Khan SR. Oxalate inducible AMBP gene and its regulatory mechanism in renal tubular epithelial cells. Biochem J 2005, (Published ahead of print as manuscript no BJ20041465).
70. Olsson MG, Olofsson T, Åkerström B. Up-regulation of α_1-microglobulin by hemoglobin and pro-oxidants in hepatoma and blood cell lines. Submitted.
71. Grubb A, Lopez C, Tejler L et al. Isolation of human complex-forming glycoprotein, heterogeneous in charge (protein HC), and its IgA complex from plasma. Physiochemical and immunochemical properties, normal plasma concentration. J Biol Chem 1983; 258(23):14698-707.
72. Grubb A, Mendez E, Fernandez-Luna JL et al. The molecular organization of the protein HC-IgA complex (HC-IgA). J Biol Chem 1986; 261(30):14313-20.
73. Calero M, Escribano J, Grubb A et al. Location of a novel type of interpolypeptide chain linkage in the human protein HC-IgA complex (HC-IgA) and identification of a heterogeneous chromophore associated with the complex. J Biol Chem 1994; 269(1):384-9.
74. Berggård T, Thelin N, Falkenberg C et al. Prothrombin, albumin and immunoglobulin A form covalent complexes with α_1-microglobulin in human plasma. Eur J Biochem 1997; 245(5):676-83.
75. Falkenberg C, Enghild JJ, Thøgersen IB et al. Isolation and characterization of a fibronectin-α_1-microglobulin complex in rat plasma. Biochem J 1994; 301(Pt 3):745-51.
76. Falkenberg C, Grubb A, Åkerström B. Isolation of rat serum α_1-microglobulin. Identification of a complex with α_1-inhibitor-3, a rat α_2-macroglobulin homologue. J Biol Chem 1990; 265(27):16150-7.
77. Wojcik EG, van der Berg M, van der Linden IK et al. Factor IX Zutphen: a Cys18—>Arg mutation results in formation of a heterodimer with α_1-microglobulin and the inability to form a calcium-induced conformation. Biochem J 1995; 311(Pt 3):753-9.
78. Kondo S, Tokunaga F, Kawano S et al. Factor XII Tenri, a novel cross-reacting material negative factor XII deficiency, occurs through a proteasome-mediated degradation. Blood 1999; 93(12):4300-8.
79. Wojcik EG, Simioni P, van der Berg M et al. Mutations which introduce free cysteine residues in the Gla-domain of vitamin K dependent proteins result in the formation of complexes with α_1-microglobulin. Thromb Haemost 1996; 75(1):70-5.

80. Fernandez-Luna JL, Moneo I, Grubb A et al. A sensitive and rapid enzyme-linked immunosorbent assay using monoclonal antibodies for simultaneous quantitation of free and IgA-complexed protein HC. J Immunol Methods 1985; 82(1):101-10.
81. Vincent C, Revillard J-P. Differential measurement by ELISA of free and IgA bound α_1-microglobulin in human serum without prior fractionation. J Immunol Methods 1985; 82(1):111-9.
82. DeMars DD, Katzmann JA, Kimlinger TK et al. Simultaneous measurement of total and IgA-conjugated α_1-microglobulin by a combined immunoenzyme/immunoradiometric assay technique. Clin Chem 1989; 35(5):766-72.
83. Ødum L, Nielsen HW. Human protein HC (α_1-microglobulin) and inter-alpha-trypsin inhibitor in connective tissue. Histochem J 1994; 26(10):799-803.
84. Ødum L, Nielsen HW. Bikunin and α_1-microglobulin in human zona pellucida and connective tissue. Histochem J 1997; 29(3):199-203.
85. Bouic P, Kanitakis J, Schmitt D et al. α_1-microglobulin: A new antigenic component of the epidermo-dermal junction in normal human skin. Br J Dermatol 1985; 112(1):35-41.
86. Bouic P, Vincent C, Revillard JP. Immunohistological localization of α_1-microglobulin in normal rat tissues. J Histochem Cytochem 1984; 32(7):717-23.
87. Bouic P, Vincent C, Revillard JP. Localization of α_1-microglobulin (protein HC) in normal human tissues: An immunohistochemical study using monoclonal antibodies. Histochem J 1984; 16(12):1311-24.
88. Santin M, Cannas M. Collagen-bound α_1-microglobulin in normal and healed tissues and its effect on immunocompetent cells. Scand J Immunol 1999; 50(3):289-95.
89. Berggård T, Enghild JJ, Badve S et al. Histological distribution and biochemical properties of α_1-microglobulin in human placenta. Am J Repr Immunol 1999; 41(1):52-60.
90. Lögdberg L, Åkerström B, Badve S. Tissue distribution of the lipocalin α_1-microglobulin in the developing human fetus. J Histochem Cytochem 2000; 48(11):1545-52.
91. Halliwell B, Gutteridge JMC. Free radicals in Biology and Medicine Oxford: Oxford University Press.
92. Lopez C, Grubb A, Mendez E. Human protein HC displays variability in its carboxyterminal amino acid sequence. Febs Lett 1982; 144(2):349-53.
93. Ishii T, Iwahashi H, Sugata R et al. Formation of hydroxanthommatin-derived radical in the oxidation of 3-hydroxykynurenine. Arch Biochem Biophys 1992; 294(2):616-22.
94. Okuda S, Nishiyama N, Saito H et al. Hydrogen-peroxide-mediated neuronal cell death induced by an endogeneous neurotoxin, 3-hydroxykynurenine. Proc Natl Acad Sci USA 1996; 93(1):12553-8.
95. Vazquez S, Garner B, Sheil MM et al. Characterization of the major autooxidation products of 3-hydroxykynurenine under physiological conditions. Free Rad Res 2000; 32(1):11-23.
96. Allhorn M, Larsson J, Olsson MG et al. Oxidative modifications on collagen and low-density lipoprotein are inhibited and reduced by the lipocalin α_1-microglobulin. Submitted.
97. Allhorn M, Klapyta A, Åkerström B. Redox properties of the lipocalin α_1-microglobulin: Reduction of cytochrome c, hemoglobin, and free iron. Free Radic Biol Med 2005; 38(5):557-67.
98. Lögdberg L, Wester L. Immunocalins: A lipocalin subfamily that modulates immune and inflammatory responses. Biochim Biophys Acta 2000; 1482(1-2):284-97.
99. Lögdberg L, Åkerström B. Immunosuppressive properties of α_1-microglobulin. Scand J Immunol 1981; 13(4):383-90.
100. Lögdberg L, Åkerström B, Shevach E. α_1-microglobulin is mitogenic for guinea pig lymphocytes. Scand J Immunol 1986; 24(5):575-81.
101. Wester L, Michaëlsson E, Holmdahl R et al. Receptor for α_1-microglobulin on T lymphocytes: Inhibition of antigen-induced interleukin-2 production. Scand J Immunol 1998; 48(1):1-7.
102. Mendez E, Fernandez-Luna JL, Grubb AO et al. Human protein HC and its IgA complex are inhibitors of neutrophil chemotaxis. Proc Natl Acad Sci USA 1986; 88(5):1472-5.
103. Babiker-Mohamed H, Olsson MO, Boketoft Å et al. α_1-microglobulin is mitogenic to human peripheral blood lymphocytes. Regulation by both enhancing and suppressive serum factors. Immunobiology 1990; 180(2-3):221-34.
104. Babiker-Mohamed H, Åkerström B, Lögdberg L. Mitogenic effect of α_1-microglobulin on mouse lymphocytes. Evidence of T- and B-cell cooperation, B-cell proliferation, and a low-affinity receptor on mononuclear cells. Scand J immunol 1990; 32(1):37-44.
105. Dröge W. Free radicals in the physiological control of cell function. Physiol Rev 2002; 82:47-95.
106. Williams MS, Kwon J. T cell receptor stimulation, reactive oxygen species, and cell signalling. Free Radic Biol Med 2004; 37(8):1144-51.
107. Fernandez-Luna JL, Levya-Cobian F, Mollinedo F. Identification of the protein HC receptor. Febs Lett 1988; 236(2):471-4.

CHAPTER 11

Glycodelin:

A Lipocalin with Diverse Glycoform-Dependent Actions

Markku Seppälä,* Hannu Koistinen, Riitta Koistinen, Philip C.N. Chiu, and William S.B. Yeung

Introduction

Glycodelin has many names in the literature, such as placental protein 14 (PP14), human placental organ-specific α2-globulin, or progesterone-dependent endometrial protein, based on electrophoretic characteristics, regulation, or tissue of first identification.[1-4] After detailed information became available on its sites of synthesis, primary structure, complex-type oligosaccharide structures and biological actions, the protein was renamed as "glycodelin" to highlight significance of its unique oligosaccharide moieties for biological activity.[5-7] Based on structural similarity with β-lactoglobulins[8] containing a retinol-binding motif[9] glycodelin is the major lipocalin protein of the reproductive axis with diverse actions in cell recognition and differentiation.[10] Due to its unique oligosaccharide moieties affecting biological functions glycodelin serves as a model in studies on functional glycomics in a number of clinical areas.

Structure

Glycodelin Gene

In the Human Genome Organisation (HUGO), the official symbol for the glycodelin gene is *PAEP* (progestogene-associated endometrial protein).[11] The glycodelin gene has been assigned to chromosome 9, band q34.[12] This is the locus in which other lipocalin genes, such as α_1-microglobulin/bikunin, prostaglandin D synthase and tear lipocalin 1 and 2 genes are also located.[13-15] The glycodelin gene is 5.05 kb long and, like many lipocalin genes, it is divided into seven exons.[16] The nucleotide sequence encoding the retinol-binding motif of β-lactoglobulins is conserved in the glycodelin gene. Four putative glucocorticoid/progesterone response elements (PRE) are found at positions –1799, -1071, -745, and –302 of the gene promoter, and two additional putative PREs are present at +1912 and +1965.[10,16]

Primary Structure

The first studies on N-terminal sequence of glycodelin uncovered 59% identity with horse β-lactoglobulin and 23% identity with human retinol-binding protein.[8,17] The amino acid sequence of glycodelin is 180 residues long, including an 18 amino acid signal sequence.[5]

*Corresponding Author: Markku Seppälä—Department of Clinical Chemistry, University of Helsinki, Biomedicum Helsinki, 4th floor, Haartmaninkatu 8, 00029 HUS, Finland. Email: mseppala@welho.com

Lipocalins, edited by Bo Åkerström, Niels Borregaard, Darren R. Flower and Jean-Philippe Salier.

Highest similarity, 91%, is with rhesus macaque glycodelin (PP14) (NCBI BLASTP search on June 8th, 2005). In β-lactoglobulins, four cysteine residues at positions 66, 106, 119, and 160 are responsible for intramolecular disulfide bridges.[18] All of them are conserved in glycodelin. But unlike β-lactoglobulins, glycodelin is a glycoprotein that contains 17.5% carbohydrate.[2] Moreover, unlike β-lactoglobulin, glycodelin-A has not been found to bind retinoic acid or retinol.[19]

Carbohydrate Moieties

Glycodelin has three potential N-linked glycosylation sites at Asn 28, Asn 63 and Asn 85 (numbering according to mature protein without signal sequence).[5] Two of them (Asn 28 and Asn 63) are glycosylated in uterine glycodelin-A and seminal plasma glycodelin-S, but in a different fashion.[6,20] In glycodelin-A, the major nonreducing epitopes in the complex-type glycans are Galβ1-4GlcNAc (lacNAc), GalNAcβ1-4GlcNAc (lacdiNAc), NeuAcα2-6Galβ1-4GlcNAc (sialylated lacNAc), NeuAcα2-6GalNAcβ1-4GlcNAc (sialylated lacdiNAc), Galβ1-4(Fucα1-3)GlcNAc (blood group Lewis^x), and GalNAcβ1-(Fucα1-3)GlcNAc (lacdiNAc analogue of Lewis^x).[6] Oligosaccharides bearing terminal sialylated lacNAc or lacdiNAc antennae may manifest immunosuppressive effects by blocking adhesive and activation-related events mediated by CD22, the human B cell receptor, and a biantennary N-linked fucosylated oligosaccharide bearing Lewis^x has been reported to inhibit E-selectin mediated adhesion. Indeed, glycodelin has been found to inhibit E-selectin mediated cell adhesion.[21] Purified glycodelin preparations isolated from amniotic fluid, endometrium, decidua and pregnancy serum are closely similar in respect of their physicochemical and immunological properties, allowing them to be categorized as glycodelin-A.[22] Glycodelin-S is a human seminal plasma glycodelin isoform that is immunologically indistinguishable from uterine glycodelin-A, but unlike the latter, it does not inhibit human sperm-zona pellucida binding, and all of its glycans are different from those of glycodelin-A. Glycodelin-S contains no sialylated glycans, its glycans are unusually fucose-rich, and the major complex-type structures are biantennary glycans with Lewis^x and Lewis^y [Fucα1-2Galβ1-4 (Fucα1-3) GlcNAc] antennae.[20] Glycodelin-F from follicular fluid also shares the protein core with glycodelin-A, but these two isoforms differ in glycosylation, as demonstrated by fluorophore-assisted carbohydrate electrophoresis and lectin binding characteristics.[23]

Folding

Differentially glycosylated glycodelin isoforms glycodelin-A and glycodelin-S share similar thermodynamic parameters of reversible denaturation. This suggests that native folding of these isoforms is not influenced by the differences in glycosylation.[19] The Swiss-Model-deduced tertiary structure of glycodelin is similar to that of bovine β-lactoglobulin and other lipocalins. The glycans may form a clustered saccharide patch,[24] in which the carbohydrates from more than one glycosylation site form a cluster. Because the folding patterns are similar in glycodelin-A and glycodelin-S they provide a model for studies on the effects of differential glycosylation on the conformational stability and function.

Tissues and Cells of Origin

Reproductive System

Many studies have shown that glycodelin-A is synthesized by glandular and luminal surface epithelium of the endometrium in response to progesterone exposure and secreted mainly to uterine fluid or amniotic fluid during pregnancy.[10] In the ovary, another glycodelin isoform, glycodelin-F, is synthesized in luteinized granulosa cells[25] and, in the fallopian tube, both glycodelin-A and –F are produced in the epithelial cells.[26-29]

Other Tissues

In normal breast tissue, glycodelin is synthesized in ductal and lobular epithelium.[30] It has also been found in sweat glands.[31] In the male reproductive tract, both glycodelin-S protein and mRNA are localized to glands of the seminal vesicle and ampullary part of the vas deferens.[32] Glycodelin mRNA is constitutively expressed in the hematopoietic tissue of the bone marrow,[33] notably in the megakaryocytic lineage.[34] Glycodelin has also been found in malignant tumors (see below).

Regulation and Biological Actions

Sperm Capacitation

Glycodelin-S from seminal plasma does not inhibit sperm-egg binding in spite of the same protein core and immunoreactivity as in glycodelin-A.[20] The biological role of glycodelin-S has been extensively elucidated by collaborative research between Hong Kong and Helsinki.[29,35] Experiments on binding kinetics have demonstrated the presence of two binding sites for glycodelin-S on human spermatozoa. These binding sites are saturable, reversible, and bind to glycodelin-S in a time- and concentration- dependent manner. Differently glycosylated glycodelin isoforms -A and –F do not compete with glycodelin-S for these binding sites. Bovine serum albumin and cyclodextrin induce cholesterol efflux from human spermatozoa. Glycodelin-S significantly reduces the cholesterol efflux induced by either of these stimulators, and it exerts this effect upstream of protein kinase activation in the adenylyl cyclase/protein kinase A/tyrosine kinase signaling pathway in spermatozoa, resulting in suppression of capacitation. Deglycosylated glycodelin-S does not bind to sperm cells and has no effect on bovine serum albumin-induced capacitation. Again, these findings demonstrate the importance of carbohydrate moieties of glycodelin-S for the biological action of this molecule,[35] particularly because the said processes are activated upon removal of glycodelin-S from the spermatozoa. In vivo, the dissociation probably takes place during passage of spermatozoa through the cervix, as suggested by in vitro experiments employing a cervical fluid surrogate.[35] In view of these observations glycodelin-S appears to play a role in maintaining an uncapacitated state in the human spermatozoa before their passage through the cervix.

Acrosome Reaction

Glycodelin-F is secreted from luteinized ovarian granulosa cells into preovulatory follicular fluid[25] and transferred with the cumulus cells into fallopian tube at ovulation. Like glycodelin-A, it binds on the sperm head. Binding kinetics studies using radiolabeled glycodelin-F have demonstrated two binding sites on human spermatozoa, and one of these binding sites also binds glycodelin-A. Thus, glycodelin-A displaces only 70% of the labeled glycodelin-F bound on human spermatozoa. Immunocytochemically glycodelin-F is localized to the acrosome region of the human spermatozoa, and sperm-bound glycodelin-F but not glycodelin-A inhibits progesterone-induced acrosome reaction and sperm-egg binding. Deglycosylation of glycodelin-F abolishes its binding to spermatozoa.[36] Studies on neoglycoproteins have shown that the binding of glycodelin-A to spermatozoa involves mannose, fucose and possibly E-selectin residues, while that of glycodelin-F involves mannose, fucose and N-acetylglucosamine, but not the selectin residue.[37] Before fertilization, during sperm passage through the cumulus/corona cell layer that surrounds the oocyte, glycodelin-F is removed from spermatozoa, and progesterone-induced acrosome reaction and sperm-egg binding capacity are restored.[25] The uptake of glycodelin-F by the cumulus cells seems to be unique among proteins of the lipocalin family, as some other lipocalin proteins do not have the same effect. Again, glycosylation may explain the difference, as the cumulus cells may partially deglycosylate glycodelin-F.[25] In view of these observations one of the biological actions of glycodelin-F appears to be in the prevention of premature acrosome reaction before the spermatozoa have penetrated through the cumulus oophorus matrix to bind to the zona pellucida.

Table 1. Biological actions of glycodelin isoforms

Isoform	Site/Biological Fluid	Suggested Function	References
Glycodelin-S	Seminal plasma	Inhibits capacitation	35
Glycodelin-A	Oocyte	Inhibits sperm-zona binding	38
	Implantation site	Immunosuppression	39-45
Glycodelin-F	Fallopian tube	Inhibits acrosome reaction	25,29
	Oocyte	Inhibits sperm-zona binding	23

Sperm Binding to the Zona Pellucida

Glycodelin-A was the first endogenous glycoprotein that was found to potently and dose-dependently inhibit binding of spermatozoa to the zona pellucida.[38] The inhibition was caused by prior binding of glycodelin-A on the spermatozoa. In the uterus, synthesis of glycodelin-A is temporally regulated by progesterone. Thus, during the estrogen-dominated fertile window, absence of glycodelin-A synthesis in the endometrium is meaningful because of its anti-fertilization activity. This property of uterine glycodelin-A during the luteal phase of the cycle is glycosylation-dependent.[20] Glycodelin-F has even stronger inhibitory activity on sperm-zona binding than glycodelin-A does.[23] This activity is reduced during passage through the cumulus oophorus matrix of the glycodelin-F covered sperm.[25] These diverse biological actions of unique isoforms make glycodelin a representative example for studies on functional glycomics during early events of the fertilization process (Table 1).

Immunosuppression, Implantation and Placentation

In addition to its anti-fertilization propensity glycodelin-A has immunosuppressive activities. These include inhibitory activity on lymphocyte proliferation,[39] NK-cell cytotoxicity,[40] T-cell proliferation and Th1-type cytokine response,[41] and induction of T-cell apoptosis.[42] The inhibition of T-cell activation is mediated by tyrosine phosphatase receptor CD45.[43] Glycodelin binds to pregnancy zone protein and alpha(2)-macroglobulin, both potentiating glycodelin's immunosuppressive activity.[44] Glycodelin also regulates B cell responses.[45] A receptor for glycodelin has been found on human monocytes.[46] While glycosylation is important for gamete interactions, the immunosuppressive properties of glycodelin are likely to be carbohydrate-independent, except in the case of apoptotic activity, where presence of sialic acid is important.[47,48] While glycodelin-A binds on spermatozoa, it remains to be proven whether glycodelin-bound sperm would have decreased immunogenicity in the female body.

Given its inhibitory activity on natural killer cells and the T cells, uterine glycodelin-A likely plays a part in fetomaternal defense mechanisms during implantation and placentation, by counteracting maternal immune cell rejection of the fetal semiallograft.[40,49] Implantation normally takes place eight days after the luteinizing hormone (LH) surge. At that time a full array of immune cells are present in the endometrium. Studies on global gene profiling have shown that, in a normal ovulatory cycle, glycodelin expression is significantly increased during the window of implantation.[50] This is translated into increased glycodelin synthesis and secretion.[10,51] Clinical observations on low glycodelin levels in uterine fluid and serum from patients with early pregnancy loss (EPL) are compatible with the role of glycodelin in placentation.[52-54]

Examples of Clinical Relevance

Controlled Ovarian Stimulation (COH)

There are many clinical situations in which glycodelin secretion may be altered because of changes in local endocrine microenvironment. COH employing ovarian suppression with gonadotropin-releasing hormone agonists in combination with controlled stimulation with follicle-stimulating hormone and LH is widely used in women undergoing in vitro fertilization. The treatment increases ovarian secretion of estrogen and androgen, and progesterone secretion may also be affected. Therefore, exogenous progesterone is commonly used to support the luteal phase. These changes are likely to affect hormone-regulated protein secretion from the endometrium in general, and glycodelin secretion in particular, interfering with endometrial receptivity. In vitro studies have shown that androgens decrease endometrial glycodelin secretion.[55] In a prospective controlled study,[56] endometrial biopsies from oocyte donors undergoing COH cycles were compared with biopsies from control women with natural cycles. Immunolocalization of glycodelin-A was demonstrated in endometrial glands and not in the endometrial stroma in all subjects throughout the implantation window. Higher and faster rise of endometrial glycodelin-A expression was noted in COH cycles compared to controls. A significant positive correlation was noted between glycodelin-A expression in the endometrium and serum estradiol levels in natural cycles, whereas neither LH nor progesterone was correlated with endometrial glycodelin-A expression. These results show that COH cycles have a significantly increased endometrial glycodelin-A expression throughout the implantation phase as compared with normal menstrual cycles and this may have an impact on implantation.

Failure of Implantation and Placentation

The polycystic ovary syndrome (PCOS) is associated with infertility and an increased rate of early pregnancy loss. This may reflect defects in ovulation, implantation and placentation. Hyperinsulinemia is an independent risk factor for EPL. Serum glycodelin correlates positively with insulin sensitivity index during weeks 3-5 of pregnancy.[54] Comparing women with PCOS who experienced EPL with those who did not, serum glycodelin was significantly lower during weeks 3-5 of pregnancy.[54] During the first trimester, both epithelial glycodelin and stromal IGFBP-1 serum concentrations are markedly decreased in PCOS, implicating endometrial epithelial and stromal dysfunction during the peri-implantation period and early pregnancy as a mechanism for EPL in PCOS. These changes may be secondary to reduced insulin sensitivity and hyperinsulinemia. Thus, glycodelin may serve as a biomarker for increased risk of early pregnancy loss due to deficient endometrial environment for maintenance of pregnancy. Interestingly, treatment with metformin, an insulin-lowering agent, increases serum glycodelin concentration in patients with PCOS, suggesting that insulin may regulate glycodelin secretion.[57] However, a study employing euglycemic hyperinsulinemic clamp has shown no acute glycodelin-lowering effect of insulin, ruling out any direct glycodelin-reducing effects of insulin.[58] However, indirect long term effects mediated by insulin through stromal factors on epithelial glycodelin secretion cannot be excluded.

Contraception

The absence of glycodelin-A in ovulatory phase endometrium is biologically meaningful because glycodelin has anti-fertilization activity.[38] But, glycodelin-A synthesis can be induced in endometrium over the fertile window. This can be achieved by sustained administration of progestagens, e.g., in the form of subdermal contraceptive implants or levonorgestrel hormone-releasing intrauterine system.[59,60] In view of the anti-fertilization effect of glycodelin-A it is possible that induction of glycodelin secretion over the fertile window may contribute to the contraceptive mechanism of these methods. Interestingly, emergency contraception with levonorgestrel brings about changes in the secretory pattern of glycodelin-A, but only in those women who take the pills before the LH surge. Importantly, in these women the serum glycodelin level rises earlier during the fertile window and glycodelin expression in endometrium is weaker during the implantation window.[61] As the great majority of human fertilizations follow from sexual intercourse during the six-day period ending in ovulation, i.e., at or before the LH surge, these results suggest that the early rise of glycodelin-A secretion may contribute to the mechanism(s) whereby pregnancy is prevented in emergency contraception.

Table 2. Glycodelin—areas of potential clinical interest

Area	Glycodelin Status	Affected Site/Function	References
Endometriosis	Increased[a]	Fertilization	70
	Decreased[b]	Implantation	71
Ovarian carcinoma	Expression	Improved survival	66
Endometrial carcinoma	Expression	Reduced growth	68
Synovial sarcoma	Expression[c]	Diagnosis	67
EPL in PCOS[d]	Reduced	Placentation	54
Recurrent miscarriage	Reduced	Placentation	52-54
Progestagen contraception	Increased[a]	Fertilization	59-61
Antiviral contraception	Chemically modified	HIV transmission	63

a, during periovulatory phase; b, during window of implantation, eutopic endometrium; c, in biphasic sarcomas; d, early pregnancy loss in polycystic ovary syndrome.

Antiviral Contraception

Like bovine β-lactoglobulin,[62] glycodelin-A can be chemically modified in such a way that it blocks the binding site on CD4 for the HIV surface glycoprotein, synthesis of viral gp 120, and infection of peripheral blood mononuclear cells by the primary HIV isolate THA/93/051, thus potentially inhibiting HIV transmission.[63] Now that a cell line producing the contraceptive isoform has been identified by recombinant technology,[64] these findings may have application for locally applied antiviral contraception.

Cancer

Glycodelin is a normal constituent of differentiated cells in the endometrium and certain other tissues. Therefore it is not frequently expressed in poorly differentiated malignant cells, whereas cancerous tissue containing both normal and malignant cells may contain glycodelin. Examples of such tumors are various histolopathological forms of breast cancer,[30] ovarian serous carcinoma,[65,66] and biphasic synovial sarcoma[67] in which epithelial components express glycodelin. Experiments on glycodelin-negative carcinoma cell lines transfected with glycodelin cDNA have demonstrated increased epithelial differentiation after transfection with the sense strand but not with the antisense strand.[31,68] Similar results have been obtained in coculture of carcinoma cells with normal stromal cells in the presence of basement membrane components.[69] Both approaches have resulted in glycodelin expression in carcinoma cells, concomitantly with reduced cell proliferation and reversion of the malignant phenotype. These results demonstrate an active role of normal stromal cells, basement membrane components and glycodelin in epithelial differentiation and glandular morphogenesis. This disposition of glycodelin is significant in patients with certain carcinomas, such as ovarian serous carcinoma in which glycodelin-expressing tumors carry better prognosis than glycodelin-negative tumors of the same clinical stage and histological grade[66] (Table 2). Indeed, glycodelin appears to have fundamental inhibitory effects on certain anti-apoptotic survival genes involved in tumor cell growth.[68]

Summary

Glycodelin is a glycosylated lipocalin whose biological actions fall into the category of functional glycomics, i.e., a glycoprotein with the same protein core has different biological actions depending on its specific glycosylation pattern. These diverse effects of glycodelin are best known in the reproductive system (Fig. 1). Glycodelin-A synthesis in secretory endometrium and glycodelin-F synthesis in luteinized granulosa cells and the oviduct are temporally related

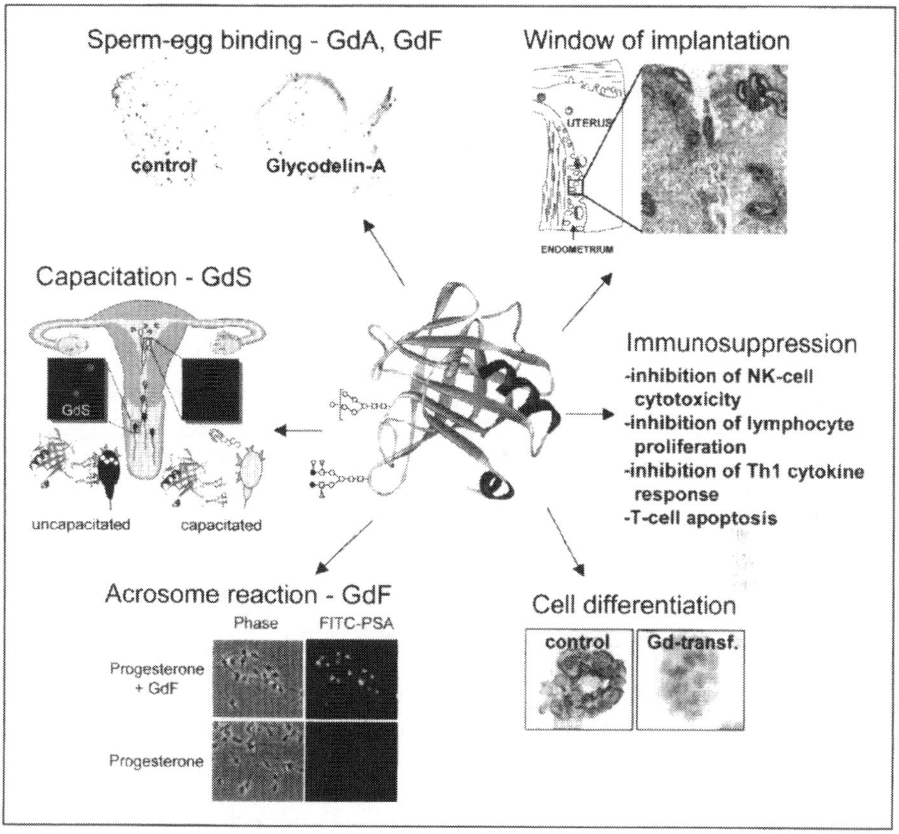

Figure 1. Biological roles of glycodelin. **Sperm-egg binding**: Glycodelin isoforms from amniotic and follicular fluids (glycodelin-A and glycodelin-F, respectively) inhibit sperm-egg binding in hemizona assay. The hemizonae with tightly bound spermatozoa, shown as black grains, are fixed and stained with Mayer's hematoxylin. **Window of implantation**: Expression of immunosuppressive glycodelin in secretory endometrial glands, demonstrated by immunohistochemical staining (dark areas in glands). **Capacitation**: Seminal fluid glycodelin-S helps maintaining spermatozoa in an uncapacitated state. **Immunosuppression**: Several immune cell functions are inhibited. **Acrosome reaction**: Follicular fluid glycodelin-F inhibits progesterone-induced acrosome reaction. Intact acrosomes (bright) are shown by FITC-PSA staining. **Cell growth and differentiation**: Glycodelin transfection to endometrial adenocarcinoma cells reduces proliferation and expression of carcinoma-associated MUC1 (black). Parts of this figure have been adapted from data in *FEBS Letters*,[19] *Journal of Biological Chemistry*,[35] and *American Journal of Obstetrics and Gynecology*.[68] Abbreviations: GdA, glycodelin-A; GdF, glycodelin-F; Gd-transf., glycodelin-transfection; GdS, glycodelin-S; FITC-PSA, fluorescein isothiocyanate-*Pisum sativum* agglutinin; phase, phase contrast microscopy.

to the action of progesterone in the female, whereas no hormonal association is known for the glycodelin-S synthesis in male seminal vesicles. These various glycodelin isoforms regulate such functions as sperm capacitation, acrosome reaction, and binding of spermatozoa to the zona pellucida of the oocyte during early events of fertilization. Examples of the mode of action and clinical relevance of these findings are provided. Unlike many other lipocalins, glycodelin does not bind retinoic acid or retinol, and experiments on its carrier functions have given negative results so far. Glycodelin has immunosuppressive activity, in part independently of glycosylation. Given that the embryo is a semiallograft in the mother, the immunosuppressive nature and abundance of glycodelin at the implantation site may have an impact on fetomaternal defence mechanisms, supported by clinical observations. Being a normal constituent of well-differentiated reproductive tissues glycodelin is rarely expressed in cancer. But where present, its expression in carcinoma tissue is associated with better prognosis. Experiments on the induction of glycodelin synthesis have demonstrated reduced proliferation and reversion of the malignant phenotype in glycodelin-expressing carcinoma cell lines. These results indicate that glycodelin has morphogen-like characteristics.

Acknowledgements

Original work in this review was supported by the Academy of Finland, the University of Helsinki, the Federation of the Finnish Life and Pension Insurance Companies, and the Cancer Foundation of Finland.

References

1. Petrunin DD, Griaznova IM, Petrunina IA et al. Immunochemical identification of human placental organ specific α2-globulin and its concentration in amniotic fluid. Biull Exp Biol Med USSR 1976; 82:803-804.
2. Bohn H, Kraus W, Winkler W. New soluble placental tissue proteins: Their isolation, characterization, localization and quantification. In: Klopper A, ed. Immunology of Human Placental Proteins. Praeger Publ Placenta, 1982:S4:67-81.
3. Bell SC, Bohn H. Immunochemical and biochemical relationship between human pregnancy-associated secreted endometrial α1- and α2-globulins (α1-and α2-PEG) and the soluble placental proteins 12 and 14 (PP12 and PP14). Placenta 1986; 7:283-294.
4. Julkunen M, Raikar RS, Joshi SG et al. Placental protein 14 and progestagen-dependent endometrial protein are immunologically indistinguishable. Hum Reprod 1986; 1:7-8.
5. Julkunen M, Seppälä M, Jänne OA. Complete amino acid sequence of human placental protein 14: A progestogerone-regulated uterine protein homologous to β-lactoglobulins. Proc Natl Acad Sci USA 1988; 85:8845-8849.
6. Dell A, Morris HR, Easton R et al. Structural analysis of the oligosaccharides derived from glycodelin, a human glycoprotein with potent immunosuppressive and contraceptive activities. J Biol Chem 1995; 270:24116-24126.
7. Seppälä M, Bohn H, Tatarinov Y. Glycodelins. Tumor Biol 1998; 19:213-220.
8. Huhtala M-L, Seppälä M, Närvänen A et al. Amino acid sequence homology between human placental protein 14 and β-lactoglobulins from various species. Endocrinology 1987; 120:2620-2622.
9. Papiz M, Sayer L, Elipoulos E et al. The structure of β-lactoglobulin and its similarity to plasma retinol-binding protein. Nature 1986; 324:383.
10. Seppälä M, Taylor RN, Koistinen H et al. Glycodelin: A major lipocalin protein of the reproductive axis with diverse actions in cell recognition and differentiation. Endocr Rev 2002; 23:401-430.
11. Kämäräinen M, Julkunen M, Seppälä M. HinfI polymorphism in the human progesterone associated endometrial protein (PAEP). Nucleic Acids Res 1991; 19:5092.
12. Van Cong N, Vaisse C, Gross MS et al. The human placentalprotein 14 (PP14) gene is localized on chromosome 9q34. Hum Genet 1991; 86:515-518.
13. White DM, Mikol DD, Espinosa R et al. Structure and chromosomal localization of the human gene for brain form of prostaglandin D2 synthase. J Biol Chem 1992; 267:23202-23208.
14. Glasgow BJ, Heinzmann C, Kojis T et al. Assignment of tear lipocalin gene to human chromosome 9q34gter. Curr Eye Res 1993; 12:1019-1023.
15. Chan P, Simon-Chazottes D, Mattei MG et al. Comparative mapping of lipocalin genes in human and mouse: The four genes for complement C8 g chain, prostaglandin-D-synthase, oncogene-24p3, and progestagen-associated endometrial protein map to HSA9 and MMU2. Genomics 1994; 23:145-150.
16. Vaisse C, Atger M, Potier B et al. Human placental protein 14 gene: Sequence and characterization of a short duplication. DNA Cell Biology 1990; 9:401-413.
17. Seppälä M, Riittinen L, Julkunen M et al. Structural studies, localization in tissue and clinical aspects of human endometrial proteins. J Reprod Fertil 1988; S36:127-141.
18. Åkerström B, Flower DR, Salier J-P. Lipocalins: Unity in diversity. Biochim Biophys Acta 2000; 1482:1-8.
19. Koistinen H, Koistinen R, Seppälä M et al. Glycodelin and β-lactoglobulin, lipocalins with a high structural similarity, differ in ligand binding properties. FEBS Lett 1999; 450:158-162.

20. Morris HR, Dell A, Easton RL et al. Gender specific glycosylation of human glycodelin affects contraceptive activity. J Biol Chem 1996; 271:32159-32167.
21. Jeschke U, Wang X, Briese V et al Glycodelin and amniotic fluid transferrin as inhibitors of E-selectin-mediated cell adhesion. Histochem Cell Biol 2003; 119:345-354.
22. Koistinen H, Easton RL, Koistinen R et al. Differences in glycosylation and sperm-egg binding inhibition of pregnancy-related glycodelin. Biol Reprod 2003; 69:1545-51.
23. Chiu PCN, Koistinen R, Koistinen H et al. Zona binding inhibitory factor-1 from human follicular fluid is an isoform of glycodelin. Biol Reprod 2003; 69:365-72.
24. Varki A. Selectin ligands. Proc Natl Acad Sci USA 1994; 91:7390-7397.
25. Tse JYM, Chiu PCN, Lee KF et al. The synthesis and fate of glycodelin in human ovary during folliculogenesis. Mol Hum Reprod 2002; 8:142-148.
26. Julkunen M, Wahlström T, Seppälä M. Human fallopian tube contains placental protein 14. Am J Obstet Gynecol 1986; 154:1076-1079.
27. Julkunen M, Koistinen R, Suikkari AM et al. Identification by hybridization histochemistry of human endometrial cells expressing mRNAs encoding a uterine β-lactoglobulin homologue and insulin-like growth factor-binding protein-1. Mol Endocrinol 1990; 4:700-707.
28. Laird SM, Hill CJ, Warren MA et al. The production of placental protein 14 by human uterine tubal epithelial cells in culture. Hum Reprod 1995; 10:1346-1351.
29. Yeung WSB, Lee K-F, Koistinen R et al. Roles of glycodelin in modulating sperm function. Mol Cell Endocrinol 2005; in press.
30. Kämäräinen M, Halttunen M, Koistinen R et al. Expression of glycodelin in human breast and breast cancer. Int J Cancer 1999; 83:738-742.
31. Kämäränen M, Seppälä M, Virtanen I et al. Expression of glycodelin in MCF-7 breast cancer cells induces differentiation into organized acinar epithelium. Lab Invest 1997; 77:565-573.
32. Koistinen H, Koistinen R, Kämäräinen M et al. Multiple forms of messenger ribonucleic acid encoding glycodelin in male genital tract. Lab Invest 1997; 76:683-690.
33. Kämäräinen M, Riittinen L, Seppälä M et al. Progesterone-associated endometrial protein (PAEP) – A constitutive marker of human erythroid precursors. Blood 1994; 84:467-473.
34. Morrow DM, Xiong N, Getty RR et al. Hematopoietic placental protein 14: An immunosuppressive factor in cells of the megakaryocytic lineage. Am J Pathol 1994; 145:1485-1495.
35. Chiu PCN, Tsang HY, Chung MK et al. Glycodelin-S in human seminal plasma reduces cholesterol efflux and capacitation of spermatozoa. J Biol Chem 2005; (Epub ahead of print PMID:15883155).
36. Chiu PCN, Koistinen R, Koistinen H et al. Binding of zona binding inhibitory factor-1 (ZIF-1) from human follicular fluid on spermatozoa. J Biol Chem 2003; 278:13570-7.
37. Chiu PCN, Tsang H-Y, Koistinen R et al. The contribution of D-mannose. L-fucose, N-acetylglucosamine and selectin residues on the binding of glycodelin isoforms to human spermatozoa. Biol Reprod 2004; 70:1710-1719.
38. Oehninger S, Coddington CC, Hodgen GD et al. Factors affecting fertilization: Endometrial placental protein 14 reduces the capacity of human spermatozoa to bind to the human zona pellucida. Fertil Steril 1995; 63:377-383.
39. Bolton AE, Pockley AG, Mowles EA et al. Identification of placental protein 14 as an immunosuppressive factor in human reproduction. Lancet 1987; 1:593-595.
40. Okamoto N, Uchida A, Takakura K et al. Suppression by human placental protein 14 of natural killer cell activity. Am J Reprod Immunol 1991; 26:137-142.
41. Mishan-Eisenberg G, Borovsky Z, Weber MC et al. Differential regulation of Th1/Th2 cytokine responses by placental protein 14. J Immunol 2004; 173:5524-5530.
42. Mukhopadhyay D, Sundereshan S, Rao C et al. Placental protein 14 (PP14) induces apoptosis in T cells, but not in monocytes. J Biol Chem 2001; 276:28268-73.
43. Rachmilewitz J, Borovsky Z, Riely GJ et al. Negative regulation of T cell activation by placental protein 14 is mediated by the tyrosine phosphatase receptor CD45. J Biol Chem 2003; 278:14059-14065.
44. Skornicka EL, Kiyatkina N, Weber MC et al. Pregnancy zone protein is a carrier and modulator of placental protein-14 in T-cell growth and cytokine production. Cell Immunol 2004; 232:144-156.
45. Yaniv E, Borovsky Z, Mishan-Eisenberg G et al. Placental protein 14 regulates selective B cell responses. Cell Immunol 2003; 222:156-63.
46. Miller RE, Fayen JD, Chakraborty S et al. A receptor for the lipocalin placental protein 14 on human monocytes. FEBS Lett 1998; 436:455-460.
47. Mukhopadhyay D, SundarRaj S, Alok A et al. Glycodelin-A, not glycodelin-S, is apoptotically active. Relevance of sialic acid modification. J Biol Chem 2004; 279:8577-2584.

48. Jayachandran R, Shaila MS, Karande AA. Analysis of the role of oligosaccharides in the apoptotic activity of glycodelin A. J Biol Chem 2004; 279:8585-8591.
49. Clark GF, Oehninger S, Patankar MS et al. A role for glycoconjugates in human development: The human feto-embryonic defense system hypothesis. Hum Reprod 1996; 11:467-473.
50. Kao LC, Tulac S, Lobo S et al. Global gene profiling in human endometrium during the window of implantation. Endocrinology 2002; 143:2119-2138.
51. Julkunen M, Koistinen R, Sjöberg J et al. Secretory endometrium synthesizes placental protein 14. Endocrinology 1986; 118:1782-1786.
52. Dalton CF, Laird SM, Estdale SE et al. Endometrial protein PP14 and CA-125 in recurrent miscarriage patients: Correlation with pregnancy outcome. Hum Reprod 1998; 13:3197-3202.
53. Tulppala M, Julkunen M, Tiitinen A et al. Habitual abortion is accompanied by low serum levels of placental protein 14 in the luteal phase of the fertile cycle. Fertil Steril 1995; 63:792-795.
54. Jakubowicz DJ, Essah PA, Seppala M et al. Reduced glycodelin and insulin-like growth factor-binding protein 1 in women with polycystic ovary syndrome during first trimester of pregnancy. J Clin Endocrinol Metab 2004; 89:833-839.
55. Tuckerman EM, Okon MA, Li T et al. Do androgens have a direct effect on endometrial function? An in vitro study. Fertil Steril 2000; 74:771-9.
56. Brown SE, Mandelin E, Oehninger S et al. Endometrial glycodelin-A expression in the luteal phase of stimulated ovarian cycles. Fertil Steril 2000; 74:130-133.
57. Jakubowicz DJ, Seppälä M, Jakubowicz S et al. Insulin reduction with metformin increases luteal phase serum glycodelin and insulin-like growth factor binding protein 1 concentrations and enhances uterine vascularity and blood flow in the polycystic ovary syndrome. J Clin Endocrinol Metab 2001; 86:1126-1133.
58. Seppälä M, Mandelin E, Koistinen R et al. Glycodelin responses to hyperinsulinaemic clamp vary according to basal glycodelin concentration. Clin Endocrinol 2005; 62:611-615.
59. Mandelin E, Koistinen H, Koistinen R et al. Endometrial expression of glycodelin in levonorgestrel-releasing subdermal implant wearing women. Fertil Steril 2001; 76:474-8.
60. Mandelin E, Koistinen H, Koistinen R et al. Levonorgestrel-releasing IUD-wearing women express contraceptive glycodelin-A in endometrium during midcycle: Another contraceptive mechanism? Hum Reprod 1997; 12:2671-2675.
61. Durand M, Seppala M, del Carmen Carvioto M et al. Late follicular phase administration of levonorgestrel as an emergency contraceptive changes the secretory pattern of glycodelin in serum and endometrium during the luteal phase of the menstrual cycle. Contraception 2005; 71:451-457.
62. Neurath AR, Jiang S, Strick N et al. Bovine β-lactoglobulin modified by 3-hydroxyphthalic anhydride blocks the CD4 cell receptor for HIV. Nat Med 1996; 2:230-234.
63. Seppälä M, Jiang S, Strick N et al. Glycodelins GdA and GdS modified by 3-hydroxyphthalic anhydride inhibit gp120-CD4 binding and HIV-1 infection in vitro. Lab Invest 1997; 77:127-130.
64. Van den Nieuwenhof IM, Koistinen H, Easton RL et al.Recombinant glycodelin carrying the same type of glycan structures as contraceptive glycodelin-A can be produced in human kidney 293 cells but not in Chinese hamster ovary cells. Eur J Biochem 2000; 267:4753-4762.
65. Kämäräinen M, Leivo I, Koistinen R et al. Normal human ovary and ovarian tumors express glycodelin, a glycoprotein with immunosuppressive and contraceptive properties. Evidence from immunohistochemical staining and in situ hybridization. Am J Pathol 1996; 148:1435-1443.
66. Mandelin E, Lassus H, Seppälä M et al. Glycodelin in ovarian serous carcinoma: Association with differentiation and survival. Cancer Res 2003; 63:6258-64.
67. Kämäräinen M, Miettinen M, Seppälä M et al. Epithelial expression of glycodelin in synovial sarcomas. Int J Cancer 1998; 76:487-490.
68. Koistinen H, Seppälä M, Nagy B et al. Glycodelin reduces carcinoma-associated gene expression in endometrial adenocarcinoma cells. Am J Obstet Gynecol 2005; 139:in press.
69. Arnold JT, Lessey BA, Seppälä M et al. Effect of normal endometrial stroma on growth and differentiation in Ishikawa endometrial adenocarcinoma cells. Cancer Res 2002; 62:79-88.
70. Cornillie FJ, Lauweryns JM, Seppälä M et al. Expression of endometrial protein 14 in pelvic and ovarian endometriotic implants. Hum Reprod 1991; 10:1411-1415.
71. Kao LC, Germeyer A, Tulac S et al. Expression profiling of endometrium from women with endometriosis reveals candidate genes for disease-based implantation failure and infertility. Endocrinology 2003; 144:2870-2881.

Chapter 12

Functional Aspects of β-Lactoglobulin, Major Urinary Protein and Odorant-Binding Protein

Andrea Cavaggioni, Paolo Pelosi, Stephen G. Edwards and Lindsay Sawyer*

Abstract

The lipocalin family contains more than 30 distinct proteins that are widely distributed throughout the living world. However, the exact physiological functions of many members of the family are unknown although several are inferred by analogy. Three of the core lipocalins, β-lactoglobulin, major urinary protein and odorant binding protein are found in three significant mammalian secretions, milk, urine and nasal mucous, respectively. Their physiological functions are discussed in the light of the available evidence. It is proposed that β-lactoglobulin is principally a nutritional protein that is closely related to glycodelin, a protein with several reported functions, including immuno-modulation. Mouse major urinary protein with its close relative, rat α_{2u}-globulin, and aphrodisin are involved in signalling amongst members of the same species. Odorant binding proteins are also signalling proteins involved in the detection of odours in the nose.

Introduction

The lipocalin family is a large and widely distributed family of small extracellular proteins many of whose physiological functions are unknown or at best inferred from their sequence and/or structural similarity to other members whose functions have been accurately assigned. Thus by analogy with serum retinol binding protein (RBP), whose function is to transport retinol from the liver to where it is required, a transport function has been assigned to β-lactoglobulin (BLG) although the evidence for this is in fact circumstantial.[1] What has become quite clear, however, is that all members of the family bind small, generally hydrophobic ligands principally in a central pocket, or calyx, formed between two anti-parallel β-sheets (see Chapter 3). The relationship between the various members has been analysed and it is possible to class members as kernel and outlier. Within the former group it is possible to group together[2] the lactoglobulins (for reviews see refs. 3-4), the odorant binding proteins (OBPs, for reviews see refs. 5-6) and the major urinary proteins (MUPs, for reviews see refs. 7-8). It is the purpose of this chapter to describe what is known about the function or functions of these proteins, including some speculation where that function is unclear.

β-Lactoglobulin

Milk is secreted by all mammals, and β-lactoglobulin (BLG) is the major milk whey protein of ruminants, horses, pigs, cats, dogs, dolphins, whales and kangaroos.[9] Although the macaque

*Corresponding Author: Lindsay Sawyer—Institute of Structural and Molecular Biology, School of Biological Sciences, The University of Edinburgh, King's Buildings, Edinburgh EH9 3JR, U.K. Email: lindsay.sawyer@ed.ac.uk

Lipocalins, edited by Bo Åkerström, Niels Borregaard, Darren R. Flower and Jean-Philippe Salier. ©2006 Landes Bioscience.

and the baboon, and perhaps all of the Old World monkeys, produce BLG, it is absent in rodent, lagomorph and human milks.[9-11]

The amino-acid sequence, and hence 3-D structure, of BLG clearly shows that it belongs to the lipocalin family, thus suggesting a transport function, although this is not necessarily its primary function in milk[2,3] (see Chapter 3). It can bind a wide range of ligands and there are reports of more than seventy distinct molecules binding to BLG, most likely in the central calyx, but there are several observations of at least two distinct binding sites.[12,13]

Various hypotheses have been put forward for the physiological function of BLG, such as hydrophobic ligand uptake and transport, enzyme regulation and neonatal passive immunity, however, these putative functions do not appear to be consistent across species[11] in consequence of which, no definite biological function has yet been identified.[14]

Ligand binding to the central cavity of BLG generally increases its stability and from the structural and sequence similarities between BLG and RBP it seemed reasonable to suppose that BLG must be involved in the transfer of retinol from mother to neonate. This role as a transporter protein is supported by the evidence of specific intestinal receptors for BLG that have been found, and also by the fact that the protein is resistant to proteolysis in the acidic milieu of the stomach.[1] However, little endogenous retinol is found bound to BLG and it is mostly fatty acids, principally palmitate, that are found associated with the cows' milk protein on initial purification.[15] Further, since retinol is largely insoluble in aqueous solution, it would be expected to be associated with the fat phase in milk. It is possible, however, that BLG may be indirectly involved in vitamin uptake by mediating a cell surface receptor's interaction with retinol.[16] An alternative but related role is that BLG may enhance fatty acid uptake, both uptake mechanisms being similar to that of retinol facilitated by serum RBP.[17]

Another modulation role has also been proposed for BLG where it sequesters free fatty acids as they are cleaved by gastric lipase, possibly improving the lipase activity.[18] However, between species there is a great deal of variability of fatty acid binding to BLG and indeed porcine BLG does not appear to bind fatty acid at all at neutral pH.[19] Therefore none of the above hypotheses probably represents the primary role of the protein which one might reasonably expect to be conserved between species.

If we assume that whatever the function is, it has been conserved through evolution, then the absence of BLG from the milks of some species presents a problem. It would appear that, apart from the useful nutritional purpose for the neonate, a specific mother-to-child transfer function, particularly of retinol, is not the reason for its presence. Indeed, the true, or original physiological function may not even lie with neonatal nutrition, but be found in the mother with continued expression in the milk during lactation being a convenient and efficient source of nutrition. However, as there appear to be more species with BLG present in their milks, it is perhaps more likely that those species that do not secrete the protein have lost the nutritional benefit of BLG. The gene structure and evolutionary relationships amongst members of the lipocalin family have been considered by Salier[20] and by Sànchez[21] (see also Chapter 2).

A comparison carried out with available, mostly published, sequences gave rise to the cladogram shown in Figure 1. The cladogram for all of the lipocalins can be found in Chapter 2. Each of the proteins considered in this chapter, together with RBP, shows up as a distinct branch and within that of the BLGs the marsupials and the ruminants form fairly distinct sub-groups. What does not show up so clearly is that, while ruminant species express significant quantities of a single BLG gene in their milk, albeit with allelic variation, the BLG of cats, dogs and horses arises from at least two distinct genes (for a review, see ref. 22), referred to as BLG-I, BLG-II etc. Further, the closest match to BLG in humans is the lipocalin glycodelin, a glycosylated protein expressed not in milk but in many tissues of the body, particularly the endometrium during early pregnancy[3,23] (see Chapter 11). A significant difference between BLG and glycodelin is that the latter does not appear to bind retinol,[24] although the actual physiological ligand, assuming the function to involve ligand-binding, is as yet unknown. Thus there is some basis for the suggestion that the function of BLG was originally associated with

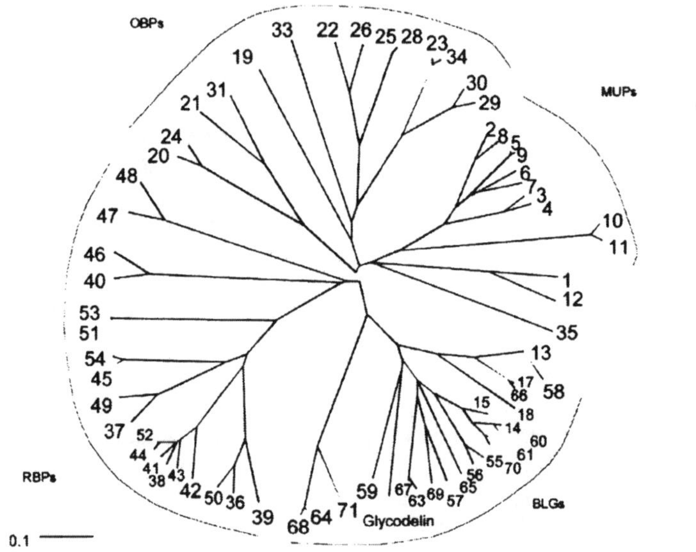

Key:

1. LCN-1 [M. musculus]
2. MUP [M. musculus]
3. MUP 5 [R. norvegicus]
4. MUP 4 [R. norvegicus]
5. MUP 2 [M. musculus]
6. MUP 4 [M. musculus]
7. MUP 5 [M. musculus]
8. MUP 3 [M. musculus]
9. MUP 1 [M. musculus]
10. [a]MUP 2 precursor [M. musculus]
11. [a]MUP 2 precursor [M. musculus]
12. [a] MUP-like LCN precursor [R. norvegicus]
13. BLG [O. aries]
14. [a]BLG II [C. familiaris]
15. [a]BLG I [C. familiaris]
16. Glycodelin [H. sapiens]
17. BLG [B. taurus]
18. BLG [S. scrofa]
19. OBP [S. scrofa]
20. OBP 2B [H. sapiens]
21. OBP 2B [R. norvegicus]
22. OBP Ia [M. musculus]
23. OBP I f [R. norvegicus]
24. OBP 2A [H. sapiens]
25. [a]OBP Ib [M. musculus]
26. [a]OBP Ia [M. musculus]
27. [a]OBP precursor [M. musculus]
28. [a]OBP Ib [M. musculus]
29. [a]OBP precursor [M. musculus]
30. [a]OBP precursor [M. musculus]
31. [a]OBP [M. musculus]
32. [a]OBP Ib [M. musculus]
33. [a]OBP Ib [R. norvegicus]
34. [a]OBP 1F [R. norvegicus]
35. P19 lipocalin [Equus caballus]
36. RBP 7, cellular [H. sapiens]
37. [a]RBP 5, cellular [C. familiaris]
38. [a]cellular RBP II [C. familiaris]
39. [a]RBP 7, cellular [C. familiaris]
40. RBP 4, plasma precursor [H. sapiens]
41. RBP 2, cellular [S. scrofa]
42. [a]RBP 2, cellular [Pan troglodytes]
43. RBP 2, cellular [H. sapiens]
44. RBP 2, cellular [M. musculus]
45. RBP 1, cellular [R. norvegicus]
46. RBP 4, plasma [M. musculus]
47. RBP 3, interstitial [B. taurus]
48. RBP 3, interstitial [M. musculus]
49. RBP 5, cellular [H. sapiens]
50. RBP 7, cellular [M. musculus]
51. cRABP I [M. musculus]
52. RBP 2, cellular [R. norvegicus]
53. cRABP 1 [H. sapiens]
54. RBP 1, cellular [H. sapiens]
55. BLG III [F. catus]
56. BLG I [F. catus]
57. BLG II [F. catus]
58. BLG [Capra hircus]
59. BLG I [P. cynocephalus]
60. BLG II [C. familiaris]
61. BLG I [C. familiaris]
62. BLG I [S. scrofa]
63. BLG II [Equus asinus]
64. BLG [Macropus giganteus]
65. BLG I [Equus asinus]
66. BLG [Bubalus bubalis]
67. BLG II [Equus caballus]
68. BLG [Macropus eugenii]
69. GENSCAN00000054793, first 7 exons [C.familiaris]
70. GENSCAN00000054793, last 8 exons
71. Glycodelin (H.sapiens)

Figure 1. Phylogenetic tree of β lactoglobulins (BLGs), retinol-binding proteins (RBPs), major urinary proteins (MUPs) and odorant-binding proteins (OBPs). The bar shows units of 0.1 nucleotide substitutions per site. The superscript 'a' denotes proteins which have been automatically identified in the genome by EnsEMBL software (www.ensembl.org).

that of glycodelin, and that gene duplication and subsequent loss of expression have led to the distribution pattern that is seen today. In this context, therefore, it is instructive to ask if there might still be some remnant of BLG detectable at the sequence level as a pseudogene.

So far, BLG pseudogenes have been sequenced in goat, sheep and cow[25,26] and these are more closely related to the monomeric BLG II forms found in some ruminant species like horse, cat and dog. The question can now be formulated in two parts: do those species that do not express BLG in their milk possess BLG pseudogenes, and conversely, do those species that do express BLG also possess a closely related endometrial protein expressed in pregnancy?[3,14]

The baboon is the only species so far examined that expresses both glycodelin and BLG. A 71 amino acid residue fragment of baboon glycodelin is extremely well conserved with 43 amino acids identical to both its own BLG and to human glycodelin (A.Fazleabas, personal communication; see ref. 14). However, endometrial lipocalins have been found in horse (uterocalin or p19[27]) and dog (cP6[28]) that are more closely related to the RBP family and both bind retinol.

It is known that many members of the lipocalin family have evolved from a single ancestor and two lipocalin pseudogenes have been identified on human chromosome 9q34 as part of a lipocalin gene family. These lie close to those coding for glycodelin and OBP[29] and it is therefore tempting to suggest that one of these pseudogenes may be the remnant of a human BLG gene, although their sequence comparisons appear to show a closer relationship to tear lipocalin than to glycodelin. There are two fragments elsewhere in the human genome, however, that do appear to bear a closer resemblance to glycodelin than those above and these might be BLG pseudogenes (Diego Sànchez, personal communication).

Thus it is certainly possible that BLG has arisen as the result of a gene duplication event from an essential, possibly endometrial, lipocalin, such as glycodelin, with a probable transport function crucial for the development of the endometrium during the early stages of pregnancy. Over-expression of BLG in the mammary gland has provided a convenient nutritional protein whose former function has been lost, although the ability to bind a variety of hydrophobic ligands has remained. This nonessential function of BLG may explain the resultant loss of functional BLG genes in some species and the lack of universal expression in milk across species.

Lipocalins with Pheromonal Activity

Pheromones are chemical substances active at low concentration as species-specific signals. They are produced and released by an animal and evoke physiological and/or behavioural responses in conspecifics, i.e., members of the same species. According to this definition, at least three classes of lipocalins, the Major Urinary Protein complex of the mouse, the urinary α_{2U}-globulin complex of the rat and the aphrodisin of the vagina of the golden hamster are pheromones, (for reviews, see refs. 7-8).

Mice and rats display a sizeable physiological lipocalin proteinuria which accounts for about 10% of nitrogen metabolism in adult male mice. The synthesis of urinary lipocalins like MUP and α_{2U}-globulin, takes place in the liver and is hormonally dependent on the presence of testosterone. Two peculiarities single out the urinary lipocalins of mice and rats among other excretory lipocalins. First, they are odorant-binding proteins that first bind in the body and then slowly release in the field, a variety of volatile odorant molecules. Second, the proteins are not excreted as a single chemical form but rather as a complex of similar but not identical forms. The urinary lipocalins become inactive as pheromones, once stripped of the bound odorants in water or saline, but become active again once dissolved in a urinary context. Thus, the combination of urinary odorants with the protein moiety seems to be needed for pheromonal activity. In nature, the volatile odorants diffuse from a urinary spot released in the field, acting rather like flags, to attract conspecifics and invite them to make a chemical assay of the proteins by licking. The protein polymorphism then provides a chemical ownership signature related to the gender, strain and individuality of the releaser[30] such that all of the essential features are present for maintaining genetic identity, avoiding inbreeding and enhancing the

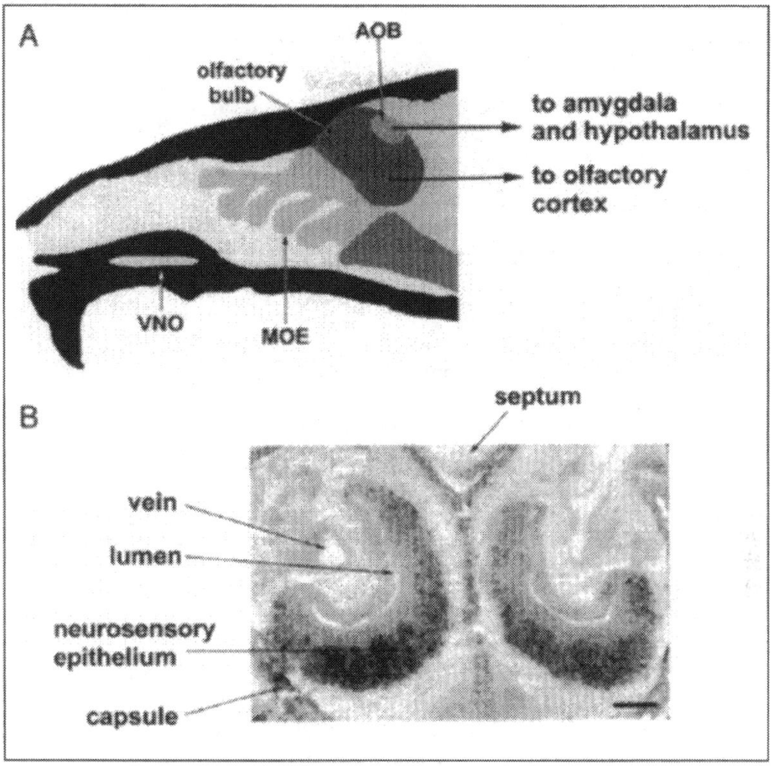

Figure 2. The vomeronasal organ of a mouse. A) Representation of a sagittal section of the head of a mouse, close to the median line. VNO: Vomeronasal organ, MOE: Main olfactory epithelium, AOB: Accessory olfactory bulb. B) Histological transverse section at the level of the vomeronasal organ. The vomeronasal lumen and, on its side, the sensory epithelium are shown. Reprinted from Tirindelli et al, Trends Neurosci 1998; 21(11):482-486, ©1998, with permission from Elsevier.[31]

choice of the best mating partner. On the basis of this detailed information, the endocrine system is primed to release responses aimed at improving the reproductive fitness. The best studied primer pheromone effects of MUP are the anticipation of the first estrus in prepubertal mice (Vandenbergh effect) and the induction of the estrus after suppression in grouped female conspecifics (Whitten effect).

The receptor for the urinary lipocalins is the vomeronasal organ, shown in Figure 2. This organ, belonging to the olfactory system, can be stimulated by nonvolatile molecules such as proteins, unlike the main olfactory epithelium which is stimulated by (volatile) odorants only. The receptor molecules for urinary lipocalins have not been identified so far, but the vomeronasal organ contains a class of receptor neurons that express receptor molecules different from the putative olfactory receptors found in the main olfactory mucosa, but homologous to the metabotropic glutamate receptor molecule.[31] Upon stimulation, the vomeronasal excitation is sent through the accessory olfactory system to the hypothalamus where the sexual cycle is controlled.[32]

The urinary proteins of mice and rats share a 65% identity of residues in amino acid sequence. The species-specificity of the protein should thus depend upon the remaining 35% which contains the epitopes that have evolved under the phylogenetic pressure for mouse vs rat differentiation. For instance, the amino-terminal sequence, which is conserved in both mouse and rat species, obviously cannot account for the species-specificity of the protein, although it displays some pheromonal activity in the mouse. The protein polymorphism may well represent a protein compatibility system involved in allorecognition among conspecifics. In the end, the binding/releasing capacity of urinary lipocalins for volatile odorants, together with the protein polymorphism, make the urinary lipocalins an efficient, clever and robust system of pheromonal communication among rodent conspecifics.

Aphrodisin is a pheromone lipocalin isolated from the vaginal secretions of estrous golden hamsters.[33] It is synthesized in situ as its gene is highly expressed in the glands of the cervix uteri. The pheromonal activity of aphrodisin is retained after organic extraction of bound ligands but is lost after proteolysis. Thus, the protein moiety seems to be responsible for the effect. Unlike MUP and α_{2U}-globulin, no protein polymorphism has been so far detected for aphrodisin. Aphrodisin activates the mounting behaviour of male hamsters, while no endocrine priming effect of aphrodisin is known.

Although the aphrodisin of golden hamsters and the urinary lipocalins of rodents are similar in the aminoacid sequence and structure (aphrodisin bears a 21% and 26% sequence identiity to MUP and α_{2U}-globulin respectively), and are all ancillary to the fitness of the species, they are pheromones tailored to carry different information and elicit different responses.

Odorant-Binding Proteins

Odorant-binding proteins were discovered for the first time in cow nasal mucosa, following a ligand-binding approach using the potent odorant 2-isobutyl-3-methoxypyrazine.[5,6,34,35] Several members of this class of proteins were purified and characterised from other mammalian species, including rat, mouse, pig, rabbit and porcupine. The only OBP isolated from a nonmammalian vertebrate is the Bowman's gland protein (BG) from frog. Each species generally expresses several OBPs, that, according to sequence identities across species, can be grouped into three sub-classes. Proteins belonging to any one of these sub-classes are most similar—in some cases identical in their amino acid sequences, although they could differ in the extent of glycosylation—to lipocalins involved in the delivery of chemical stimuli, such as the urinary proteins of mouse and rat or the pig salivary proteins (SALs). For instance, the pig SALs—two sequences differing by only three substitutions—are strictly male specific and loaded with the two components of the pig sex pheromone (androstenone and androstenol) when expressed in the salivary glands, while they are equally present in the nasal area of both sexes, in such case devoid of any ligand.[36-38] Similarly, the mouse expresses in the nose of both sexes proteins identical in amino acid sequence, to MUPs. Given such high similarity between proteins synthesised in different parts of the body and performing different functions, it has been agreed to restrict the term of OBP to those secreted in the nasal area.

An important characteristic of OBPs is their property of reversibly binding, with dissociation constants in the micromolar range, a broad spectrum of organic molecules of medium size and rather hydrophobic nature, including some pheromones. Three-dimensional structures of complexes of OBPs with various ligands have shown that different organic molecules can be accommodated within the same binding pockets, but assuming different orientations.[39-42]

OBPs are synthesised by glands of the nasal epitheliun located in the respiratory or vomeronasal regions, but not in the olfactory area (see Fig. 2). The site of expression, their high affinities for pheromones and their high similarity to urinary and salivary proteins strongly suggest that OBPs play a role in pheromone perception in the vomeronasal organ, rather than being involved with detection of general odorants.[43] Another element supporting this view is the observation that at least one of the two OBPs secreted in the mouse vomeronasal organ is expressed soon after birth and then again at puberty, two periods when pheromones play a role recognising the mother and finding the partner, respectively.[44]

As for BLG, the main open question concerning OBPs is their physiological function. A role of passive carrier of hydrophobic ligands across the aqueous mucus does not fully justify the large amounts produced and the number of genes involved. On the other hand, the wide spectrum of binding observed with different ligands seems to exclude any specific function. The available data cannot reasonably support any other hypothesis. Some suggestions, however, come from the insect world, where other families of soluble binding proteins are active in the chemoreception. At least two major classes of proteins (each comprising several sub-classes) called OBPs and CSPs (chemosensory proteins), are secreted in high concentrations in the proximity of chemosensory neurons.[45] Insects' OBPs, unlike those of vertebrates, as well as CSPs, are mainly structured in α-helical domains, arranged in a very compact fold with internal binding pockets. Interestingly, conformational changes have been observed as a consequence of ligand-binding and as a function of pH,[46] in this latter case rather like BLG.[47,48] It has also been suggested that such changes could be involved in the efficient release of the bound ligand in the proximity of the membrane-bound receptor. Alternatively, such conformational changes may generate new surface features, capable of interacting with membrane-bound receptors.

No major conformational effects have yet been observed with mammalian OBPs, although even minor modifications in the shape of these proteins, such as the exposure of a single amino acid residue, could trigger a specific interaction with a receptor, as is the case of retinol-binding protein and transthyretin.[49]

Conclusions

By considering the species distribution and binding properties of BLG, it is apparent that the presence of the protein in the milk of most, but not all, species provides most probably a nutritionally useful component possibly left over from the evolutionary stage before the separation of the *Mammalia* into *Eutheria* and *Metatheria*. As glycodelin is the human protein most closely related to β-lactoglobulin, it may hold the key to the origins of the physiological role of β-lactoglobulin in ruminant and other mammalian species.

The pheromonal proteins MUP and α_{2U}-globulin are closely related to the OBPs but function externally as a means of communicating between members of a species. Not only do the proteins promote the slow release of volatile signalling molecules but also the significant polymorphism amongst the proteins themselves serves to provide information about gender and individuality.

Finally, the OBPs function within the nasal mucous in a manner that is not yet understood. Their binding behaviour allows them to form complexes with a variety of odorant molecules, in this respect not unlike BLG, and as such perhaps to act as mediators or modulators between the volatile odours and the nasal receptor.

In conclusion, these three core members of the lipocalin family have provided a great wealth of information about their properties that lead to hypotheses regarding their physiological function but in the case of both BLG and OBP it is still necessary to validate these ideas.

Acknowledgement

L. Sawyer is most grateful to Diego Sanchez for stimulating discussions about pseudogenes.

References

1. Papiz MZ, Sawyer L, Eliopoulos EE et al. The structure of β-lactoglobulin and its similarity to plasma retinol binding protein. Nature 1986; 324:383-385.
2. Flower DR, North ACT, Sansom CE. The lipocalin protein family: Structural and sequence overview. Biochim Biophys Acta 2000; 1482:9-24.
3. Sawyer L, Kontopidis G. The core lipocalin, bovine β-Lactoglobulin. Biochim Biophys Acta 2000; 1482:136-148.
4. Sawyer L. β-Lactoglobulin. In: Fox PF, McSweeney PLH, eds. Advanced Dairy Chemistry - I, Part A, 3rd edn. New York: Kluwer Academic/Plenum Publishers, 2003:319-386.
5. Pelosi P. Odorant binding proteins. Crit Rev Biochem Mol Biol 1994; 29:99-228.

6. Tegoni M, Pelosi P, Vincent F et al. Mammalian odorant binding proteins. Biochim Biophys Acta 2000; 1482:229-240.
7. Cavaggioni A, Mucignat-Caretta C. Major urinary proteins, α_{2U}-globulins and aphrodisin. Biochim Biophys Acta 2000; 1482(1-2):218-228.
8. Beynon RJ, Hurst JL. Urinary proteins and the modulation of chemical scents in mice and rats. Peptides 2004; 25(9):1553-63.
9. Pérez MD, Calvo M. Interaction of β-lactoglobulin with retinol and fatty acids and its role as a possible biological function for this protein: A review. J Dairy Sci 1995; 78:978-988.
10. Azuma N, Yamauchi K. Identification of α-lactalbumin and β-lactoglobulin in cynomolgus monkey (Macaca fascicularis) milk. Comp Biochem Phys B 1991; 99(4):917-921.
11. Kontopidis G, Holt C, Sawyer L. β-Lactoglobulin: Binding properties, structure, and function. J Dairy Sci 2004; 87(4):785-796.
12. Narayan M, Berliner LJ. Fatty acids and retinoids bind independently and simultaneously to β-lactoglobulin. Biochemistry 1997; 36(7):1906-1911.
13. Lübke M, Guichard E, Tromelin A et al. Nuclear magnetic resonance spectroscopic study of β-lactoglobulin interactions with two flavor compounds, γ-decalactone and β-ionone J. Agric Food Chem 2002; 50(24):7094-7099.
14. Kontopidis G, Holt C, Sawyer L. The ligand-binding site of bovine β-lactoglobulin: Evidence for a function? J Mol Biol 2002; 318(4):1043-1055.
15. Perez MD, Diaz de Villegas C, Sanchez L et al. Interaction of fatty acids with β-lactoglobulin and albumin from ruminant milk. J Biochem (Tokyo) 1989; 106(6):1094-1097.
16. Kushibiki S, Hodate K, Kurisaki J et al. Effect of beta-lactoglobulin on plasma retinol and triglyceride concentrations, and fatty acid composition in calves. J Dairy Res 2001; 68(4):579-586.
17. Perez MD, Puyol P, Ena JM et al. Comparison of the ability to bind lipids of β-lactoglobulin and serum albumin of milk from ruminant and nonruminant species. J Dairy Res 1993; 60(1):55-63.
18. Perez MD, Sanchez L, Aranda P et al. Effect of β-lactoglobulin on the activity of pregastric lipase. A possible role for this protein in ruminant milk. Biochim Biophys Acta 1992; 1123(2):151-155.
19. Frapin D, Dufour E, Haertlé T. Probing the fatty-acid-binding site of beta-lactoglobulins. J Prot Chem 1993; 12(4):443-449.
20. Salier J-P. Chromosomal location, exon/intron organization and evolution of lipocalin genes. Biochim Biophys Acta 2000; 1482:25-34.
21. Gabriel Gutiérrez G, Ganfornina MD, Sánchez D. Evolution of the lipocalin family as inferred from a protein sequence phylogeny. Biochim Biophys Acta 2000; 1482:35-45.
22. Ng-Kwai-Hang KF, Grosclaude F. Genetic polymorphism of milk proteins. In: Fox PF, McSweeney PLH, eds. Advanced Dairy Chemistry - I, Part A, 3rd edn. New York: Kluwer Academic/Plenum Publishers, 2003:739-816.
23. Seppala M, Taylor RN, Koistinen H et al. Glycodelin: A major lipocalin protein of the reproductive axis with diverse actions in cell recognition and differentiation. Endocr Rev 2002; 23(4):401-430.
24. Koistinen H, Koistinen R, Seppala M et al. Glycodelin and beta-lactoglobulin, lipocalins with a high structural similarity, differ in ligand binding properties. FEBS Lett 1999; 450(1-2):158-162.
25. Passey RJ, Mackinlay AG. Characterisation of a second, apparently inactive, copy of the bovine β-lactoglobulin gene. Eur J Biochem 1995; 233(3):736-743.
26. Folch JM, Coll A, Hayes HC et al. Characterization of a caprine β-lactoglobulin pseudogene, identification and chromosomal localization by in situ hybridization in goat, sheep and cow. Gene 1996; 177(1-2):87-91.
27. Suire S, Stewart F, Beauchamp J et al. Uterocalin, a lipocalin provisioning the preattachment equine conceptus: Fatty acid and retinol binding properties, and structural characterization. Biochem J 2001; 356:369-376.
28. Buhi WC, Alvarez IM, Shille VM et al. A secreted uterine retinol-binding protein in the bitch. Biochem J 1995; 311:407-415.
29. Lacazette E, Gachon AM, Pitiot G. A novel human odorant-binding protein gene family resulting from genomic duplicons at 9q34: Differential expression in the oral and genital spheres. Hum Mol Genet 2000; 9(2):289-301.
30. Hurst JL, Payne CE, Nevison CM et al. Individual recognition in mice mediated by major urinary proteins. Nature 2001; 414(6864):631-634.
31. Tirindelli R, Mucignat-Caretta C, Ryba NJ. Molecular aspects of pheromonal communication via the vomeronasal organ of mammals. Trends Neurosci 1998; 21(11):482-486.
32. Brennan PA, Keverne EB. Something in the air? New insights into mammalian pheromones. Curr Biol 2004; 14(2):R81-89.

33. Briand L, Trotier D, Pernollet JC. Aphrodisin, an aphrodisiac lipocalin secreted in hamster vaginal secretions. Peptides 2004; 25(9):1545-1552.
34. Pelosi P, Baldaccini NE, Pisanelli AM. Identification of a specific olfactory receptor for 2-isobutyl-3-methoxypyrazine. Biochem J 1982; 201:245-248.
35. Pelosi P. Perireceptor events in olfaction. J Neurobiol 1996; 30:3-19.
36. Marchese S, Pes D, Scaloni A et al. Lipocalins of boar salivary glands binding odours and pheromones. Eur J Biochem 1998; 252:563-568.
37. Scaloni A, Paolini S, Brandazza A et al. Purification, cloning and characterisation of odorant- and pheromone-binding proteins from pig nasal epithelium. Cell Mol Life Sci 2001; 58:823-834.
38. Spinelli S, Vincent F, Pelosi P et al. Boar salivary lipocalin. Three-dimensional X-ray structure and androsterol/androstenone docking simulations. Eur J Biochem 2002; 269:2449-2456.
39. Bianchet MA, Bains G, Pelosi P et al. The three dimensional structure of bovine odorant-binding protein and its mechanism of odor recognition. Nat Struct Biol 1996; 3:934-939.
40. Tegoni M, Ramoni R, Bignetti E et al. Domain swapping creates a third putative combining site in bovine odorant binding protein dimer. Nat Struct Biol 1996; 3:863-867.
41. Spinelli S, Ramoni R, Grolli S et al. The structure of the monomeric porcine odorant binding protein sheds light on the domain swapping mechanism. Biochemistry 1998; 37:7913-7818.
42. Vincent F, Spinelli S, Ramoni R et al. Complexes of porcine odorant binding protein with odorant molecules belonging to different chemical classes. J Mol Biol 2000; 300:127-139.
43. Miyawaki A, Matsushita F, Ryo Y et al. Possible pheromone-carrier function of two lipocalin proteins in the vomeronasal organ. EMBO J 1994; 13:5835-5842.
44. Pelosi P. The role of perireceptor events in vertebrate olfaction. Cell Mol Life Sc 2001; 58:503-509.
45. Vogt RG. Biochemical diversity of odor detection: OBPs, ODEs and SNMPs. In: Blomquist GJ, Vogt RG, eds. Insect Pheromone Biochemistry and Molecular Biology. London: Elsevier Academic Press, 2003:391-446.
46. Horst R, Damberger F, Luginbuhl P et al. NMR structure reveals intramolecular regulation mechanism for pheromone binding and release. Proc Natl Acad Sci USA 2001; 98:14374-14379.
47. Tanford C, Bunville LG, Nozaki Y. The reversible transformation of β-lactoglobulin at pH 7.5. J Am Chem Soc 1959; 81(15):4032-4036.
48. Qin BY, Bewley MC, Creamer LK et al. Structural basis of the Tanford transition of bovine β-lactoglobulin. Biochemistry 1998; 37(40):14014-14023.
49. Monaco HL, Rizzi M, Coda A. Structure of a complex of two plasma proteins: Transthyretin and retinol-binding protein. Science 1995; 268:1039-1041.

Chapter 13

The Plasma Lipocalins α_1-Acid Glycoprotein, Apolipoprotein D, Apolipoprotein M and Complement Protein C8γ

Willem van Dijk,* Sonia Do Carmo, Eric Rassart, Björn Dahlbäck and James M. Sodetz

General Introduction

A variety of molecules have been identified in blood plasma that exhibit lipocalin-like properties, but they do not seem to be functionally related. This review is restricted to four of these lipocalins: α_1-acid glycoprotein (AGP), apolipoprotein D (apoD), apolipoprotein M (apoM) and complement protein C8γ (C8γ).

AGP belongs to the so-called outlier group of lipocalins. It is one of the major acute-phase proteins of human blood. This places it functionally in the group of molecules that are supposed to dampen excessive inflammatory reactions. Although both drug-binding and anti-inflammatory or immunomodulatory properties have been described for AGP, its exact physiological function is still not clarified. AGP contains five N-linked complex type glycans that are strongly sialylated, which makes it one of the most acidic glycoproteins of human plasma. The composition of these glycans changes during inflammatory processes. It is of great interest that a number of the anti-inflammatory and immunomodulatory properties are affected by the composition of the glycans, and thus may be influenced by the state of the inflamed condition.

ApoD is a 29-kDa glycoprotein that was isolated for the first time in 1973 from a human plasma HDL fraction. It is an atypical apolipoprotein since its synthesis is not restricted to the liver and/or the intestine. In addition, it is expressed at different levels in a wide range of tissues of several mammalian species and chicken. The wide distribution of the apoD gene in many species supports its evolutionary importance in the chordate lineage. Structural analyses revealed that the protein is part of the lipocalin family. Several candidate hydrophobic molecules were identified as potential ligand for apoD, such as progesterone, pregnenolone, bilirubin, cholesterol, arachidonic acid and E-3-methyl-2-hexenoic acid. Recent detailed ligand binding studies revealed that apoD discriminates well in its binding function between closely related compounds. Due to this apparent heterogeneity, a role as a multi-ligand:multi-function protein has been proposed.

ApoM is a novel apolipoprotein that is predominantly present in HDL and to a minor extent in chylomicrons, VLDL and LDL. An unusual feature of apoM is the retention of the

*Corresponding Author: Willem van Dijk—Glycoimmunology Group, Department of Molecular Cell Biology and Immunology, VU University Medical Center, Amsterdam, The Netherlands. Email: w.vandijk@vumc.nl

Lipocalins, edited by Bo Åkerström, Niels Borregaard, Darren R. Flower and Jean-Philippe Salier.

signal peptide in the mature protein, which serves as a hydrophobic anchor in the phospholipid layer of the lipoproteins. The apoM gene is expressed in liver and kidney and is driven by the transcription factor HNF-1α. ApoM is a member of the lipocalin superfamily: molecular modelling suggests a typical lipocalin fold with one α-helix and eight β-strands forming a coffee filter-like shape. ApoM contains a predicted hydrophobic ligand-binding pocket and preliminary results suggest that it can bind retinol and retinoic acid. However, the physiological ligand and function of apoM is unknown. It can bind to megalin, a member of the LDL-receptor family, which interacts with many lipocalins, e.g., in tubular endothelium of kidneys. ApoM is present in the urine of megalin-deficient mice but not those of normal mice, suggesting a megalin-mediated uptake of apoM in the tubular epithelium of normal mice.

C8γ is a subunit of the eighth component of complement (C8), which is one of five components (C5b, C6, C7, C8, C9) that assemble on the surface of pathogenic organisms to form C5b-9, a cytolytic macromolecular complex referred to as the membrane attack complex, or MAC. C8γ is unique in that it is the only lipocalin among the approximately thirty-five proteins in the human complement system. Among the lipocalins it is one of only a few that form a covalent complex with another protein, i.e., C8α. While many complement components can undergo protein-protein interactions, none are specifically designed to bind small molecules. C8γ is capable of doing both and this dual function suggests it has an as yet unrecognized role in the formation and/or function of the MAC. The structure of C8γ suggests its natural ligand has a long and narrow hydrophobic tail with a negatively charged moiety at one end. This and the fact C8γ associates with cell membranes as part of the MAC raises the possibility that it may bind the acyl chain of a phospholipid or similar fatty acid-like molecule.

Human α_1-Acid Glycoprotein, a Drug Binding and Immunomodulatory Protein

α_1-Acid glycoprotein (AGP), also known as orosomucoid, belongs to the so-called outlier group of lipocalins. This is based on its theoretical three-dimensional structure as well as on its ability to bind and transport a large number of basic lipophilic compounds. In addition to these protein-born properties, AGP also can express a variety of immunological properties that are dependent on or influenced by the composition of its five N-linked glycans. Both types of properties will be reviewed below.

Biochemical Properties of AGP

AGP, is an important constituent of human plasma in being one of the major positive acute-phase proteins.[1] Its plasma level can rise from about 0.8 to 3 g/L within 24 h following the unset of an acute-phase response. About one-third of its number-average molecular weight of 35-37 kD (MW_n, determined by MALDI-ToF/MS[2]) is due to the presence of its five asparagine-linked complex-type glycans.[3,4] These glycans are strongly sialylated and make AGP to one of the most acidic plasma proteins with a pI of 3.2-3.6.[5] Desialylation reduces the number of isoforms detectable within this pH range from ten to three or four with a pI of 4.2-4.7.[5,6] The desialylated isoforms represent the polypeptide variants of AGP that are determined by three adjacent genes, AGP-A, AGP-B and AGP-B', of which the latter two are identical.[7] These genes are clustered in one locus of chromosome 9, q31-q34.1, of the human genome and encode for the two molecular forms, AGP-1 (also known as ORM1, AGP-A) and AGP-2 (ORM2, AGP-B/B'); they differ in only 21 of the 181 amino acids. Polymorphism exists, especially in AGP-1, which can be detected by isoelectric focussing of desialylated AGP.[6,8] The AGP-1 isoforms predominate in a ratio of 3:1 over AGP-2 in normal human plasma[7,9] but variations in ratios have been reported under various pathological conditions.[9-11] AGP-1 contains four and AGP-2 five cysteine residues.[7] Two disulfide bonds have been described, i.e., Cys5-Cys165 and Cys72-Cys147,[4] for a total AGP preparation in which AGP-1 is the major

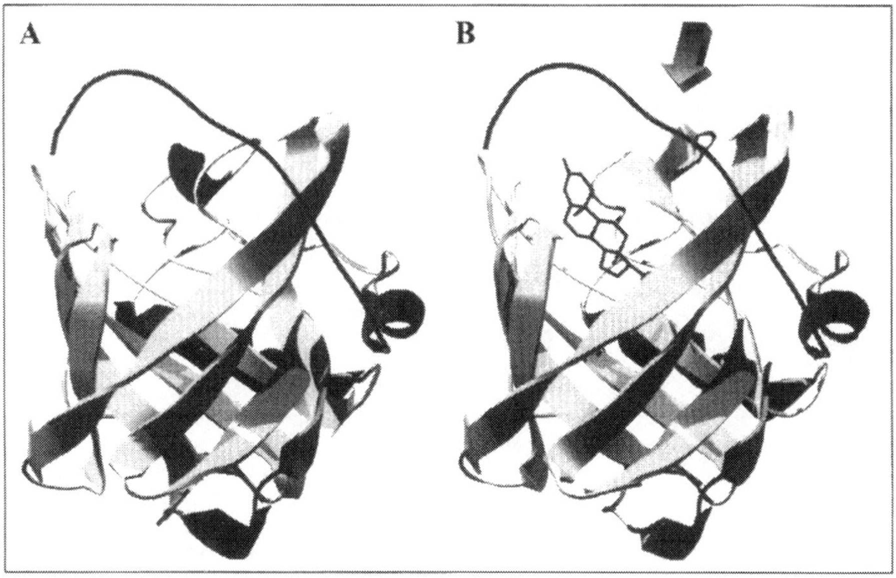

Figure 1. The ribbon molecular model of AGP in the native state (A) shows an eight-stranded antiparallel β-barrel. It can be seen that during progesterone binding (B) the α-helix in the first loop above the β-barrel was transformed into antiparallel β-sheet (marked with an arrow). Reprinted with permission from Kopecky Jr V et al; Biochem Biophys Res Commun 2003; 300:41-46, ©2003 with permission from Elsevier.[15]

isoform. It is not know whether in AGP-2 the additional fifth cysteine residue, Cys149, plays a role in disulfide formation.

The amino acid sequence of AGP and the intron locations in the three genes show close similarity with those of lipocalins.[12,13] In addition, a partial similarity with the EGF receptor has been described by Toh et al.[12] The protein has been crystallized,[14] but the three-dimensional structure could not be analyzed because of the high carbohydrate content and the genetic polymorphism of the AGP preparations. Nevertheless, Kopecky et al[15] have proposed a model of the secondary and tertiary structure of AGP that was based on infrared and Raman spectroscopies (Fig. 1). The lipocalin-like character of AGP is clearly visible in this model by the eight β-strands and the "coffee-filter-like cone" with a hydrophobic interior, that was originally proposed by Brian Halsall.[2] Vibrational spectroscopy confirmed details of the secondary structure and the structure content predicted by homology modeling of the protein moiety, i.e., 15% α-helices, 41% β-sheets, 12% β-turns, 8% bands, and 24% unordered structure at pH 7.4.[15]

Carrier Functions of AGP

The exact physiological function of AGP is not known, but it has been shown that it can bind and transport a large number of basic as well as neutral lopophilic compounds. An extensive overview was published in 1988 by Kremer et al[16] and more recently by Israili and Dayton.[17] The latter review enlist more than 300 drugs that can bind to AGP. A number of basic drugs binding more strongly to AGP than to albumin; affinity ratios of AGP to albumin range from about 11 for imiprazole to 1000 for thioridazine.[17] AGP-1 and AGP-2 may differ in binding specificity for various drugs.[18-21] A great number of studies emphasize that increases in plasma levels of AGP, as induced by inflammatory conditions, will influence the free plasma concentrations of drugs for which AGP is the main carrier. Various reports have indicated that this can give a serious threat for the patients because the efficacies of the drugs are affected.[16] For example, the increased plasma levels of AGP in depressed patients resulted in lowering of the effective concentration of imipramine in these patients.[22] AGP appeared to act in this way as an endogenous inhibitor for the binding of imipramine to the serotonin receptor and may have influenced the plasma-to-brain transport of this drug and comparable ones.[23] Furthermore, high affinity binding to AGP of plasticizers, like di-(2-ethylhexyl) phthalate and tris-(2-butoxyethyl) phosphate, that are present in rubber stoppers of some blood collection tubes and blood-storage backs can leak into the blood. This has been shown to increase the free concentration of many drugs that have lower binding affinities for AGP, in blood samples as well as in the circulation after blood transfusion.[16,17,24]

Recent studies, employing a biomembrane model,[25,26] have indicated that AGP, like other lipocalins,[27] can undergo structural and functional changes under mild acidic conditions like occurring on membrane surfaces. A unique conformational transition of a β-sheet to an α-helix was detected under these conditions, providing in a mechanism for ligand uncoupling upon interaction with cells. Indeed, it was shown that such a conformational change promoted an interaction with the membrane under ligand release, either by a change in affinity or closer membrane association.[26]

Glycosylation-Dependent Properties of AGP

Human AGP contains five N-linked complex-type glycans. Glycan structures, in general, are very flexible and therefore can hide a relatively large area of the protein. Indeed, studies with monoclonal antibodies raised against AGP have indicated that the glycans cover most of the surface of the AGP molecule.[28] The determinants for the antibodies were located almost exclusively in the drug-binding site; no antibodies were detected in close vicinity of the five glycosylation sites.

Analysis of the glycans composition of human plasma AGP has revealed a substantial heterogeneity in their structures.[3,29] This heterogeneity results from the presence in plasma of various distinct AGP-glycoforms of which the plasma levels, and thus their relative occurrence, are dependent on the (patho)physiological condition.[30-33] Examples of two major AGP-glycoforms that are present in normal human plasma are depicted in Figure 2. They differ in extent of branching of the glycans as well in extent of fucosylation. The presence of a fucose residue on a sialylated branch of a glycan results in the expression of the sialyl Lewis^x (sLe^x) determinant, a sugar structure of highly biological significance (e.g., see ref. 34). This blood group is low expressed on normal human plasma AGP, but increases strongly during acute and chronic inflammatory processes[32,35-37] (Fig. 2). Most probably this is due to increased synthesis by the liver of sLe^x-expressing AGP-glycoforms under the influence of inflammatory cytokines.[38]

It is of great interest that a number of immunomodulatory properties of AGP are affected by differences in branching of the glycans as well by the extent of sialylation and fucosylation, i.e., the presence and number of sLe^x groups per AGP molecule. An extensive review of these effects was recently presented by Hochepied et al.[39] So, AGP-glycoforms expressing a high amount of sLe^X groups (cf. Fig. 2) have been shown to be able to ameliorate neutrophil-mediated injuries in lung and intestine in a rat reperfusion model.[40] AGP-glycoforms lacking sLe^X groups were clearly less active.[40] The sLe^X expressing AGP-glycoforms may have affected the primary interaction of neutrophils between endothelial selectins, because we found that such AGP-glycoforms, but not nonfucosylated AGP, can bind in a concentration- and Ca^{2+}-dependent manner to human E-selectin in vitro (Fig. 3). The reperfusion studies, in addition, indicated that the sLe^X-containing AGP-glycoforms inhibited complement-mediated injuries in lung and intestine. In vitro studies of the group of Hechtmann[40] and of our group[41,42] have shown that AGP can inhibit both the alternative and the classical complement activation route and that a high amount of sLe^X groups on AGP strongly enhanced these effects. Other glycosylation-dependent immunomodulatory properties of AGP are dependent on the extent of branching of the glycans, like the ability to inhibit the proliferation of lymphocytes[43,44] and to induce an inhibitor of IL-1 comitogenic activity in macrophages.[45] The degree of sialylation of AGP has been shown to be essential for its inhibitory effect on platelet aggregation.[46] For

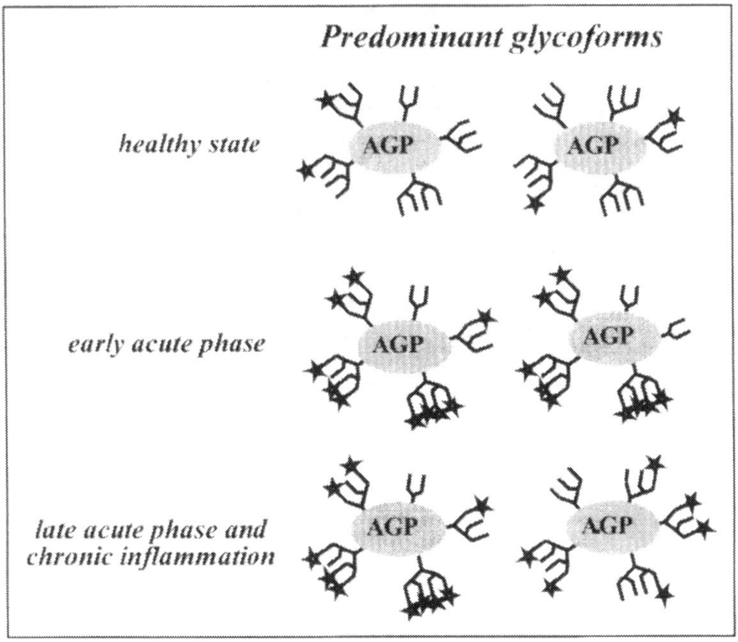

Figure 2. sLe^x-dependent binding of AGP-glycoforms to E-selectin-IgG chimera immobilized on a carboxymethylated dextran CM5 sensor chip, a Biacore 2000 study. AGP was isolated from the serum of a patient suffering from acute trauma; strongly fucosylated and nonfucosylated subfractions were prepared by chromatography on AAL-agarose (Brinkman-van der Linden ECW, Havenaar EC and Van Dijk W, unpublished observations).

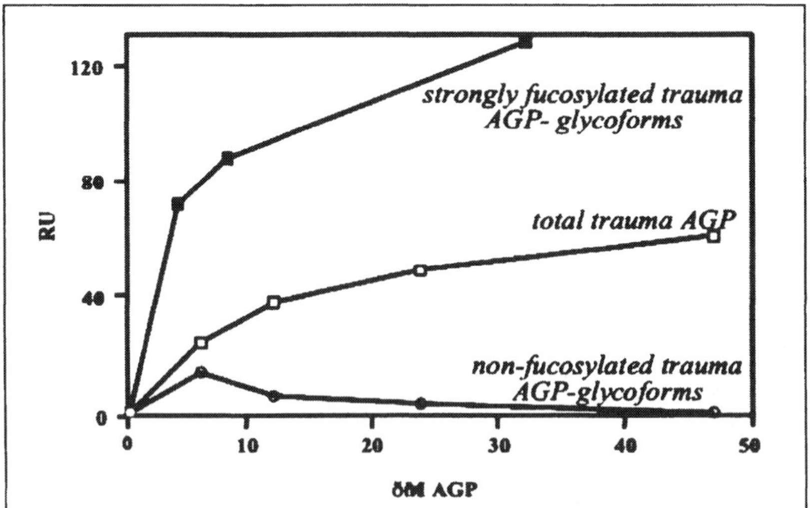

Figure 3. Schematic representations of the predominant or most increased glycoforms of AGP in plasma under various conditions. Di-, tri-, and tetra-branched N-linked glycans are represented by the 2-, 3- and 4-forked structures; diamonds represent sLe^x groups that are present on branches (lactosamine units) that express both α2,3-linked sialic acid and α1,3-linked fucose residues.

more details for these and other glycosylation dependent effects is referred to the review of Hochepied et al.[39]

Anti-Inflammatory Properties of AGP Thus far Unknown to Be Dependent on Glycosylation

AGP also can exert a number of very interesting anti-inflammatory activities of which it is not yet studied or fully excluded whether its glycans are involved or not. So, AGP can protect against TNF-α-induced lethality in mice when injected 3-4 h previous to the lethal challenge.[47] TNF-α is considered to be one of the most important mediators of the shock-inducing activities of endotoxins. It has been suggested that AGP can exert this protective effect by inducing an increased capillary barrier, maintaining perfusion of vital organs.[39] This is based on the observation that AGP is able to antagonize the capillary leakage induced by e.g., histamine and platelet-activating factor.[48] It has been reported that AGP can bind to the microvascular endothelium in a manner that adheres to the rules of receptor-ligand interactions.[49-51] Finally, a number of studies have described beneficial effects of AGP on wound healing and related collagen formation after severe burns as well as peripheral nerve outgrowth after injury.[52-54] It would be interesting to know whether inflammation-induced changes in glycosylation of AGP (cf. Fig. 2) contribute to these effects.

Concluding Remarks about AGP

Although the exact physiological role is not known, AGP appears to be a multi-functional protein under normal and inflammatory conditions. On the one hand, it can bind and transport a large number of particularly therapeutic drugs as well as endogenous compounds, and thus influence their free plasma concentrations. On the other hand, it can exert a variety of immunomodulatory and anti-inflammatory effects that, at least partly, are dependent on the composition of its glycans. The available literature clearly shows that the inflammation-induced changes in concentration as well as in glycosylation of AGP can be of great help to battle excessive inflammation.

Apolipoprotein D

Introduction

Apolipoprotein D (apoD), so called because of its association with plasma lipoproteins, belongs to the lipocalin superfamily. In human, apoD is present in many fluids, expressed in several tissues although poorly expressed in the liver and intestine, which are the major sites of synthesis of other apolipoproteins. In serum of adult healthy subjects, apoD levels vary from 50-200 mg/L plasma. Plasma apoD is mainly associated with both apoA-I and apoA-II in HDL and VHDL particles. It is also present as apoD/apoB-100 heterodimers in LDL and VLDL lipoproteins.

The gene and protein structures, the tissue distribution and known ligands of apoD have been reviewed in reference 55. Here, the role of apoD is discussed with respect to its modulation (1) by biological factors, (2) in various types of cancer, and (3) in the normal and injured nervous systems.

Modulation of ApoD Levels

The regulation of apoD expression is complex and many authors have shown the importance of biological factors in the modulation of this protein.

Cellular Growth and Differentiation

Many lines of evidence suggest an inverse relationship between apoD expression and cell proliferation.[56-62] Furthermore, apoD selectively suppresses the proliferative response of vascular smooth muscle cells to growth factors by a mechanism related to nuclear translocation of ERK1/2.[63]

However, apoD expression might also be related to the differentiation period that is attained following growth arrest. Both retinoic acid[64] and 1,25-dihydroxyvitamin D3,[65] well known for their differentiation properties, were able to induce apoD expression in breast cancer cells. This induction was shown later to be mediated by the nuclear retinoic acid receptors (RAR) leading to a significant inhibition of cell proliferation suggesting that apoD could be a biochemical marker of RAR-mediated growth arrest and cell differenciation.[66]

There is also evidence for such a role in vivo as apoD is expressed mostly in nonproliferating or terminally differentiated prostate glandular epithelial cells.[67] However, this observation can not be generalized since not all nonproliferating or terminally differentiated cells do express apoD.[62]

Metabolic Studies and Energy Homeostasis

ApoD modulation is found in pathologies such as Tangier disease,[68] familial LCAT deficiency,[69] mutations in the apoA-I gene,[70] type 2 diabetes,[71-73] alcoholism,[74,75] corneal Fuch's dystrophy,[76] renal dysfunction[77,78] and ischemic tissue injuries.[79,80] This association with such diseases may be fortuitous and a reflection of lipid metabolism imbalance. The apoD correlation with type 2 diabetes, obesity and hyperinsulinemia may be liver X receptor-dependent and the activation of inflammatory and pro-angiogenic pathways.[73,81]

In the hypothalamus, apoD interacts specifically with the cytoplasmic portion of the long form of the leptin receptor, Ob-Rb, known to play a key function in regulating food intake and body weight. Moreover, the hypothalamic level of apoD mRNA is stimulated by dietary fat, and is strongly, positively correlated with body fat mass and circulating leptin levels. This positive association with body fat, however, is lost in obese ob-/ob- and db-/db- mice, which exhibit markedly reduced levels of hypothalamic apoD mRNA compared to that of wild-type mice. These results suggest that apoD in the hypothalamus is involved in the leptin/Ob-Rb signal transduction pathway that controls body fat accumulation on a high-fat diet.[82]

Development

ApoD is also involved in gestation[83] and foetus development.[84] In mouse, apoD is selectively modulated from E9 to birth in mesenchyme and neuroepithelium.[85] In rat brain, during development and in the early neonatal period, maturation-associated induction of apoD gene expression coincides with the period of active myelination as well as synaptogenesis in rodent brain.[86]

ApoD is also present in the yolk of the rapidly growing chicken oocyte. First proposed to play a role in the transport and/or mobilization of lipids during embryogenesis in oviparous species,[87] it is presumed to transport regulatory molecules such as vitamin A and thyroid hormones.[88] In the chicken foetus, apoD is expressed in ectodermal derivatives in the developing feather follicles and the nervous system. Both neurons and glia express chicken apoD in a subset-dependent form, although very dynamic temporal changes of expression cannot be ruled out. By contrast with the mouse, chicken apoD was not found in pericytes and meningeal cells. It was suggested that the common ancestor of birds and mammals might have expressed apoD in both mesenchymal and neuroectodermal derivatives. The expression profiles possibly changed after the split of mammals and birds, with chicken apoD expression being restricted to neuroectodermal derivatives.[89]

Cancer

ApoD is overexpressed in several types of cancer such as breast, ovarian, endometrium, prostate, retinal, skin, pancreatic, and central nervous system (CNS) carcinomas. It is underexpressed in thyroid oncocytomas.[90] It is silenced by DNA methylation in esophageal squamous cell carcinoma and was identified as a candidate of tumor suppression.[91] However, the correlation between the degree of tumoral differentiation and apoD expression remains ambiguous. This association can either be positive or negative depending on the type of tumor analysed. It can also vary depending upon the sample analysed and there are even discrepancies within the same tumor type. Whether apoD expression is a cause or a consequence of these cellular transformations remains unclear.

In breast and central nervous system cancers, high apoD expression is mostly correlated with highly differentiated, noninvasive/metastatic carcinomas and it is associated with a longer relapse-free and better survival. The presence of apoD in these tumors serves as a good prognostic indicator but could simply be a reflection of growth arrest due to cellular differentiation.[92-96]

In contrast, the inverse relationship between apoD and cell proliferation appears to be lost during malignant transformation and elevated apoD staining is associated with advanced invasive prostate,[67,97-99] skin[100,101] and pancreatic carcinomas[102,103] and may therefore be a prognostic of unfavorable evolution. It is not known if this is due to an increased rate of cellular proliferation or a decrease in cell death.[99] Moreover, apoD protease activity,[104] its implication in cell motility in response to growth factors[105] and its association with the progression to a more differentiated tumoral stage suggests a potential role for apoD in tumoral invasiveness.

In other cancers, like endometrial and ovarian carcinomas and retinoblastomas, there is no significant relationship between apoD immunostaining and either age of the patients, or the stage and histologic grade of the tumors.[106-108] There are also contradictions in the same cancer type. Some studies of breast and prostate cancers demonstrate no significant association between apoD expression and the degree of differentiation although an association with patient survival remains.[109,110]

So far, it appears that the utility of apoD as a prognostic remains uncertain and more studies are required. It was shown that apoD could be useful in the choice and follow-up of hormonal therapy to treat breast cancers.[111-113] This observed discrepancy among a wide range of cancer could be due to the fact that some cancers show hormonal dependance for their growth and the correlation may be restricted to some specific cancer cell types. For example, differences in apoD expression between male and female breast cancer, the latter having a higher level of apoD and a better outcome, opens the field of a more selective hormonal therapy.[114] There are numerous sterol responsive elements in the apoD promoter[62] and one would expect a fine regulation of apoD expression by steroids.

Nervous System

Peripheral Nervous System

The first report showing an implication of apoD in the nervous system was the study of Boyles[115] that showed an increased expression in the regenerating sciatic nerve of the rat after a crush injury. The level of apoD mRNA and protein transiently increased by 40- and 500-fold, respectively, at the time when axons from the proximal stump grow into the distal nerve segment.[115,116] ApoE, another apolipoprotein, increased 250-fold after crush injury. ApoD also accumulated in the regenerating sciatic nerve of two other species, the rabbit and the marmoset monkey.[115] Only small amounts of apoD mRNA are detected in rat noninjured mature nerve and modest expression was detected in transected nerves which were prevented from regeneration by ligation. Since peripheral neural tissue is capable of local synthesis of some apolipoproteins such as apoD and apoE, it seems probable that it would have its own independent system of lipoproteins that may serve as the vehicles for lipid movement between cells. This lipid transport system should be most active in the injured peripheral nerves, in which massive quantities of lipid are freed due to myelin degradation and are subsequently stored and reused during regeneration.

Central Nervous Sytem

In human, apoD is produced by astrocytes and oligodendrocytes in the white matter and in some scattered neurons and protoplasmic astrocytes in the grey matter.[117] It is also present in perivascular cells, and pericytes in the walls of blood vessels suggesting a role in the transport of sterols and small hydrophobic molecules to, or from, blood vessels to the cortex.[118-120]

All the apoD ligands, cholesterol, progesterone, pregnenolone, arachidonic acid and bilirubin are important molecules in the CNS. ApoD also binds androgens and oestrogens to a lower extent. Two ligands, progesterone and pregnenolone, are synthesized by astrocytes and oligodendrocytes. Pregnenolone was shown to accumulate as pregnenolone sulfate in the brains and sciatic nerves of humans and rats.[121] ApoD could, therefore, be implicated in the local transport of steroid hormones and may participate in reinnervation process.

Both apoD protein and mRNA increase in aged cerebral cortex probably due to the increased number of total and reactive astrocytes.[122-124] The same observation is also reported in microglia of mice that lack cystatin B, a model for Unverricht-Lundborg disease, a human disorder characterized by progressive neurological dysfunction and seizures.[125] Also, an even larger increase of apoD is observed in aged PDAPP transgenic mice, a model for Alzheimer's disease that express the mutated human amyloid precursor protein.[126] This may represent a glial cell compensatory response to beta-amyloid deposition in Alzheimer's disease.

ApoD is increased in the cerebrospinal fluid (CSF) and hippocampus of patients with Alzheimer disease.[127] This increase is correlated with the Braak stage of neurodegeneration.[128] Although independent of apoE protein concentrations, it is correlated with the apoE genotype depending on the brain region analysed.[127-129] The differential association with apoE genotype may be explained by disease progression. In early stages, an apoD increase in the presence of E4 allele could be a compensatory mechanism and be indicative of ongoing reinnervation rather than cell injury and death. However, this correlation is lost during disease progression.[123,129] The apoD increase correlates with the number of neurofibrillary tangles (NFT) but not with senile plaques.[128] Furthermore, whereas apoE is always located overlapping the amyloid core, apoD seems preferably around and near the amyloid.[130] Colocalization of apoD and NFT in the same neuron was rare and apoD transcription is impaired during NFT formation,[123] suggesting that apoD is present in stressed neurons before they possibly accumulate NFT.

ApoD levels are also elevated in the CSF of patients with stroke, meningoencephalitis, motor neuron disease, dementia,[127] chronic inflammatory demyelinating polyneuropathy, Guillain-Barre Syndrome, multiple sclerosis,[131] and cerebrotendinous xanthomatosis.[132] Increased apoD transcripts were identified in glaucoma[133] and in the nervous system of virus-infected mice.[134-136]

ApoD decreases significantly in serum samples from schizophrenic patients.[137] This supports recent hypotheses involving systemic insufficiencies in lipid metabolism /signaling in schizophrenia. Deficiencies in arachidonic acid, a primary apoD ligand, were reported in schizophrenic patients.[138,139] In contrast, apoD levels were significantly and selectively increased both in schizophrenic and bipolar subjects depending on the brain region analysed allowing the discrimination between the two disorders.[140] This apoD elevation in affected CNS regions is emphasized by neuroleptic drugs suggesting a focal compensatory response.[137] Indeed, elevated apoD levels were reported in post-mortem brains, as well as plasma, of schizophrenic patients and in rodent brains after chronic treatment with clozapine and other atypical antipsychotics.[141-143] This increase was even more significant in clozapine-treated chronic patients.[144]

The Niemann-Pick disease (NPC) is another human genetic disorder affecting lipid homeostasis that shows altered apoD expression. It is an inherited, lysosomal cholesterol disorder whose major phenotypic feature is a progressive neurodegeneration. The animal model of human NPC also shows abnormalities in cholesterol metabolism and increased apoD levels in plasma, brain, adipose tissue, heart, thymus and cultured astrocytes.[145,146] In particular, strongly apoD-immunolabeled cells are present in the brain regions previously shown to display the most significant neurodegenerative changes.[147] Thus, increased synthesis of apoD, a cholesterol transporter, could reflect an attempt to improve the cellular cholesterol trafficking.

ApoD is normally produced by astrocytes and oligodendrocytes.[117,148] Following an acute brain injury in rodents, its expression is upregulated in astrocytes and is also found in neurons. It is the case following kainic acid injection,[149,150] entorhinal cortex lesion,[151] traumatic brain injury[152] and chronic treatments with MK-801, a noncompetitive antagonist of the NMDA glutamate receptor.[153] However, apoD expression appears as a distinct event in glial and neuronal cells. Astrocyte activation is a general response after a brain lesion. The recovery of neural tissue after an injury is probably in part the result of an equilibrium between the negative effects of reactive astroglia on axon growth and the beneficial properties of the substances released by these reactive astroglia, such as sex steroids.[154] The latter could potentially be responsible for the local increase in apoD mRNA and protein by either a direct action on the apoD gene promoter or, indirectly, by causing a growth arrest of astrocytes that usually appears 3 to 4 days post-lesion. The production of apoD by neurons, however, could be a desperate attempt to retrieve essential survival factors such as steroids or growth factors. Alternatively, apoD expression could be devoted to the release of toxic molecules. ApoD may be part of the antioxidant defense system and act as a scavenger to remove heme-related molecules such as bilirubin.[108,155] ApoD clearly has an important function in the nervous system in both normal and pathological situations.

Conclusions about ApoD

In spite of the quantity of studies on apoD, little is known about its physiological functions. It seems possible that because of its multi-ligand properties, apoD acts through different pathways in each tissue or organ where its expression was reported. Its implication in cellular processes and pathologies highlights the importance of understanding the molecular mechanisms controlling its expression.

Apolipoprotein M, a Lipocalin with Unusual Phospholipid-Binding Properties

Apolipoprotein M (apoM) is a novel apolipoprotein that is predominantly present in high-density lipoproteins (HDL) and to a minor extent in chylomicrons, very low-density lipoproteins (VLDL), and low-density lipoproteins (LDL). An unusual feature of apoM is the retention of the signal peptide in the mature protein due to a missing signal peptidase cleavage site. This signal peptide binds to the phospholipid of the lipoproteins, thus serving as a hydrophobic anchor for apoM. The apoM gene is located in the major histocompatibility complex class III locus and contains six exons. The gene is expressed in liver and kidney and expression is driven by the transcription factor HNF-1α. The apoM protein contains 188 amino acid residues and is a member of the lipocalin superfamily. Molecular modelling suggests a typical lipocalin fold with one α -helix and eight β-strands forming a coffee filter-like shape. ApoM contains a predicted hydrophobic ligand-binding pocket and preliminary results suggest that it can bind retinol and retinoic acid. However, the physiological ligand of apoM is unknown and so is the function of apoM. ApoM can bind to megalin, a member of the LDL-receptor family, which interacts with many lipocalins, e.g., in tubular endothelium of kidneys. ApoM is present in the urine of megalin-deficient mice but not in the urine of normal mice, suggesting megalin-mediated uptake of apoM in the tubular epithelium of normal mice.

ApoM Associated with Plasma Lipoproteins

Apolipoprotein M (apoM) is a plasma protein predominantly associated with high-density lipoproteins (HDL).[156] To a minor extent, it is also present in triglyceride (TG) rich lipoproteins such as chylomicrons and very low-density lipoproteins (VLDL) and in low-density lipoproteins (LDL). The mature human apoM protein contains 188 amino acid residues, predicting a molecular weight of 21 kDa. The majority of apoM in human plasma contains an N-linked carbohydrate side chain. The exact plasma concentration of apoM is not known. Estimates from different laboratories range from 20 to 150 mg/l, which is much lower than the concentration of the dominating protein in HDL, apoAI (1-2 g/l, Mw 25 kDa).[156-158] Therefore, it can be concluded that apoM is only present in a subpopulation of HDL particles in plasma. HDL is important for reverse cholesterol transport, i.e., the transport of cholesterol from periferal cells to the liver. ApoAI has a key role in this reverse cholesterol transport, activating lecithin-cholesterol acyl transferas (LCAT) and interacting with ATP-binding cassette transporter A1 (ABCA1).[159-163] HDL particles may contain other apolipoproteins, e.g., apoAII, apoAIV, apoAV, apoCI-III, and apoE and the protein composition of HDL in plasma is highly heterogeneous.[164-168] The other apolipoproteins serve different functions, e.g., apoAV and apoCII stimulates lipoprotein lipase, and apoE interacts with lipoprotein receptors. The physiological significance of apoM being present in a minor fraction of the HDL particles is unknown.

An unusual property of apoM is that the signal peptide is retained in the mature circulating protein.[156] During synthesis of secreted proteins, the signal peptide directs the nascent peptide chain through the phospholipid of the endoplasmic reticulum (ER). In almost all proteins, the signal peptide is cleaved off by a signal peptidase that reacts with a conserved recognition site at the end of the signal peptide. In contrast, the signal peptidase cleavage site is lacking in apoM and therefore the signal peptide remains present in the mature protein. It is likely that the retained signal peptide serves as a hydrophobic anchor, which serves to localize apoM to the phospholipid layer of the lipoproteins. Another HDL-associated protein, paraoxonase-1 (PON-1), also retains the signal peptide, which anchors the protein to the lipoprotein phospholipid.[169] PON-1 is an important anti-oxidant component of a subpopulation of HDL. The phospholipid membranes of lipoproteins are single layer, which makes them distinct from cell membranes that comprise a double layer phospholipid. The molecular explanation for the preferential localization of the apoM and PON-1 signal peptides to the single layer lipoprotein phospholipid has not yet been elucidated.

The plasma distribution of apoM has been characterized in normal mice and in mice where different lipoprotein genes have been genetically modified.[158] In normal mice, the majority of apoM is recovered in HDL. In contrast, apoM is distributed in the TG-rich VLDL lipoproteins in apoE-deficient mice, particularly after high fat diet. In LDL-receptor deficient mice on the other hand, apoM is present in the LDL fraction in addition to the HDL particles. Taken together these results suggest that apoM is not only able to bind to HDL but can be transferred to other lipoprotein particles as well. In ApoAI-deficient mice, the concentration of apoM is decreased to 33% of normal, the apoM being localized in lipoprotein particles having the same size as normal HDL. It is well established that HDL-like particles are present in the plasma of apoAI deficient mice, other apolipoproteins playing the role of apoAI.[170] The results obtained with the apoAI-deficient mice suggest that the secretion or turnover of apoM to a certain extent depend on apoAI or other lipoproteins with similar functions. To date, apoM knockout mice or apoM transgenic mice have not been described.

Lipocalin Fold of ApoM

Sensitive sequence searches, threading and comparative model building have revealed that the region following the signal peptide in apoM (residues 20-188) is structurally related to the lipocalin protein family.[171] Like many other lipocalins, apoM has very limited sequence identity with other proteins of the lipocalin family. This explains that linear sequence comparisons performed with programs such as BLAST and FASTA were unsuccessful in identifying apoM as a lipocalin. Rather, the initial identification of apoM as a lipocalin was based on secondary structure predictions and protein motif searches. Many lipocalins share three conserved motifs and constitute the kernel group of lipocalins. ApoM can be classified into the more divergent group of lipocalins, the outlier lipocalins, having only two conserved sequence motifs. The three-dimensional (3D) structures of several lipocalins have been determined and the most closely related 3D structure to apoM was MUP, major urinary protein.[172] It is interesting to note that apoM and MUP only shares 19% linear sequence identity despite having similar lipocalin fold (Fig. 4A). MUP is predominately found in the urine of male rats and the hydrophobic pocket of MUP binds pheromones that affect the behaviour and sexual response of female rats.

The 3D model of apoM is characterized by an eight-stranded anti-parallel β-barrel.[171] The overall shape resembles a coffee filter holder and on the outside, an α -helix runs like a handle (Fig. 4B). Like other lipocalins, the β-barrel surrounds a hydrophobic pocket. The

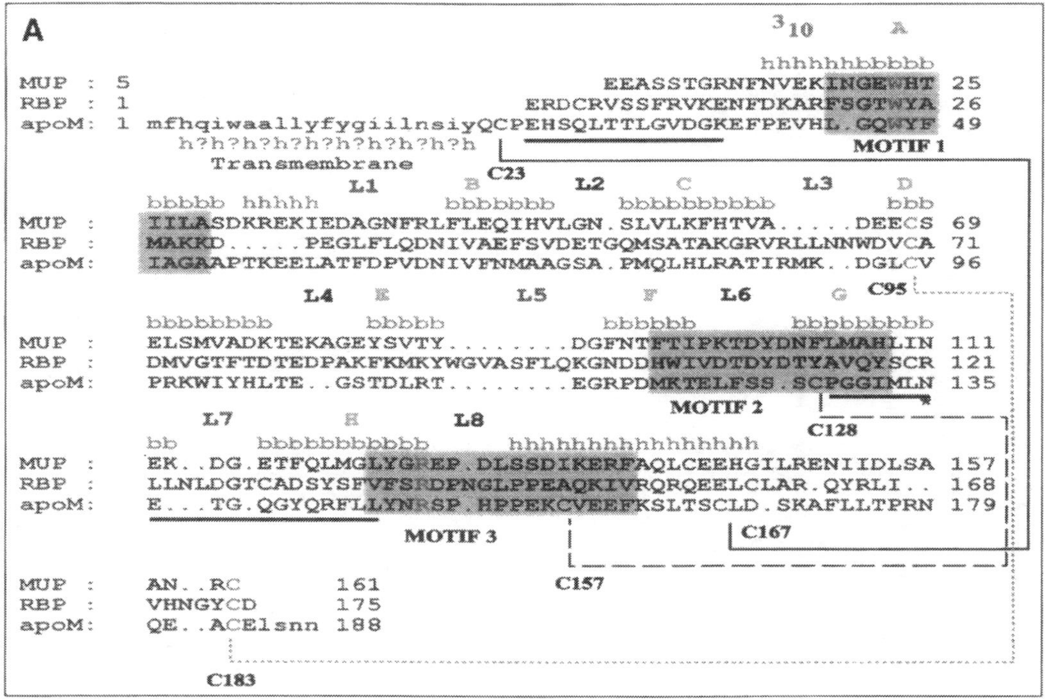

Figure 4. A) Sequence alignments of apoM. Sequences of MUP, RP and apoM were aligned even though the sequences were extremely divergent. The amino acid numbering follows that of the apoM sequence, whereas those of MUP and RBP were taken from the respective PDB files. The three conserved sequence motifs of kernel lipocalins are shadowed, only motifs 1 and 2 being present in apoM. The glycosylated Asn135 is marked with a star. The MUP structure was used as template to build the model with exception of the two underlined areas, which were built using RBP as template. The lower case letters represent the signal peptide, which was not part of the model. Dashed lines denote the three proposed disulphide bridges. Secondary structure elements labelled A-H are taken from the MUP structure, hhhh defining α-helices and bbb β-strands. A conservative contact in the lipocalins involves packing of Arg149 against TRP47 together with hydrogen bonds between the guanido group of Arg149 with the main chain carbonyl groups of the N-terminal 3_{10} helix. The loops are labelled L1-L8. Figure reprinted with permission from Duan et al.[171] Figure continued on next page.

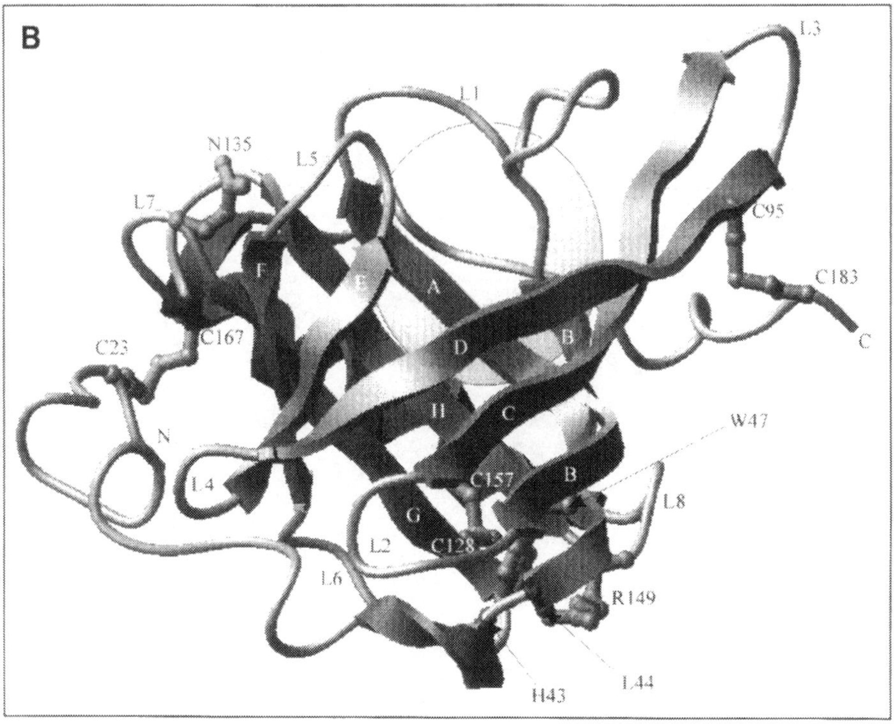

Figure 4, continued. B) Molecular model of apoM. Ribbon drawing of apoM, the loops and predicted secondary structure elements being labelled as in Figure 4A. Loops L1, L3, L5, and L7 are located near the opening of the binding pocket, the remaining loops being on the opposite end. The glycosylated Asn135 and the predicted disulphide bridges are indicated. The sphere highlights the proposed hydrophobic binding cavity. The figure is from Duan et al.[171]

physiological ligand binding to the hydrophobic pocket in apoM is unknown. However, we have preliminary results suggesting that recombinantly expressed mouse apoM can bind retinol and retinoic acid in a manner similar to that of another lipocalin, retinol-binding protein.[173] Whether the retinol binding to apoM has any physiological implications remains to be determined. Other potential ligands such a cholesterol, prostaglandins, fatty acids did no bind to apoM. Human apoM contains three cysteines and the model suggests a disulfide-bond pairing of Cys23-Cys167, Cys95-Cys183, and Cys128-Cys157.[171] In human apoM, Asn135 located near the opening of the hydrophobic pocket carries an N-linked carbohydrate. In contrast, mouse apoM is not glycosylated. Other structural features noticed during the evaluation of the 3D model were two electronegative regions located around the N-terminus and the opening of the binding pocket. In addition, an exposed hydrophobic patch at the surface of apoM that could potentially be involved in intermolecular interactions was observed.

Genomic Structure and Expression Profile of ApoM

In the human genome, the apoM gene is located on chromosome 6 position p21.33.[156] This is part of the major histocompatibility complex class III region, which is a chromosomal section very rich in genes. Close neighbours are HLA-B-associated transcripts BATS and BAT4 and an open reading frame called C6orf47 (G4 protein) (http://www.ncbi.nlm.nih.gov/entrez). These are genes of unknown function. Other genes located in the vicinity are the genes for TNF and lymphotoxin alpha and beta. In the mouse, the apoM gene is located in a corresponding HLA region on chromosome 17 position 17 B2. The rat chromosome 20 harbours the rat apoM gene (location 20p12). The human apoM gene contains 6 exons and spans a genomic region of approximately 2.3 kb and the apoM genes in other species are very similar in structure. By searching different public databases, it was observed that other mammals have identified apoM genes, e.g., the orangutan (*Pongo pygmaeus*), the chimpanzee (*Pan troglodytes*), the cow (*Bos taurus*), the pig (*Sus scrofa*), and the dog (*Canine familiaris*). ApoM genes appear to be present also in zebra fish (*Danio rerio*), the African clowed frog (*Xenopus laevis*), and the puffer fish (*Tetraodon nigroviridis*). This widespread presence in very distant species suggests a fundamental biological role of apoM. It is interesting to note that also apoAI and apoE have been identified in e.g., the Zebra fish, demonstrating the presence of a lipoprotein system in these animals.

ApoM gene expression is strongest in the liver followed by the kidney, whereas expression levels in other organs are very low.[156,158,174,175] In the liver, the hepatocytes express apoM and in the kidneys, the apoM is expressed in the tubular epithelial cells. During embryogenesis, apoM gene expression is detectable from days 7-10 in mice and in humans, apoM is expressed from the third month of gestation. The expression of apoM in both liver and kidney is dependent on stimulation by HNF-1α, a conserved HNF-1α binding site being present in the promotor region of the apoM gene.[157] In mice lacking HNF-1α, no apoM mRNA is detectable on Northern blotting and no apoM is expressed in the liver and kidney. In these mice, apoM is not detectable in the circulation. In mice with heterozygous HNF-1α deficiency, the apoM level in plasma is approximately 50% of the wild-type apoM level. In humans, HNF-1α gene mutations are associated with MODY3 (maturity-onset diabetes of the young). In patients with MODY3, the plasma levels of apoM were found to be decreased as compared to controls.[157]

Leptin has been suggested to upregulate expression of the apoM gene.[176] Leptin-deficient ***ob/ob*** mice and leptin-receptor deficient ***db/db*** mice demonstrate significantly reduced apoM expression and apoM concentration in plasma.[177] In the ***ob/ob*** mice, administration of leptin resulted in increased apoM expression. Similar leptin dependence of gene expression in ***ob/ob*** mice and normalization by leptin administration was reported for several of the genes involved in lipoprotein metabolism, e.g., apoAI, apoAII, apoAIV, and hepatic lipase.[176] Several in vitro expression studies using HepG2 cells have been reported. Platelet activating factor (PAF) was found to stimulate expression in these cells, whereas a PAF-receptor antagonist (Lexipafant) strongly inhibited apoM gene expression and apoM synthesis.[178] TNF-α and IL-1α had no effects and transforming growth factor-β decreased apoM synthesis.[179]

Function of ApoM

The physiological significance of apoM expression in the kidney is unknown. The expression is confined to the tubular epithelium but it is not known if the apoM is secreted into the urine or to plasma. No apoM is detected in normal human or mouse urine.[158] However, in mice specifically lacking megalin in the kidneys, apoM is found in the urine.[173] In these mice, apoAI is not present in the urine arguing against glomerular filtration of HDL-associated apoM. On gelfiltration chromatography of urine from the megalin-deficient animals, apoM elutes smaller than albumin, whereas apoM in plasma is recovered in the HDL fraction. In direct binding experiments using Biacore, megalin was found to bind apoM. Moreover, apoM was found to bind to cells expressing megalin and binding was followed by endocytosis. Megalin is a high molecular weight lipoprotein receptor-like protein present in the membrane of tubular epithelial cells on the side facing the urine. It is a multi-potent receptor for low molecular weight proteins being filtered in the glomeruli, many of the proteins belonging to the lipocalin group, e.g., retinol binding protein. Taken together the results suggest apoM to be synthesized in the tubular epithelium of the kidney, secreted to the urine and then under normal circumstances being reabsorbed by megalin. Whether urinary apoM binds a hydrophobic substance in its binding pocket before being reabsorbed remains to be determined.

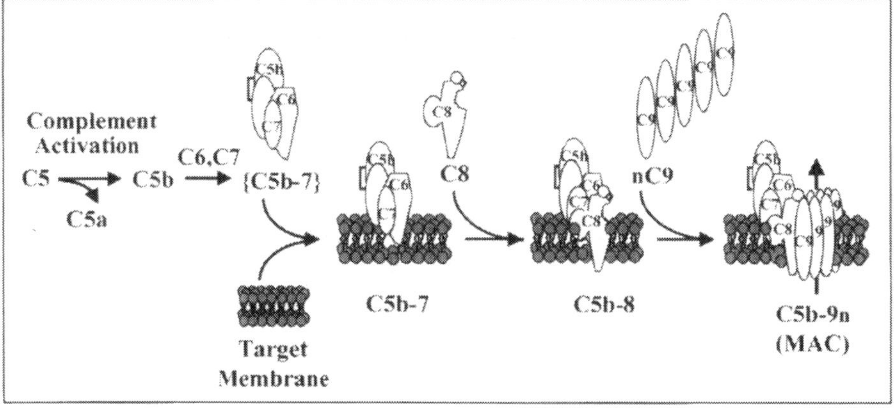

Figure 5. Formation of the membrane attack complex (MAC). MAC assembly is initiated by activation of the complement system and cleavage of C5 by the proteolytic enzyme C5 convertase. The product C5b then binds C6 and C7 to form C5b-7, a trimeric complex that expresses a transient, high-affinity lipid binding site. C5b-7 physically associates with the target cell membrane and subsequently binds C8 to form tetrameric C5b-8. C5b-8 functions as a C9 receptor by promoting the binding and self-polymerization of multiple C9 molecules to produce the MAC (C5b-9). Within this complex, poly C9 forms a cylindrical porelike structure that is typical of a lytically active MAC. In this depiction, C5b is shown as a disulfide-linked dimer. C6, C7 and C9 are single chain proteins, and C8 is an oligomer composed of three subunits. C8 and C9 are the principal components inserted into the bilayer.

Human Complement Protein C8γ

C8γ is a subunit of the eighth component of complement (C8), which is one of five components (C5b, C6, C7, C8, C9) that assemble on the surface of pathogenic organisms to form C5b-9, a cytolytically active macromolecular complex referred to as the membrane attack complex, or MAC (Fig. 5).[180] Individually, these components circulate as hydrophilic proteins but when combined they form an amphiphilic complex capable of inserting into cell membranes.[181,182] Upon insertion, lipids in the target membrane undergo a disruptive rearrangement. On gram-negative bacteria this rearrangement increases outer membrane permeability, which in turn induces lethal changes in the inner membrane.[181,183]

Human C8

C8 is an oligomeric protein composed of an α (64 kDa), β (64 kDa) and γ (22 kDA) subunit.[184,185] The subunits are synthesized independently from different genes and are secreted in the form of a disulfide-linked C8α–γ dimer that is noncovalently associated with C8β.[186] C8α and C8β are homologous to each other and to C6, C7 and C9. Together they form the "MAC family" of proteins whose common features include the presence of tandemly arranged modules at their N- and C-termini.[182,187] By contrast, C8γ is unique in that it is the only lipocalin among the approximately thirty-five proteins in the human complement system. Among the lipocalins it is one of only a few that form a covalent complex with another protein, i.e., C8α.

Studies using serum-derived and recombinant forms of C8α and C8β have identified specific roles for each in the formation and function of the MAC.[188-193] This is in contrast to C8γ, which is the most structurally well-characterized MAC protein but is one whose function is still unknown (reviewed in ref. 194). C8γ is not required for the synthesis and secretion of C8α; C8α can be expressed independently as a recombinant protein in mammalian and insect cells.[190,195] Binding between C8α–γ and C8β is likewise not dependent on C8γ; purified C8α and C8β can form a noncovalent 1:1 complex in the absence of C8γ.[195] C8γ is

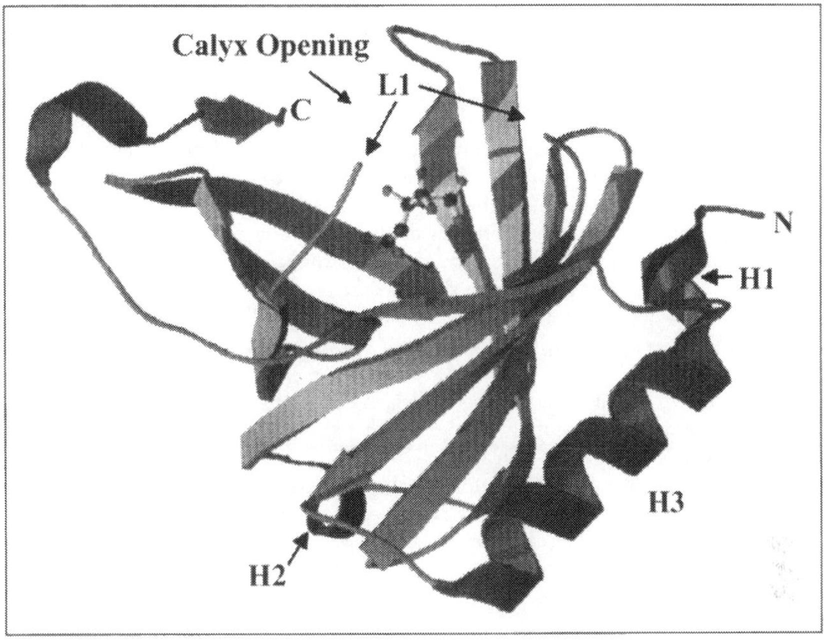

Figure 6. Ribbon diagram of C8γ. The opening of the calyx is at the top. The structure consists of an eight-stranded continuously hydrogen bonded β-barrel with an additional ninth and tenth β-strand completing the barrel. Included is a short 3_{10} helix at the N-terminus (H1), a second 3_{10} helix (H2) near the closed end of the barrel and an α-helix (H3) that flanks the calyx. Loop 1 (L1) is incomplete due to partial disorder. A citrate ion from the crystallization buffer is shown in the calyx opening.

also not essential for incorporation of C8 into the MAC nor is it required for expression of MAC cytolytic activity. Although not as efficient, a complex of C8α + C8β is an effective substitute for C8 in the MAC-mediated lysis of simple cells such as erythrocytes and in the killing of gram-negative bacteria.[195,196]

Structure of C8γ

Recombinant human C8γ containing a C40 → G40 substitution has been produced in insect cells, crystallized, and the structure determined to 1.2 Å resolution.[197] C8γ displays a typical lipocalin fold forming a calyx with a distinct binding pocket for a small molecule (Fig. 6). Loop 1 (residues 38-52), which spans the open end of the calyx and restricts access to the binding pocket in some lipocalins, is partially disordered. It contains C40, which forms the disulfide bond to C164 in C8α. During intracellular processing, C8α and C8γ must associate prior to disulfide bond formation, therefore each must contain a complementary binding site for the other. In C8α, the C8γ binding site resides on a 17-residue segment that includes C164.[191] A synthetic peptide containing this sequence was shown to specifically bind to C8γ in solution. C8γ crystals soaked with the peptide revealed additional electron density near loop 1, thus suggesting that one or more loops at the calyx opening are involved in binding C8α. It may be that during assembly of the MAC, conformational changes in C8α alter the position of loop 1 so as to allow access to the binding pocket.

Several positively charged residues surround the entrance to the calyx. In the initial crystal structure, well-resolved electron density corresponding to a citrate ion from the crystallization buffer was observed within hydrogen bonding distance from the ε-NH_3^+ group of K129 at the opening of the calyx.[197] The citrate ion is anchored by this lysine and stabilized through additional hydrogen bonds to the guanidinium group of R70 and three water molecules. This suggests the natural ligand for C8γ may contain an anionic moiety such as a carboxylate or phosphate group at one end.

C8γ differs from other lipocalins in that its ligand binding pocket is much deeper. Smaller side chains at the base of the pocket create a hydrophobic cavity at the bottom of the calyx with an opening of sufficient size to allow penetration of a single hydrocarbon chain (Fig. 7). Side chains on Y83 and Y131 appear to partially restrict access to the bottom of the calyx. The cavity itself is comparatively large (volume of ~100 $Å^3$). When crystals were exposed to Xe gas to produce heavy atom derivatives, strong density corresponding to two Xe atoms, and a third weaker peak, was observed within the cavity.[197] The relatively large Xe atom has been used as a probe to study cavities and hydrophobic sites in other proteins.[198] Such regions are usually devoid of ordered water and normally accommodate a single Xe atom. In C8γ, the deepest portion of the C8γ calyx is hydrophobic and devoid of ordered water, but it is much larger than a typical protein cavity.

The structure of C8γ suggests its natural ligand has a long and narrow hydrophobic tail with a negatively charged moiety at one end. This and the fact C8γ associates with cell membranes as part of the MAC raised the possibility that it may bind the acyl chain of a phospholipid or similar fatty acid-like molecule. To investigate this, crystals were soaked with a mixture of saturated fatty acids (C12:0 to C20:0) and analyzed by X-ray diffraction (L. Lebioda and J.M. Sodetz, unpublished results). A C8γ-laurate complex was observed in which the alkyl chain penetrates into the hydrophobic cavity and the negative carboxylate group interacts with positive charges at the calyx entrance (Fig. 8). As predicted by modeling and shown by Xe binding, the hydrophobic cavity can accommodate a larger moiety than the alkyl chain of laurate. In these experiments, laurate may have bound preferentially over longer chain fatty acids due to solubility differences, or other variables that were not controlled. Experiments are currently focused on individual fatty acids of different lengths as well as glycerophospholipids as potential ligands. Also significant is the movement of Y83 and Y131 in complexes with laurate or Xe. Movement of these residues may regulate access to the lower cavity.

Function of C8γ

One can still only speculate on the function of C8γ and the identity of its natural ligand. At one time it was reported that C8α–γ binds radiolabeled retinoids and C8γ may be a retinol-transport protein.[199] Later studies used spectrophotometric methods to detect retinol binding to recombinant human C8γ and could not corroborate these findings.[195] When C8γ was first identified as a lipocalin based on sequence similarity, it was thought the putative binding pocket might recognize and bind to a small hydrophobic region on C8α. Such binding would serve to "dock" C8α and C8γ prior to intracellular formation of C8α–γ. The C8γ crystal structure now suggests this is unlikely. The size and features of the binding pocket are not conducive to binding multiple or even single amino acid side chains as ligands.

If the ligand is indeed a fatty acid-like molecule, one potential source is the target membrane on which the MAC is assembled. Exposure of membrane lipid, i.e., phospholipid or lipid A on gram-negative bacteria, may render fatty acid side chains accessible to C8γ. Such a binding function was not considered previously because membrane photolabeling experiments indicated no direct interaction between C8γ and the membrane bilayer.[200] C8γ is not absolutely required for MAC-mediated lysis of membranes but it does enhance this activity. A complex of C8α + C8β is effective, however the activity of this complex is lower compared to C8. Interestingly, the addition of purified C8γ to a C8α + C8β complex increases both hemolytic and bactericidal activities to near normal levels.[195,196] Thus, C8γ functions in some manner to increase MAC activity. During MAC assembly it may bind to C8α and either increase C8 affinity for C5b-7 or C5b-8 affinity for C9. Another possibility to now be considered is whether C8γ participates in the cytolytic mechanism by binding directly to membrane phospholipid.

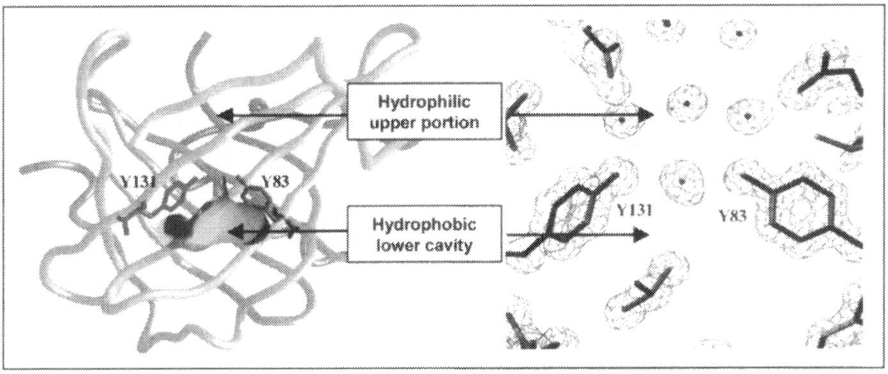

Figure 7. Side view of the C8γ calyx. Left: GRASP rendering of the hydrophobic lower cavity colored according to the polarity of residues lining the cavity (nonpolar = yellow, polar = blue). The surface is predominantly hydrophobic with some hydrophilic character present where the surface contacts the hydroxyls of Y83 and Y131. The cavity has a flat bottom that extends horizontally to fill the space created by the hydrophobic residues. Right: Cross-sectional side view of the middle portion of the C8γ binding pocket. Y83 and Y131 appear to form a gate to the hydrophobic lower cavity. Above the gate and out of view is the positively charged hydrophilic entrance to the calyx. The tyrosine hydroxyls are hydrogen-bonded to four water molecules to form a hexagonal arrangement that separates the upper hydrophilic and lower hydrophobic portions of the binding pocket.

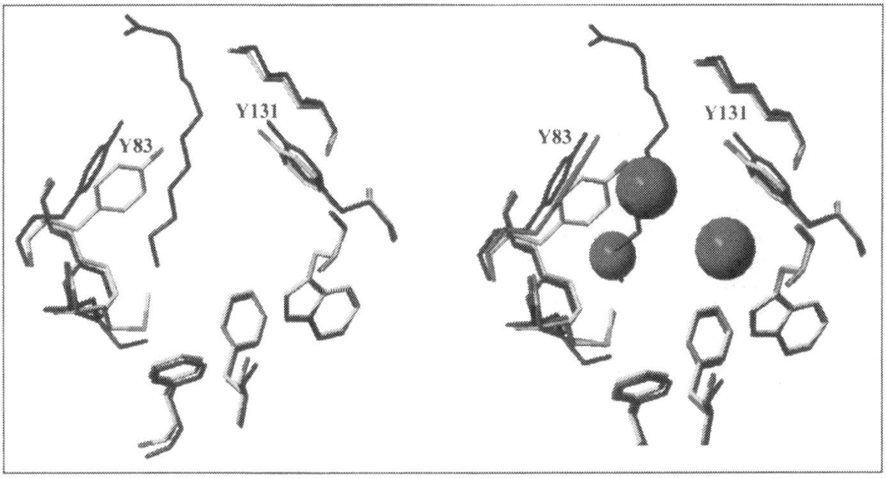

Figure 8. Model of the C8γ-laurate complex. Left: Superposition of the C8γ-laurate structure (blue) on native C8γ (standard atomic colors). The alkyl chain of laurate extends into the lower cavity. Not shown are several positively charged residues located in the upper portion of the binding pocket and in close proximity to the carboxyl group. In this rendering, Y83 and Y131 are on opposite sides from the rendering in (Fig. 7). Both side chains shift upon binding with Y83 undergoing the greatest movement. Right: Superposition of the C8γ-laurate (blue), C8γ-Xe (red), and native C8γ structures. The lower cavity is only partially occupied by laurate while Xe occupies a larger portion of the available volume. Distances between the phenolic oxygens of Y83 and Y131 differ in each structure. In native C8γ this distance is 5.1 Å, while in the Xe derivatized and laurate structures these distances are 6.6 Å and 7.4 Å, respectively. It is clear that Y83 and Y131 are capable of moving to allow for ligand penetration into the lower cavity.

If the natural ligand is not membrane-associated but instead is a small molecule than one must consider the source of such a molecule. The MAC is assembled on the outer membrane of gram-negative bacteria; thus C8γ is exposed in vivo to small inflammatory mediators, e.g., prostaglandins, eicosanoids, etc. C8γ could in theory regulate the immune response by binding and neutralizing such molecules at the site of MAC deposition. Features of the binding pocket suggest this is unlikely; the upper portion appears too shallow and the size and polarity of such molecules are not compatible with the lower cavity.

Also to be considered is the possibility that C8γ binds microbial siderophores. Siderophores are iron chelators synthesized by microbes and are essential for growth under iron-limiting conditions.[201] Human siderocalin (neutrophil gelatinase-associated lipocalin or NGAL, lipocalin 2, Lcn2) and tear lipocalin (Lcn 1) exhibit antimicrobial activity because of their ability to bind siderophores and interfere with iron acquisition.[202-205] Crystallographic studies of siderocalin-siderophore complexes identified a triad of positively-charged residues at the calyx opening as being critical for binding.[203,206] This triad is conserved in C8γ, however the shape of the C8γ binding pocket is not compatible with any known siderocalin. Indeed, studies using a panel of siderophores including ones that bind to siderocalin failed to convincingly demonstrate specific binding to C8γ.[206] Whether C8γ binds other siderophores cannot be excluded. Importantly, C8γ differs from siderocalin in that it has a lower cavity that can accommodate a ligand of considerable length. Whether this distinction has functional significance remains to be determined.

C8γ is a unique complement protein because of its potential to bind a small ligand. While many complement components can undergo protein-protein interactions, none are specifically designed to bind small molecules. C8γ is capable of doing both and this dual function suggests it has an as yet unrecognized role in the formation and/or function of the MAC.

References

1. Fournier T, Medjoubi N, Porquet D. Alpha-1-acid glycoprotein. Biochim Biophys Acta 2000; 1482:157-71.
2. Halsall HB, Austin RC, Dage H et al. Structural aspects of alpha-1-acid glycoprotein and its interactions. In: Otagari M, Sugiyama Y, Testa B, Tillement J-P, eds. Proc. Intern. Symp. Serum Albumin and Alpha-1-acid Glycoprotein: From Basic Science to Clinical Applications. Kumamato, Japan: 2001.
3. Schmid K, Nimberg RB, Kimura A et al. The carbohydrate units of human plasma alpha1-acid glycoprotein. Biochim Biophys Acta 1977; 492:291-302.
4. Schmid K, Burgi W, Collins JH et al. The disulfide bonds of alpha1-acid glycoprotein. Biochemistry 1974; 13:2694-7.
5. Iijima S, Shiba K, Kimura M et al. Changes of alpha1-acid glycoprotein microheterogeneity in acute inflammation stages analyzed by isoelectric focusing using serum obtained postoperatively. Electrophoresis 2000; 21:753-9.
6. Eap CB, Cuendet C, Baumann P. Orosomucoid (alpha1-acid glycoprotein) phenotyping by use of immobilized pH gradients with 8 M urea and immunoblotting. A new variant encountered in immunoblotting. Hum Genet 1988; 80:183-5.
7. Dente L, Pizza MG, Metspalu A et al. Structure and expression of the genes coding for human alpha1-acid glycoprotein. EMBO J 1987; 6:2289-96.
8. Nakamura H, Yuasa I, Umetsu K et al. The rearrangement of the human alpha(1)-acid glycoprotein/orosomucoid gene: Evidence for tandemly triplicated genes consisting of two AGP1 and one AGP2. Biochem Biophys Res Commun 2000; 276:779-84.
9. van Dijk W, Pos O, Van der Stelt ME et al. Inflammation-induced changes in expression and glycosylation of genetic variants of alpha1-acid glycoprotein. Studies with human sera, primary cultures of human hepatocytes and transgenic mice. Biochem J 1991; 276:343-7.
10. Eap CB, Fischer J-F, Baumann P. Variations in relative concetrations of variants of human alpha1-acid glycoprotein after acute-phase conditions. Clin Chim Acta 1991; 203:379-86.
11. Duche JC, Urien S, Simon N et al. Expression of the genetic variants of human alpha-1-acid glycorprotein in cancer. Clin Biochem 2000; 33:197-202.
12. Toh H, Hayashida H, Kikuno R et al. Sequence similarity between EGF receptor and alpha-1-acid glycoprotein. Nature 1985; 314:199.

13. Pervaiz S, Brew K. Homology and structurefuction correlations between alpha1-acid glycoprotein and serum retinol-binding protein and its relatives. FASEB J 1987; 1:209-14.
14. McPherson A, Friedman ML, Halsall HB. Crystallization of alpha 1-acid glycoprotein. Biochem Biophys Res Commun 1984; 124:619-24.
15. Kopecky Jr V, Ettrich R, Hofbauerova K et al. Structure of human alpha1-acid glycoprotein and its high-affinity binding site. Biochem Biophys Res Commun 2003; 300:41-6.
16. Kremer JMH, Wilting J, Janssen LHM. Drug binding to human alpha1-acid glycoprotein in health and disease. Pharmacol Rev 1988; 40:1-46.
17. Israili ZH, Dayton PG. Human alpha-1-glycoprotein and its interactions with drugs. Drug Metab Rev 2001; 33:161-235.
18. Herve F, Caron G, Duche JC et al. Ligand specificity of the genetic variants of human alpha1-acid glycoprotein: Generation of a three-dimensional quantitative structureactivity relationship model for drug binding to the A variant. Mol Pharmacol 1998; 54:129-38.
19. Fitos I, Visy J, Zsila F et al. Specific ligand binding on genetic variants of human alpha1-acid glycoprotein studied by circular dichroism spectroscopy. Biochem Pharmacol 2004; 67:679-88.
20. Kuroda Y, Matsumoto S, Shibukawa A et al. Capillary electrophoretic study on pH dependence of enantioselective disopyramide binding to genetic variants of human alpha1-acid glycoprotein. Analyst 2003; 128:1023-7.
21. Kuroda Y, Shibukawa A, Nakagawa T. The role of branching glycan of human alpha1-acid glycoprotein in enantioselective binding to basic drugs as studied by capillary electrophoresis. Anal Biochem 1999; 268:9-14.
22. Nieto E, Vieta E, Alvarez L et al. Alpha1-acid glycoprotein in major depressive disorder. Relationships to severity, response to treatment and imipramine plasma levels. J Affec Disord 2000; 59:159-64.
23. Nemeroff CB, Krishnan KR, Blazer DG et al. Elevated plasma concentrations of alpha 1-acid glycoprotein, a putative endogenous inhibitor of the tritiated imipramine binding site, in depressed patients. Arch Gen Psychiatry 1990; 47:337-40.
24. Sager G, Little C. The effect of the plasticizers TBEP (Tris-(2-butoxyethyl)-phosphate) and DEHP (di-(2-ethylhexyl)phthalate) on beta-adrenergic ligand binding to alpha1acid glycoprotein and mononuclear leukocytes. Biochem Pharmacol 1989; 38:2551-7.
25. Nishi K, Sakai N, Komine Y et al. Structural and drug-binding properties of alpha(1)-acid glycoprotein in reverse micelles. Biochim Biophys Acta 2002; 1601:185-91.
26. Nishi K, Maruyama T, Halsall HB et al. Binding of alpha1-acid glycoprotein to membrane results in a unique structural change and ligand release. Biochemistry 2004; 43:10513-9.
27. Gasymov OK, Abduragimov AR, Gasimov EO et al. Tear lipocalin: Potential for selective delivery of rifampin. Biochim Biophys Acta 2004; 1688:102-11.
28. Halsall HB, Villalobos AP, Ivancic JS et al. Monoclonal antibodies against human orosomucoid: Tools for the exploration of structure, function and interactions. In: Baumann P, Eap CB, Testa B, Tillement J-P, eds. Alpha-1-Acid Glycoprotein: Genetics, Biochemistry, Physiological Functions and Pharmacology. New York: Liss, 2002:67-84.
29. Yoshima H, Matsumoto A, Mizuochi T et al. Comparative study of the carbohydrate moieties of rat and human plasma alpha1-acid glycoproteins. J Biol Chem 1981; 256:8476-84.
30. Mackiewicz A, Mackiewicz K. Glycoforms of serum alpha1-acid glycoprotein as markers for disease. Glycoconjugate J 1995; 12:241-7.
31. Mackiewicz A, Marcinkowska-Pieta R, Ballou S et al. Microheterogeneity of alpha1-acid glycoprotein in the detection of intercurrent infection in systemic lupus erythemathosis. Arthritis Rheumatism 1987; 30:513-8.
32. De Graaf TW, Van der Stelt ME, Anbergen MG et al. Inflammation-induced expression of sialyl Lewis X-containing glycan structures on alpha 1-acid glycoprotein (orosomucoid) in human sera. J Exp Med 1993; 177:657-66.
33. Schalkwijk CG, Poland DC, van Dijk W et al. Plasma concentration of C-reactive protein is increased in type I diabetic patients without clinical macroangiopathy and correlates with markers of endothelial dysfunction: Evidence for chronic inflammation. Diabetologia 1999; 42:351-7.
34. Lowe JB. Glycan-dependent leukocyte adhesion and recruitment in inflammation. Curr Opin Cell Biol 2003; 15:531-8.
35. Brinkman-Van der Linden CM, Havenaar EC, Van Ommen CR et al. Oral estrogen treatment induces a decrease in expression of sialyl Lewis x on alpha 1-acid glycoprotein in females and male-to-female transsexuals. Glycobiology 1996; 6:407-12.
36. Havenaar EC, Axford JS, Brinkman-van der Linden EC et al. Severe rheumatoid arthritis prohibits the pregnancy-induced decrease in alpha3-fucosylation of alpha1-acid glycoprotein. Glycoconj J 1998; 15:723-9.

37. Hashimoto S, Asao T, Takahashi J et al. alpha1-acid glycoprotein fucosylation as a marker of carcinoma progression and prognosis. Cancer 2004; 101:2825-36.
38. Poland DC, Kulik W, van Dijk W et al. Distinct glycoforms of human alpha(1)-acid glycoprotein have comparable synthesis rates: A [(13)C]valine-labelling study in healthy humans. Glycoconj J 2003; 20:99-105.
39. Hochepied T, Berger FG, Baumann H et al. alpha(1)-Acid glycoprotein: An acute phase protein with inflammatory and immunomodulating properties. Cytokine Growth Factor Rev 2003; 14:25-34.
40. Williams JP, Weiser MR, Pechet TTV et al. alpha1-Acid glycoprotein reduces local and remote injuries after intestinal ischemia in the rat. Am J Physiol 1997; 273:G1031-G1035, (Gastrointest. Liver Physiol 36).
41. Poland DC, Vallejo JJ, Niessen HW et al. Activated human PMN synthesize and release a strongly fucosylated glycoform of {alpha}1-acid glycoprotein, which is transiently deposited in human myocardial infarction. J Leukoc Biol 2005; 78(2):453-61.
42. Poland DCW. Neutrophills synthesize and can release a specifically fucosylated glycoform of alpha-1-acid glycoprotein which is a very potent inhibitor of the classical route of complement. Tissue specific fucosylation of alpha-1-acid glycoprotein and its potency to inhibit the classical route of complement. Amsterdam: Ph.D thesis, Vrije Universiteit, 2002:71-88.
43. Pos O, Oostendorp RA, Van der Stelt ME et al. Con A-nonreactive human alpha 1-acid glycoprotein (AGP) is more effective in modulation of lymphocyte proliferation than Con A-reactive AGP serum variants. Inflammation 1990; 14:133-41.
44. Shiyan SD, Bovin NV. Carbohydrate composition and immunomodulatory activity of different glycoforms of alpha1-acid glycoprotein. Glycoconj J 1997; 14:631-8.
45. Bories PN, Guenounou M, Feger J et al. Human alpha 1-acid glycoprotein-exposed macrophages release interleukin 1 inhibitory activity. Biochem Biophys Res Comm 1987; 147:710-5.
46. Costello M, Fiedel BA, Gewurz H. Inhibition of platelet aggregation by native and desialised alpha1-acid glycoprotein. Nature 1979; 281:677-8.
47. Libert C, Brouckaert P, Fiers W. Protection by alpha1-acid glycoprotein against tumor necrosis factor-induced lethality. J Exp Med 1994; 180:1571-5.
48. Muchitsch EM, Teschner W, Linnau Y et al. In vivo effect of alpha 1-acid glycoprotein on experimentally enhanced capillary permeability in guinea-pig skin. Arch Int Pharmacodyn Ther 1996; 331:313-21.
49. Schnitzer JE, Pinney E. Quantitation of specific binding of orosomucoid to cultured microvascular endothelium: Role in capillary permeability. Am J Physiol 1992; 263:H48-H55.
50. Johnsson E, Haraldsson B. Addition of purified orosomucoid preserves the glomerular permeability for albumin in isolated perfused rat kidneys. Acta Physiol Scand 1993; 147:1-8.
51. Huxley VH, Curry FE. Differential actions of albumin and plasma on capillary solute permeability. Am J Physiol 1991; 260:H1645-H1654.
52. Liu HM, Lei HY, Schmid K. Alpha1-acid glycoprotein and peripheral nerve injury, studied with a wound chamber. Lab Invest 1993; 68:577-83.
53. Salomatin VV, Liutov AG, Eneekiva SA et al. Effect of alpha1-acid glycoprotein on chemiluminescenc and lipid peroxidation in experimental thermal trauma. Voprosy Med Khimii 1993; 39:24-6.
54. Franzblau C, Schmid K, Faris B et al. The interaction of collagen with alpha1-acid glycoprotein. Biochim Biophys Acta 1976; 427:302-14.
55. Rassart E, Bedirian A, Do Carmo S et al. Apolipoprotein D. Biochim Biophys Acta 2000; 1482(1-2):185-98.
56. Simard J, Dauvois S, Haagensen DE et al. Regulation of progesterone-binding breast cyst protein GCDFP-24 secretion by estrogens and androgens in human breast cancer cells: A new marker of steroid action in breast cancer. Endocrinology 1990; 126(6):3223-31.
57. Provost PR, Marcel YL, Milne RW et al. Apolipoprotein D transcription occurs specifically in nonproliferating quiescent and senescent fibroblast cultures. FEBS Lett 1991; 290(1-2):139-41.
58. Simard J, Veilleux R, de Launoit Y et al. Stimulation of apolipoprotein D secretion by steroids coincides with inhibition of cell proliferation in human LNCaP prostate cancer cells. Cancer Res 1991; 51(16):4336-41.
59. Simard J, de Launoit Y, Haagensen DE et al. Additive stimulatory action of glucocorticoids and androgens on basal and estrogen-repressed apolipoprotein-D messenger ribonucleic acid levels and secretion in human breast cancer cells. Endocrinology 1992; 130(3):1115-21.
60. Blais Y, Sugimoto K, Carriere MC et al. Potent stimulatory effect of interleukin-1 alpha on apolipoprotein D and gross cystic disease fluid protein-15 expression in human breast-cancer cells. Int J Cancer 1994; 59(3):400-7.

61. Blais Y, Sugimoto K, Carriere MC et al. Interleukin-6 inhibits the potent stimulatory action of androgens, glucocorticoids and interleukin-1 alpha on apolipoprotein D and GCDFP-15 expression in human breast cancer cells. Int J Cancer 1995; 62(6):732-7.
62. Do Carmo S, Seguin D, Milne R et al. Modulation of apolipoprotein D and apolipoprotein E mRNA expression by growth arrest and identification of key elements in the promoter. J Biol Chem 2002; 277(7):5514-23.
63. Sarjeant JM, Lawrie A, Kinnear C et al. Apolipoprotein D inhibits platelet-derived growth factor-BB-induced vascular smooth muscle cell proliferated by preventing translocation of phosphorylated extracellular signal regulated kinase 1/2 to the nucleus. Arterioscler Thromb Vasc Biol 2003; 23(12):2172-7.
64. Lopez-Boado YS, Tolivia J, Lopez-Otin C. Apolipoprotein D gene induction by retinoic acid is concomitant with growth arrest and cell differentiation in human breast cancer cells. J Biol Chem 1994; 269(43):26871-8.
65. Lopez-Boado YS, Puente XS, Alvarez S et al. Growth inhibition of human breast cancer cells by 1,25-dihydroxyvitamin D3 is accompanied by induction of apolipoprotein D expression. Cancer Res 1997; 57(18):4091-7.
66. Lopez-Boado YS, Klaus M, Dawson MI et al. Retinoic acid-induced expression of apolipoprotein D and concomitant growth arrest in human breast cancer cells are mediated through a retinoic acid receptor RARalpha-dependent signaling pathway. J Biol Chem 1996; 271(50):32105-11.
67. Aspinall JO, Bentel JM, Horsfall DJ et al. Differential expression of apolipoprotein-D and prostate specific antigen in benign and malignant prostate tissues. J Urol 1995; 154(2 Pt 1):622-8.
68. Alaupovic P, Schaefer EJ, McConathy WJ et al. Plasma apolipoprotein concentrations in familial apolipoprotein A-I and A-II deficiency (Tangier disease). Metabolism 1981; 30(8):805-9.
69. Albers JJ, Adolphson J, Chen CH et al. Defective enzyme causes lecithin-cholesterol acyltransferase deficiency in a Japanese kindred. Biochim Biophys Acta 1985; 835(2):253-7.
70. Deeb SS, Cheung MC, Peng RL et al. A mutation in the human apolipoprotein A-I gene. Dominant effect on the level and characteristics of plasma high density lipoproteins. J Biol Chem 1991; 266(21):13654-60.
71. Baker WA, Hitman GA, Hawrami K et al. Apolipoprotein D gene polymorphism: A new genetic marker for type 2 diabetic subjects in Nauru and south India. Diabet Med 1994; 11(10):947-52.
72. Vijayaraghavan S, Hitman GA, Kopelman PG. Apolipoprotein-D polymorphism: A genetic marker for obesity and hyperinsulinemia. J Clin Endocrinol Metab 1994; 79(2):568-70.
73. Hansen L, Gaster M, Oakeley EJ et al. Expression profiling of insulin action in human myotubes: Induction of inflammatory and pro-angiogenic pathways in relationship with glycogen synthesis and type 2 diabetes. Biochem Biophys Res Commun 2004; 323(2):685-95.
74. Lewohl JM, Wang L, Miles MF et al. Gene expression in human alcoholism: Microarray analysis of frontal cortex. Alcohol Clin Exp Res 2000; 24(12):1873-82.
75. Saito M, Smiley J, Toth R et al. Microarray analysis of gene expression in rat hippocampus after chronic ethanol treatment. Neurochem Res 2002; 27(10):1221-9.
76. Gottsch JD, Bowers AL, Margulies EH et al. Serial analysis of gene expression in the corneal endothelium of Fuchs' dystrophy. Invest Ophthalmol Vis Sci 2003; 44(2):594-9.
77. Dieplinger H, Schoenfeld PY, Fielding CJ. Plasma cholesterol metabolism in end-stage renal disease. Difference between treatment by hemodialysis or peritoneal dialysis. J Clin Invest 1986; 77(4):1071-83.
78. DeWan AT, Arnett DK, Atwood LD et al. A genome scan for renal function among hypertensives: The HyperGEN study. Am J Hum Genet 2001; 68(1):136-144.
79. James RW, Martin B, Pometta D et al. Apoprotein D in a healthy, male population and in male myocardial infarction patients and their male, first-degree relatives. Atherosclerosis 1986; 60(1):49-53.
80. Lin CS, Ho HC, Gholami S et al. Gene expression profiling of an arteriogenic impotence model. Biochem Biophys Res Commun 2001; 285(2):565-9.
81. Hummasti S, Laffitte BA, Watson MA et al. Liver X receptors are regulators of adipocyte gene expression but not differentiation: Identification of apoD as a direct target. J Lipid Res 2004; 45(4):616-25.
82. Liu Z, Chang GQ, Leibowitz SF. Apolipoprotein D interacts with the long-form leptin receptor: A hypothalamic function in the control of energy homeostasis. FASEB J 2001; 15(7):1329-31.
83. Provost PR, Tremblay Y, el-Amine M et al. Guinea pig apolipoprotein D RNA diversity, and developmental and gestational modulation of mRNA levels. Mol Cell Endocrinol 1995; 109(2):225-36.
84. McConathy WJ, Lane DM. Studies on the apolipoproteins and lipoproteins of cord serum. Pediatr Res 1980; 14(5):757-61.

85. Sanchez D, Ganfornina MD, Martinez S. Expression pattern of the lipocalin apolipoprotein D during mouse embryogenesis. Mech Dev 2002; 110(1-2):225-9.
86. Ong WY, Lau CP, Leong SK et al. Apolipoprotein D gene expression in the rat brain and light and electron microscopic immunocytochemistry of apolipoprotein D expression in the cerebellum of neonatal, immature and adult rats. Neuroscience 1999; 90(3):913-22.
87. Vieira AV, Lindstedt K, Schneider WJ et al. Identification of a circulatory and oocytic avian apolipoprotein D. Mol Reprod Dev 1995; 42(4):443-6.
88. Yao Y, Vieira A. Comparative 17beta-estradiol response and lipoprotein interactions of an avian apolipoprotein. Gen Comp Endocrinol 2002; 127(1):89-93.
89. Ganfornina MD, Sanchez D, Pagano A et al. Molecular characterization and developmental expression pattern of the chicken apolipoprotein D gene: Implications for the evolution of vertebrate lipocalins. Dev Dyn 2005; 232(1):191-9.
90. Baris O, Savagner F, Nasser V et al. Transcriptional profiling reveals coordinated up-regulation of oxidative metabolism genes in thyroid oncocytic tumors. J Clin Endocrinol Metab 2004; 89(2):994-1005.
91. Yamashita K, Upadhyay S, Osada M et al. Pharmacologic unmasking of epigenetically silenced tumor suppressor genes in esophageal squamous cell carcinoma. Cancer Cell 2002; 2(6):485-95.
92. Diez-Itza I, Vizoso F, Merino AM et al. Expression and prognostic significance of apolipoprotein D in breast cancer. Am J Pathol 1994; 144(2):310-20.
93. Serra Diaz C, Vizoso F, Lamelas ML et al. Expression and clinical significance of apolipoprotein D in male breast cancer and gynaecomastia. Br J Surg 1999; 86(9):1190-7.
94. Lamelas ML, Vazquez J, Enguita MI et al. Apolipoprotein D expression in metastasic lymph nodes of breast cancer. Int J Surg Investig 2000; 2(4):285-93.
95. Hunter S, Young A, Olson J et al. Differential expression between pilocytic and anaplastic astrocytomas: Identification of apolipoprotein D as a marker for low-grade, noninfiltrating primary CNS neoplasms. J Neuropathol Exp Neurol 2002; 61(3):275-81.
96. Porter D, Lahti-Domenici J, Keshaviah A et al. Molecular markers in ductal carcinoma in situ of the breast. Mol Cancer Res 2003; 1(5):362-75.
97. Zhang SX, Bentel JM, Ricciardelli C et al. Immunolocalization of apolipoprotein D, androgen receptor and prostate specific antigen in early stage prostate cancers. J Urol 1998; 159(2):548-54.
98. Ashida S, Nakagawa H, Katagiri T et al. Molecular features of the transition from prostatic intraepithelial neoplasia (PIN) to prostate cancer: Genome-wide gene-expression profiles of prostate cancers and PINs. Cancer Res 2004; 64(17):5963-72.
99. Hall RE, Horsfall DJ, Stahl J et al. Apolipoprotein-D: A novel cellular marker for HGPIN and prostate cancer. Prostate 2004; 58(2):103-8.
100. Miranda E, Vizoso F, Martin A et al. Apolipoprotein D expression in cutaneous malignant melanoma. J Surg Oncol 2003; 83(2):99-105.
101. West RB, Harvell J, Linn SC et al. Apo D in soft tissue tumors: A novel marker for dermatofibrosarcoma protuberans. Am J Surg Pathol 2004; 28(8):1063-9.
102. Iacobuzio-Donahue CA, Maitra A, Shen-Ong GL et al. Discovery of novel tumor markers of pancreatic cancer using global gene expression technology. Am J Pathol 2002; 160(4):1239-49.
103. Ryu B, Jones J, Hollingsworth MA et al. Invasion-specific genes in malignancy: Serial analysis of gene expression comparisons of primary and passaged cancers. Cancer Res 2001; 61(5):1833-8.
104. Kesner L, Yu WS, Bradlow HL et al. Proteases in cyst fluid from human gross cyst breast disease. Cancer Res 1988; 48(22):6379-83.
105. Leung WC, Lawrie A, Demaries S et al. Apolipoprotein D and platelet-derived growth factor-BB synergism mediates vascular smooth muscle cell migration. Circ Res 2004; 95(2):179-86.
106. Vazquez J, Gonzalez L, Merino A et al. Expression and clinical significance of apolipoprotein D in epithelial ovarian carcinomas. Gynecol Oncol 2000; 76(3):340-7.
107. Rojo JV, Gonzalez LO, Lamelas ML et al. Apolipoprotein D expression in endometrial carcinomas. Acta Obstet Gynecol Scand 2001; 80(2):158-61.
108. Alvarez ML, Barbon JJ, Gonzalez LO et al. Apolipoprotein D expression in retinoblastoma. Ophthalmic Res 2003; 35(2):111-6.
109. Rodriguez JC, Diaz M, Gonzalez LO et al. Apolipoprotein D expression in benign and malignant prostate tissues. Int J Surg Investig 2000; 2(4):319-26.
110. Selim AA, El-Ayat G, Wells CA. Immunohistochemical localization of gross cystic disease fluid protein-15, -24 and -44 in ductal carcinoma in situ of the breast: Relationship to the degree of differentiation. Histopathology 2001; 39(2):198-202.
111. Hall RE, Aspinall JO, Horsfall DJ et al. Expression of the androgen receptor and an androgen-responsive protein, apolipoprotein D, in human breast cancer. Br J Cancer 1996; 74(8):1175-80.

112. Weber-Chappuis K, Bieri-Burger S, Hurlimann J. Comparison of prognostic markers detected by immunohistochemistry in male and female breast carcinomas. Eur J Cancer 1996; 32A(10):1686-92.
113. Harding C, Osundeko O, Tetlow L et al. Hormonally-regulated proteins in breast secretions are markers of target organ sensitivity. Br J Cancer 2000; 82(2):354-60.
114. Serra C, Vizoso F, Lamelas ML et al. Comparative study of two androgen-induced markers (apolipoprotein D and pepsinogen C) in female and male breast carcinoma. Int J Surg Investig 2000; 2(3):183-92.
115. Boyles JK, Notterpek LM, Anderson LJ. Accumulation of apolipoproteins in the regenerating and remyelinating mammalian peripheral nerve. Identification of apolipoprotein D, apolipoprotein A-IV, apolipoprotein E, and apolipoprotein A-I. J Biol Chem 1990; 265(29):17805-15.
116. Spreyer P, Schaal H, Kuhn G et al. Regeneration-associated high level expression of apolipoprotein D mRNA in endoneurial fibroblasts of peripheral nerve. EMBO J 1990; 9(8):2479-84.
117. Patel SC, Asotra K, Patel YC et al. Astrocytes synthesize and secrete the lipophilic ligand carrier apolipoprotein D. Neuroreport 1995; 6(4):653-7.
118. Navarro A, Tolivia J, Astudillo A et al. Pattern of apolipoprotein D immunoreactivity in human brain. Neurosci Lett 1998; 254(1):17-20.
119. Hu CY, Ong WY, Sundaram RK et al. Immunocytochemical localization of apolipoprotein D in oligodendrocyte precursor-like cells, perivascular cells, and pericytes in the human cerebral cortex. J Neurocytol 2001; 30(3):209-18.
120. Navarro A, Del Valle E, Tolivia J. Differential expression of apolipoprotein d in human astroglial and oligodendroglial cells. J Histochem Cytochem 2004; 52(8):1031-6.
121. Baulieu EE. Neurosteroids: A novel function of the brain. Psychoneuroendocrinology 1998; 23(8):963-87.
122. Kalman J, McConathy W, Araoz C et al. Apolipoprotein D in the aging brain and in Alzheimer's dementia. Neurol Res 2000; 22(4):330-6.
123. Belloir B, Kovari E, Surini-Demiri M et al. Altered apolipoprotein D expression in the brain of patients with Alzheimer disease. J Neurosci Res 2001; 64(1):61-9.
124. del Valle E, Navarro A, Astudillo A et al. Apolipoprotein D expression in human brain reactive astrocytes. J Histochem Cytochem 2003; 51(10):1285-90.
125. Lieuallen K, Pennacchio LA, Park M et al. Cystatin B-deficient mice have increased expression of apoptosis and glial activation genes. Hum Mol Genet 2001; 10(18):1867-71.
126. Thomas EA, Sautkulis LN, Criado JR et al. Apolipoprotein D mRNA expression is elevated in PDAPP transgenic mice. J Neurochem 2001; 79(5):1059-64.
127. Terrisse L, Poirier J, Bertrand P et al. Increased levels of apolipoprotein D in cerebrospinal fluid and hippocampus of Alzheimer's patients. J Neurochem 1998; 71(4):1643-50
128. Glockner F, Ohm TG. Hippocampal apolipoprotein D level depends on Braak stage and APOE genotype. Neuroscience 2003; 122(1):103-10.
129. Thomas EA, Laws SM, Sutcliffe JG et al. Apolipoprotein D levels are elevated in prefrontal cortex of subjects with Alzheimer's disease: No relation to apolipoprotein E expression or genotype. Biol Psychiatry 2003; 54(2):136-41.
130. Navarro A, Del Valle E, Astudillo A et al. Immunohistochemical study of distribution of apolipoproteins E and D in human cerebral beta amyloid deposits. Exp Neurol 2003; 184(2):697-704.
131. Reindl M, Knipping G, Wicher I et al. Increased intrathecal production of apolipoprotein D in multiple sclerosis. J Neuroimmunol 2001; 119(2):327-32.
132. Salen G, Berginer V, Shore V et al. Increased concentrations of cholestanol and apolipoprotein B in the cerebrospinal fluid of patients with cerebrotendinous xanthomatosis. Effect of chenodeoxycholic acid. N Engl J Med 1987; 316(20):1233-8.
133. Tomarev SI, Wistow G, Raymond V et al. Gene expression profile of the human trabecular meshwork: NEIBank sequence tag analysis. Invest Ophthalmol Vis Sci 2003; 44(6):2588-96.
134. Dandoy-Dron F, Guillo F, Benboudjema L et al. Gene expression in scrapie. Cloning of a new scrapie-responsive gene and the identification of increased levels of seven other mRNA transcripts. J Biol Chem 1998; 273(13):7691-7.
135. Kang S, Seo S, Hill J et al. Changes in gene expression in latent HSV-1-infected rabbit trigeminal ganglia following epinephrine iontophoresis. Curr Eye Res 2003; 26(3-4):225-9.
136. Saha S, Rangarajan PN. Common host genes are activated in mouse brain by Japanese encephalitis and rabies viruses. J Gen Virol 2003; 84(Pt 7):1729-35.
137. Thomas EA, Dean B, Pavey G et al. Increased CNS levels of apolipoprotein D in schizophrenic and bipolar subjects: Implications for the pathophysiology of psychiatric disorders. Proc Natl Acad Sci USA 2001; 98(7):4066-71.

138. Thomas EA, George RC, Sutcliffe JG. Apolipoprotein D modulates arachidonic acid signaling in cultured cells: Implications for psychiatric disorders. Prostaglandins Leukot Essent Fatty Acids 2003; 69(6):421-7.
139. Yao JK, Thomas EA, Reddy RD et al. Association of plasma apolipoproteins D with RBC membrane arachidonic acid levels in schizophrenia. Schizophr Res 2005; 72(2-3):259-266.
140. Thomas EA, Dean B, Scarr E et al. Differences in neuroanatomical sites of apoD elevation discriminate between schizophrenia and bipolar disorder. Mol Psychiatry 2003; 8(2):167-75.
141. Thomas EA, Danielson PE, Nelson PA et al. Clozapine increases apolipoprotein D expression in rodent brain: Towards a mechanism for neuroleptic pharmacotherapy. J Neurochem 2001; 76(3):789-96.
142. Khan MM, Parikh VV, Mahadik SP. Antipsychotic drugs differentially modulate apolipoprotein D in rat brain. J Neurochem 2003; 86(5):1089-100.
143. Thomas EA, George RC, Danielson PE et al. Antipsychotic drug treatment alters expression of mRNAs encoding lipid metabolism-related proteins. Mol Psychiatry 2003; 8(12):983-93, 950.
144. Mahadik SP, Khan MM, Evans DR et al. Elevated plasma level of apolipoprotein D in schizophrenia and its treatment and outcome. Schizophr Res 2002; 58(1):55-62.
145. Yoshida K, Cleaveland ES, Nagle JW et al. Molecular cloning of the mouse apolipoprotein D gene and its upregulated expression in Niemann-Pick disease type C mouse model. DNA Cell Biol 1996; 15(10):873-82.
146. Suresh S, Yan Z, Patel RC et al. Cellular cholesterol storage in the Niemann-Pick disease type C mouse is associated with increased expression and defective processing of apolipoprotein D. J Neurochem 1998; 70(1):242-51.
147. Ong WY, Hu CY, Patel SC. Apolipoprotein D in the Niemann-Pick type C disease mouse brain: An ultrastructural immunocytochemical analysis. J Neurocytol 2002; 31(2):121-9.
148. Boyles JK, Notterpek LM, Wardell MR et al. Identification, characterization, and tissue distribution of apolipoprotein D in the rat. J Lipid Res 1990; 31(12):2243-56.
149. Ong WY, He Y, Suresh S et al. Differential expression of apolipoprotein D and apolipoprotein E in the kainic acid-lesioned rat hippocampus. Neuroscience 1997; 79(2):359-67.
150. Montpied P, de Bock F, Lerner-Natoli M et al. Hippocampal alterations of apolipoprotein E and D mRNA levels in vivo and in vitro following kainate excitotoxicity. Epilepsy Res 1999; 35(2):135-46.
151. Terrisse L, Seguin D, Bertrand P et al. Modulation of apolipoprotein D and apolipoprotein E expression in rat hippocampus after entorhinal cortex lesion. Brain Res Mol Brain Res 1999; 70(1):26-35.
152. Franz G, Reindl M, Patel SC et al. Increased expression of apolipoprotein D following experimental traumatic brain injury. J Neurochem 1999; 73(4):1615-25.
153. O'Donnell J, Stemmelin J, Nitta A et al. Gene expression profiling following chronic NMDA receptor blockade-induced learning deficits in rats. Synapse 2003; 50(3):171-80.
154. Garcia-Segura LM, Naftolin F, Hutchison JB et al. Role of astroglia in estrogen regulation of synaptic plasticity and brain repair. J Neurobiol 1999; 40(4):574-84.
155. Peitsch MC, Boguski MS. Is apolipoprotein D a mammalian bilin-binding protein? New Biol 1990; 2(2):197-206.
156. Xu N, Dahlback B. A novel human apolipoprotein (apoM). J Biol Chem 1999; 274(44):31286-31290.
157. Richter S, Shih DQ, Pearson ER et al. Regulation of apolipoprotein M gene expression by MODY3 gene hepatocyte nuclear factor-1alpha: Haploinsufficiency is associated with reduced serum apolipoprotein M levels. Diabetes 2003; 52(12):2989-2995.
158. Faber K, Axler O, Dahlback B et al. Characterization of apoM in normal and genetically modified mice. J Lipid Res 2004; 45(7):1272-1278.
159. Brewer Jr HB. Increasing HDL cholesterol levels. N Engl J Med 2004; 350(15):1491-1494.
160. Brewer Jr HB. High-density lipoproteins: A new potential therapeutic target for the prevention of cardiovascular disease. Arterioscler Thromb Vasc Biol 2004; 24(3):387-391.
161. Rader DJ. High-density lipoproteins as an emerging therapeutic target for atherosclerosis. Jama 2003; 290(17):2322-2324.
162. Rye KA, Barter PJ. Formation and metabolism of prebeta-migrating, lipid-poor apolipoprotein A-I. Arterioscler Thromb Vasc Biol 2004; 24(3):421-428.
163. Barter PJ, Nicholls S, Rye KA et al. Antiinflammatory properties of HDL. Circ Res 2004; 95(8):764-772.
164. Barter PJ, Brewer Jr HB, Chapman MJ et al. Cholesteryl ester transfer protein: A novel target for raising HDL and inhibiting atherosclerosis. Arterioscler Thromb Vasc Biol 2003; 23(2):160-167.
165. Blanco-Vaca F, Escola-Gil JC, Martin-Campos JM et al. Role of apoA-II in lipid metabolism and atherosclerosis: Advances in the study of an enigmatic protein. J Lipid Res 2001; 42(11):1727-1739.

166. Meir KS, Leitersdorf E. Atherosclerosis in the apolipoprotein-E-deficient mouse: A decade of progress. Arterioscler Thromb Vasc Biol 2004; 24(6):1006-1014.
167. Castellani LW, Lusis AJ. ApoA-II versus ApoA-I: Two for one is not always a good deal. Arterioscler Thromb Vasc Biol 2001; 21(12):1870-1872.
168. Schaap FG, Rensen PC, Voshol PJ et al. ApoAV reduces plasma triglycerides by inhibiting very low density lipoprotein-triglyceride (VLDL-TG) production and stimulating lipoprotein lipase-mediated VLDL-TG hydrolysis. J Biol Chem 2004; 279(27):27941-27947.
169. Sorenson RC, Bisgaier CL, Aviram M et al. Human serum Paraoxonase/Arylesterase's retained hydrophobic N-terminal leader sequence associates with HDLs by binding phospholipids: Apolipoprotein A-I stabilizes activity. Arterioscler Thromb Vasc Biol 1999; 19(9):2214-2225.
170. Plump AS, Azrolan N, Odaka H et al. ApoA-I knockout mice: Characterization of HDL metabolism in homozygotes and identification of a post-RNA mechanism of apoA-I up-regulation in heterozygotes. J Lipid Res 1997; 38(5):1033-1047.
171. Duan J, Dahlback B, Villoutreix BO. Proposed lipocalin fold for apolipoprotein M based on bioinformatics and site-directed mutagenesis. FEBS Lett 2001; 499(1-2):127-132.
172. Bocskei Z, Groom CR, Flower DR et al. Pheromone binding to two rodent urinary proteins revealed by X-ray crystallography. Nature 1992; 360(6400):186-188.
173. Faber K. ApoM. Structural and functional studies of a novel apolipoprotein in mice. Ph.D. Thesis. Malmo: Lund University, 2005.
174. Zhang XY, Dong X, Zheng L et al. Specific tissue expression and cellular localization of human apolipoprotein M as determined by in situ hybridization. Acta Histochem 2003; 105(1):67-72.
175. Zhang XY, Jiao GQ, Hurtig M et al. Expression pattern of apolipoprotein M during mouse and human embryogenesis. Acta Histochem 2004; 106(2):123-128.
176. Liang CP, Tall AR. Transcriptional profiling reveals global defects in energy metabolism, lipoprotein, and bile acid synthesis and transport with reversal by leptin treatment in ob/ob mouse liver. J Biol Chem 2001; 276(52):49066-49076.
177. Xu N, Nilsson-Ehle P, Hurtig M et al. Both leptin and leptin-receptor are essential for apolipoprotein M expression in vivo. Biochem Biophys Res Commun 2004; 321(4):916-921.
178. Xu N, Zhang XY, Dong X et al. Effects of platelet-activating factor, tumor necrosis factor, and interleukin-1alpha on the expression of apolipoprotein M in HepG2 cells. Biochem Biophys Res Commun 2002; 292(4):944-950.
179. Xu N, Hurtig M, Zhang XY et al. Transforming growth factor-beta down-regulates apolipoprotein M in HepG2 cells. Biochim Biophys Acta 2004; 1683(1-3):33-37.
180. Müller-Eberhard HJ. Molecular organization and function of the complement system. Annu Rev Biochem 1988; 57:321-347.
181. Esser AF. The membrane attack complex of complement: Assembly, structure and cytotoxic activity. Toxicology 1994; 87:229-247.
182. Plumb ME, Sodetz JM. Proteins of the membrane attack complex. In: Volanakis JE, Frank MM, eds. The Human Complement System in Health and Disease. New York: Marcel Dekker, 1998:119-148.
183. Wang Y, Bjes ES, Esser AF. Molecular aspects of complement-mediated bacterial killing: Periplasmic conversion of C9 from a protoxin to a toxin. J Biol Chem 2000; 275:4687-4692.
184. Steckel EW, York RG, Monahan JB et al. The eighth component of human complement: Purification and physiochemical characterization of its unusual subunit structure. J Biol Chem 1980; 255:11997-12005.
185. Lebioda L, Sodetz JM. Complement protein C8. In: Morikis D, Lambris JD, eds. Structural Biology of the Complement System. Boca Raton: CRC/Taylor & Francis, 2005:233-250.
186. Ng SC, Rao AG, Howard OMZ et al. The eighth component of human complement (C8): Evidence that it is an oligomeric serum protein assembled from products of three different genes. Biochemistry 1987; 26:5229-5233.
187. Hobart MJ, Fernie BA, DiScipio RG. Structure of the human C7 gene and comparison with the C6, C8A, C8B, and C9 genes. J Immunol 1995; 154:5188-5194.
188. Monahan JB, Sodetz JM. Binding of the eighth component of human complement to the soluble cytolytic complex is mediated by its β-subunit. J Biol Chem 1980; 255:10579-10582.
189. Lockert DH, Kaufman KM, Chang C-P et al. Identity of the segment of human complement C8 recognized by complement regulatory protein CD59. J Biol Chem 1995; 270:19723-19728.
190. Plumb ME, Scibek JJ, Barber, TD et al. Chimeric and truncated forms of human complement protein C8α reveal binding sites for C8β and C8γ within the membrane attack complex/perforin region. Biochemistry 1999; 38:8478-8484.
191. Plumb ME, Sodetz JM. An indel within the C8α subunit of human complement C8 mediates intracellular binding of C8γ and formation of C8α–γ. Biochemistry 2000; 39:13078-13083.

192. Scibek JJ, Plumb ME, Sodetz JM. Binding of human complement C8 to C9: Role of the N-terminal modules in the C8α subunit. Biochemistry 2002; 41:14546-14551.
193. Musingarimi P, Plumb ME, Sodetz JM. Interaction between the C8α–γ and C8β subunits of human complement C8: Role of the C8β N-terminal thrombospondin type 1 module and membrane attack/perforin domain. Biochemistry 2002; 41:11255-11260.
194. Schreck SF, Parker CL, Plumb ME et al. Human complement protein C8γ. Biochim Biophys Acta 2000; 1482:199-208.
195. Schreck SF, Plumb ME, Platteborze PL et al. Expression and characterization of recombinant subunits of human complement component C8: Further analysis of the function of C8α and C8γ. J Immunol 1998; 161:311-318.
196. Parker CL, Sodetz JM. Role of the human C8 subunits in complement-mediated bacterial killing: Evidence that C8γ is not essential. Mol Immunol 2002; 39:453-458.
197. Ortlund E, Parker CL, Schreck SF et al. Crystal structure of human complement protein C8γ at 1.2Å resolution reveals a lipocalin fold and a distinct ligand binding site. Biochemistry 2002; 41:7030-7037.
198. Prange T, Schiltz M, Pernot L et al. Exploring hydrophobic sites in proteins with xenon or krypton. Proteins 1998; 30:61-73.
199. Haefliger JA, Peitsch MC, Jenne DE et al. Structural and functional characterization of complement C8γ, a member of the lipocalin protein family. Mol Immunol 1991; 28:123-131.
200. Steckel EW, Welbaum BE, Sodetz JM. Evidence of direct insertion of terminal complement proteins into cell membrane bilayers during cytolysis: Labeling by a photosensitive membrane probe reveals a major role for the eighth and ninth components. J Biol Chem 1983; 258:4318-4324.
201. Ratledge C, Dover LG. Iron metabolism in pathogenic bacteria. Annu Rev Microbiol 2000; 54:881-941.
202. Goetz DH, Willie ST, Armen RS et al. Ligand preference inferred from the structure of neutrophil gelatinase associated lipocalin. Biochemistry 2000; 39:1935-1941.
203. Goetz DH, Holmes MA, Borregaard N et al. The neutrophil lipocalin NGAL is a bacteriostatic agent that interferes with siderophoremediated iron aquisition. Mol Cell 2002; 10:1033-1043.
204. Flo TH, Smith KD, Sato S et al. Lipocalin 2 mediates an innate immune response to bacterial infection by sequestering iron. Nature 2004; 432:917-921.
205. Fluckinger M, Haas H, Mershak P et al. Human tear lipocalin exhibits antimicrobial activity by scavenging microbial siderophores. Antimicrob Agents Chem 2004; 48:3367-3372.
206. Holmes MA, Paulsene W, Jide X et al. Siderocalin (Lcn2) also binds carboxymycobactins, potentially defending against mycobacterial infections through iron sequestration. Structure 2005; 13:29-41.

Chapter 14

Lipocalin Receptors:
Into the Spotlight

Brian J. Burke, Clara Redondo, Bernhard Redl and John B.C. Findlay*

Abstract

Evidence has been steadily accruing over time that a significant number of lipocalins interact with specific membrane receptors. The transfer of RBP:retinol across the cell membrane, faciliated by the elusive RBP receptor was for many years the archetypal example of a lipocalin:receptor mediated process. Megalin was subsequently identified as a low affinity/high capacity endocytic membrane receptor capable of binding a range of lipocalins including RBP.

Progress in the field has accelerated in recent years. A receptor for glycodelin has been found on T-cells and is thought to be involved in the downregulation of inflammatory processes. A receptor for lipocalin-1 has been cloned and shown to be a member of an undefined gene family of membrane proteins potentially possessing multiple transmembrane segments.

It is hoped that continuing advances in studying molecular interactions will allow the identification and cloning of other lipocalin receptors leading to a more complete understanding of how lipocalins participate in biological processes.

Introduction

Over the last three decades, there has been growing experimental evidence that an increasing number of lipocalins bind to specific cell surface receptors. To date and chronologically, the first reports in the literature included receptors for retinol-binding protein (RBP),[1] α-1-microglobulin (α-1M),[2] purpurin,[3] β-lactoglobulin (Blg),[4] epididymal retinoic acid binding protein (ERABP),[5] insecticyanin,[6] α-1-acid glycoprotein (AGP),[7] odorant-binding protein (OBP),[8] glycodelin,[9] and lipocalin-1 (Lcn-1).[10]

The aim of this review is twofold; firstly to produce an up to date summary of work in the lipocalin receptor field; secondly to encourage innovative approaches to answering the many questions that remain on this topic. At this stage it is not known how many lipocalin receptors there are or whether they are related and how many families they represent. Thus the rest of the decade should be a period of excitement in the lipocalin receptor field.

The review will focus on the few systems for which there is a body of structural and functional data, which authenticate the presence of a receptor.

The RBP Receptor Story

RBP is perhaps the most widely understood of the lipocalins in terms of its structure and function (see Chapter 7). Its roles in the release of retinol from liver, and transport of this

*Corresponding Author: John B.C. Findlay—School of Biochemistry and Microbiology, University of Leeds, Mount Preston Road, Leeds LS2 9JT, U.K. Email: J.B.C.Findlay@leeds.ac.uk

Lipocalins, edited by Bo Åkerström, Niels Borregaard, Darren R. Flower and Jean-Philippe Salier. ©2006 Landes Bioscience.

vitamin in blood are important for growth, differentiation, reproduction and normal vision. In healthy humans, RBP serum levels are typically in the region of 2 μM, with RBP:retinol complexed with another protein, transthyretin (TTR).[11] This 76 kDa complex, consisting usually of 4 molecules of TTR and 2 of RBP, is of sufficient size not be excreted in the glomeruli.[12] Once RBP releases its retinol, the resultant *apo*protein also has a much lower affinity for both receptor and TTR and is subsequently filtered in the kidneys and degraded in the urine. In the meantime, the retinol transferred into cells is bound to cellular retinol binding protein (CRBP), a 15 kDa protein, prior to metabolism and utilization.[13-15]

Several schools of thought have developed on the exact mechanism of retinol release and entry into the cell. Some argue that retinol, being in equilibrium with RBP and extremely hydrophobic, readily partitions into the lipid phase of the membrane and thence to CRBP.[12,16-22] During studies on RBP gene-disrupted mice, a reduction in levels of serum retinol was observed and additionally in the first five months of life, animals exhibited visual impairment.[23] The same study showed that an enhanced vitamin A diet could restore vision and that the mice were viable, suggesting that other routes can exist for introducing vitamin A into the cellular metabolism. Later work by the same group demonstrated that the visual defects could be reversed by engineered expression of human RBP in the background of RBP-null mice,[24] indicating a role for RBP. The authors further suggest that at least in the visual system there appears to be the need for a membrane receptor.

Around the same time, two human siblings were reported with apparently undetectable levels of serum RBP which was put down to point mutations in their RBP gene.[25] Both suffered night blindness and mild retinal dystrophy but none of the other complications associated with clinical vitamin A deficiency.

Perhaps of some importance is the observation that mice obtain sufficient retinol during weaning to sustain them throughout their entire life cycle calling into question interpretations of data from mice models.[26] Additionally, there are indications that age and disease status may influence retinol metabolism. There is also evidence to suggest that very low intakes of retinol reduce retinol utilisation, in a manner consistent with maintaining adequate reserves for key processes (see ref. 27). Thus, one should bear in mind the different living conditions of native and laboratory animals, together with the evolutionary pressures which pertain to animals living in the wild.

For such reasons, a greater number of researchers were persuaded by the regulatory logic of an RBP:receptor mediated uptake system. This was borne out by indications of a specific receptor for RBP in the plasma membrane of many eukaryotic cells including the visual photoreceptor cells and placenta. The RBP receptor has been confirmed in Sertoli,[28] stellate,[29,30] peritubular pigment epithelial,[31] embryonic carcinoma,[31] choroid plexus[32] and a variety of tissue culture cells.[33,34] The placental receptor showed high affinity for the *holo*-form of RBP but not for the *apo*-form.[35] Purification studies suggested it had a molecular weight in the region of 55kDa.[36] More recent work with RBP- knockout mice has supported a natural receptor-mediated retinol uptake system, which does not involve endocytosis or free diffusion of retinol.[23,24,37] Other lipocalins were unable to bind to this receptor indicating specificity. Since then it has been found that there are also specific membrane receptors for other lipocalins.[4,38,39]

The subsequent fate of RBP upon binding to the cell surface has been a source of some debate. Some reports show that the retinol is released and the *apo*protein remains in the extracellular compartment.[40-42] Other reports say that the RBP:retinol complex is internalized.[43-46] Evidence has been presented to suggest that cells which endocytose the RBP:retinol complex do so to create vitamin A stores that can be rereleased into the blood stream when required.[47,48] For that reason, that may be the same explanation for evidence that RBP can be synthesised in tissues other than the liver.[49] Another possibility is that the difference is due to the various biological systems and methods employed. Heller[1] used intact, isolated pigment epithelium cells recovered from bovine eyes and the Findlay group[42] used placental vesicles. Groups characterising the RBP:receptor interaction as an endocytic process were all using cultured cells. It is possible that both routes occur, defined by tissue type.

During these two decades, the research focus has shifted towards a more detailed understanding of the RBP:receptor interaction and the process has reinforced the evidence for a receptor.

Using site-directed mutagenesis, Sivaprasadarao and Findlay[50] delineated the roles of the loop regions present around the mouth of the RBP beta barrel in recognising both the RBP membrane receptor and TTR. Their results demonstrated that RBP interacts with both TTR and the receptor via loops "CD" and "EF." These binding sites, however, were overlapping rather than identical and there was the possibility also of an additional contact with TTR via loop "AB." One of the implications of these results is that RBP, when bound to TTR, cannot bind simultaneously to the receptor. Confirming these observations, Melhus and colleagues[51] subsequently used monoclonal antibodies to try and block both the TTR and RBP receptor binding sites. Epitope mapping using synthetic peptides corresponding to RBP loop regions reported the TTR and RBP binding sites in agreement with the earlier study.[50] Structural data, produced afterwards, confirmed the nature of the RBP:TTR interaction as derived from the protein engineering studies.[52,53] The interaction of RBP and TTR from a structural perspective is discussed in more detail in Chapter 7.

Further studies focussed on the "CD" loop of RBP (comprising amino acids G59-D68 of the sequence), which is crucial for receptor recognition and interaction. Sundaram et al[50] demonstrated that this loop and the ability to recognise the RBP receptor, could be transferred successfully to the equivalent position in ERABP[54] and in a quite different lipocalin such as MUP (Redondo, Vouropoulou and Findlay, in preparation).

They further extended the notion of Sivaprasadarao and Findlay that the RBP receptor is essential for transfer of retinol from extracellular RBP to cellular retinol-binding protein (CRBP) inside the cell (see Fig. 1). Membranes lacking the receptor (e.g., erythrocytes) and heat-denatured systems are ineffective. Whether this occurs via the receptor itself or a separate protein system is recruited for the transfer, remains to be determined. CRBP is a member of the fatty acid-binding protein family (FABP) which is structurally related to the lipocalins. Intracellular levels of retinoids are regulated by CRBP which has been suggested to exert in a buffer-like role mediating further metabolism and utilization.

Recent observations[55] suggest that CRBP itself also exhibits specific membrane-association characteristics indicating the existence of a protein receptor, which may or may not be related to retinol metabolising enzymes, supporting the early results obtained by Sundaram et al.[56]

It is worth pointing out that this is a unique delivery/uptake system not yet seen elsewhere in eukaryotic biology but the principle does exist in prokaryotes, although the extracellular and intracellular binding proteins and the membrane receptor/transporter are quite different.[57]

Megalin

Megalin, a member of the low density lipoprotein receptor family, was the first, but nonspecific, lipocalin receptor identified. It is a 600-kDa Type–1 cell surface protein, classically consisting of a single transmembrane domain, a large amino-terminal extracellular domain and a short carboxy-terminal cytoplasmatic tail.[58] The detailed structure, expression, and physiological properties of megalin have been studied in great detail over the past few years and there are numerous recent reviews (refs. 59-65). Megalin mediates endocytosis of a large variety of ligands including vitamin-binding proteins and other carrier proteins, lipoproteins, hormones and hormone receptors, drugs and toxins, enzymes and enzyme inhibitors, immune- and stress-response-related proteins and some other proteins, eg. cytochrome C (for a review see ref. 58). Therefore, it is not unexpected to find megalin involved in the cellular uptake of lipocalins. First evidence for a role of megalin in cellular uptake of RBP came from megalin deficient mice. These mice showed a lack of RBP production in their renal proximal tubules and instead revealed highly increased urinary excretion of RBP and retinol, indicating that glomerular filtered RBP-retinol complexes escape uptake by proximal tubules.[66] A direct interaction of megalin and RBP was confirmed by surface plasmon resonance analysis and partial inhibition of RBP uptake by a polyclonal megalin antibody in Brown Norway rat yolk sac epithelial cells.[66]

Further investigations on the megalin knock-out mice revealed that megalin binds and mediates uptake of a number of other lipocalins present in mouse plasma and urine, including alpha-1 microglobulin, major urinary protein 6 and odorant-binding protein IA.[67] Because in the mouse system, many of the proteins that were found to bind megalin were plasma carriers for lipophilic compounds (e.g., fatty acids, odors, vitamin A and D metabolites), it was suggested that megalin might be essential to prevent urinary loss of lipids bound to small carrier proteins. This more general function is further supported by the fact, that the affinities of megalin for lipocalins and other binding proteins is only 0.1 to 1.8 μM, indicating a low affinity but high capacity receptor.[68]

LIMR

In a search for proteins that interact with the lipocalin Lcn-1 using phage-display technology, the Redl group identified a novel receptor , which was termed LIMR (lipocalin-1 interacting membrane receptor[10]). Lcn-1 is a lipocalin with broad ligand binding specificity,[69] which structurally differs from other lipocalin members by its extremely wide ligand cavity framed by a set of four loops at the open end. This cavity extends deeply into the barrel structure and ends with two distinct lobes.[70] Lcn-1 is thought to act as a physiological scavenger of potentially harmful hydrophobic molecules and was recently found to be another member of the lipocalin subgroup of siderocalins, since it binds bacterial and fungal siderophores.[71,72]

LIMR is essential for the internalization of Lcn-1 in human NT2 cells,[73] and thus functions as a endocytic receptor similar to megalin. However, the structural composition of LIMR appears to be completely different from that of megalin and other endocytic receptors. LIMR is a 55 kDa protein, which consists of a short extracellular domain, nine putative transmembrane domains interrupted by a large intracellular loop and an intermediate length cytoplasmatic tail.[73,74] Phage-display experiments and interaction analyses of purified recombinant peptides in solution revealed that LIMR binds Lcn-1 via the N-terminal region.[10] These results are supported by the fact that a polyvalent antiserum raised against the LIMR N-terminus abrogates uptake of FITC-Lcn-1 in NT2 cells (Redl, unpublished).

LIMR appears to constitute a novel family of endocytic receptors, whose members can be found in a wide variety of organisms including ***Mus musculus, Fugu rubripes, Caenorhabditis elegans, Drosophila melanogaster, Anopheles gambiae, Dictyostelium discoideum*** and ***Arabidopsis thaliana.*** Humans and mice, each have two closely related proteins in addition to up to 7 other putative homologues. In humans there is a LIMR orthologous protein called Difl4, which is encoded on chromosome 7q36, whereas LIMR is encoded on chromosome 12p11. In mouse there are also two closely related proteins, one is highly similar to human LIMR, whereas the other, Lmbr1, is more similar to human Difl4. It is interesting to note, that sequence comparison of human and mouse LIMR-related proteins reveals a relatively lower degree of conservation in the N-terminal region of the proteins. For example, whereas the overall sequence conservation of human LIMR and Difl4 is 58.3%, amino acid identity in the N-terminal region of the respective proteins is only 12.5%. Assuming that the N-terminal parts of these proteins are responsible for receptor-ligand interaction, as shown with LIMR, these results might indicate that the orthologous proteins found within one species interact with different target proteins.

A very recent study demonstrated that the LIMR gene is one of the genes that are highly induced in response to dietary iron-deprivation in rat duodenum.[75] An enhanced cellular uptake of siderocalins might indeed play a role in iron homeostasis in the gut since they are known to participate in the iron-depletion strategy of the immune system and are able to

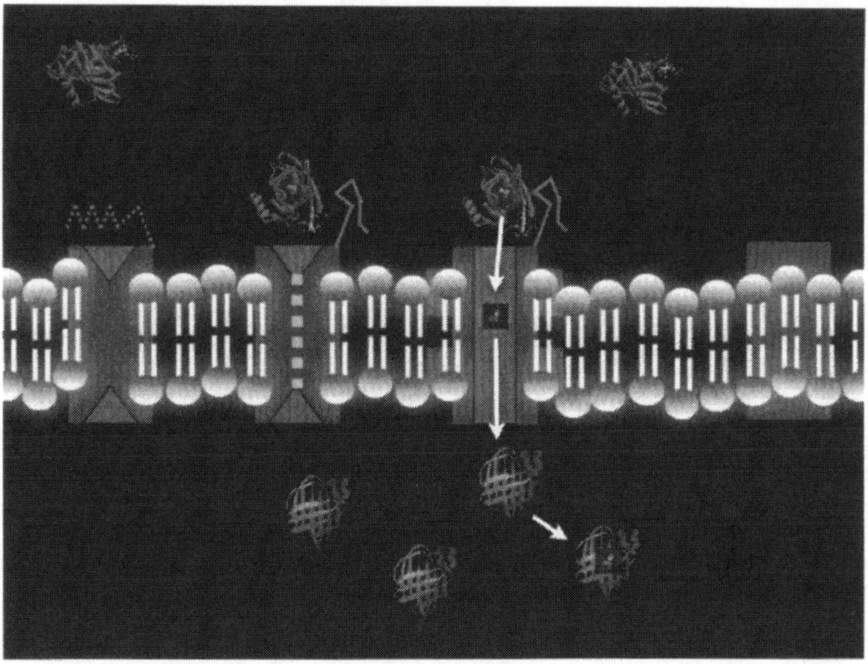

Figure 1. Receptor-mediated uptake of retinol from retinol binding protein. Schematic representation of the events proposed to facilitate transfer of retinol across the cell membrane. Retinol binding protein (shown at the top of the bilayer) complexes with retinol and adopts a *holo* conformation. The receptor binding site on the lipocalin becomes accessible and is able to induce a conformational change in the membrane receptor. This allows transfer of retinol to an intracellular protein such as cellular retinol binding protein (shown beneath the bilayer) in its *apo* form.

mediate iron absorption by some cell types[75,76]). Thus, it is possible, that LIMR or its orthologues have a function in the internalization of siderocalins, in general. However, the exact role of these receptors in binding of additional lipocalins or other transport proteins remains to be determined.

Glycodelin

Glycodelin is implicated in a variety of processes within the reproductive axis (reported in detail in Chapter 11). Specifically, in the case of females, at least one isoform of glycodelin is capable of binding to sperm in order to inhibit egg-binding.[77] More generally, it is implicated in regulation of inflammation, specifically by downregulation of T cells. This is achieved by the binding of glycodelin to CD45 (a major protein tyrosine phosphatase receptor) and other glycosylated entities on the surface of T-cells in a lectin-like manner.[78] The glycosylation state of glycodelin is critical to the varying roles of the three isoforms found.[79,80] Researchers have proposed that glycodelin alters the local balance between tyrosine kinases and phosphatases thereby attenuating T cell receptor signalling. The net effect is to damp down the immune response. The exact role of CD45 in T cell activation is unclear at this stage. Interestingly, glycodelin has been observed in the cumulus/corona cells which accompany the ovum during ovulation (which themselves do not contain mRNA for glycodelin[81]) implying the presence of an uptake pathway.[81] Which receptor is potentially involved remains to be determined.

Lipocalin-2

The functional characterisation of lipocalin-2 (Lcn2) has advanced significantly in recent years. Proposed roles include that of an antibacterial agent[76] as well as involvement in cellular differentiation.[82] Additionally, both transcriptomic and proteomic methodologies have detected significant increases in the levels of Lcn2 related to various cancers.[83,84]

As is the case with many lipocalins, Lcn2 is taken up by the megalin scavenging receptor.[85] Interestingly, the tightness of interaction is much greater than for other lipocalins so far examined, with Lcn2 displaying a dissociation constant that is approximately three orders of magnitude lower when compared to lipocalins such as RBP and MUP. It has been suggested that this might be explained in part by the net positive charge displayed by Lcn2.[85]

However, interaction with megalin is not sufficient to explain some of the functional behaviours of Lcn2, especially given that Lcn2 is present in a broader range of tissues than the endocytic receptor. A recently published paper documented the involvement of Lcn2 in regulation of cell phenotype by acting as an inhibitor of the intracellular pathway responsible for E-cadherin degradation.[82] The effect is enhanced by the presence of siderophore:iron complexes but not when the iron is substituted by gallium.[82] Furthermore, supplementation of cell systems with free iron does not facilitate this effect and at higher levels increased degradation of E-cadherin is observed.[82] Given that megalin does not distinguish between *apo* and *holo* forms of Lcn2, nor is it specific as to which proteins it endocytoses, one would anticipate another mode of modulating this intracellular pathway. Further work is required to delineate the precise role of this lipocalin and whether a membrane receptor other than megalin is implicated in mediating its observed effects.

The Future of Lipocalins and Their Receptors

Perhaps the most consistent and difficult challenge facing researchers working in the field of lipocalin receptors is the very nature of the receptors themselves. Membrane proteins, whilst accounting for as much as 30% of the genome, are more difficult to detect, purify and characterize than soluble proteins. This has delayed the molecular biology substantially. It is therefore not surprising that lipocalin receptors have proved so elusive in revealing their identities.

As greater understanding of the evolution of the lipocalins is developed (see Chapter 2), it will be interesting to observe the broader evolution of functional systems.

The eye is an excellent example of this. The more complex eyes found in chordates rely on a sophisticated system, requiring the delivery of retinol by RBP to the RPE by a receptor-mediated process. The chromophore can only be regenerated in the RPE. It would be reasonable to infer that rhodopsin, RBP, iRBP, CRBP and the RBP receptor have all evolved within a similar time frame to help facilitate chordate vision in a way that offers clear regulatory advantages. The same is likely to be even truer of the reproductive and foetal systems.

Some researchers have speculated as to the existence of a structurally related family of membrane receptors, specific for the lipocalins.[68] This maybe likely for lipocalins that are similar in terms of sequence and function. However, the scope of lipocalin function is so diverse that one cannot rule out more than one class of specific lipocalin receptors. Moreover, it seems likely that the roles of these receptors will vary to include signal transduction, endocytosis, and transport. The presence of additional subunits would not be unexpected, nor would be the presence of a substantial quaternary structure. The one definitive receptor class now identified, raises a number of interesting fundamental issues because its putative structure is so atypical of an endocytic receptor.

Artificial evolution of the lipocalin fold gives further indication of the potential for lipocalin:receptor interaction. An engineered lipocalin was recently produced with novel specificity for CTLA-4, a receptor on the surface of T cells which in principle could allow lipocalins to behave has receptor agonists/antagonists.[86]

As our knowledge of lipocalins and their receptors continues to evolve, we anticipate a greater understanding of the functional significance of certain lipocalins.

Acknowledgements

The authors wish to acknowledge and thank Prof. Markku Seppälä for helpful discussions regarding glycodelin.

References

1. Heller J. Interactions of plasma retinol-binding protein with its receptor. J Biol Chem 1975; 380(10):3613-3619.
2. Pearlstein E, Turesson I, Tejler L et al. Expression of protein-Hc on plasma-membrane of different human cell-types. J Immunol 1977; 119(3):824-829.
3. Schubert D, Lacorbiere M. Isolation of a cell-surface receptor from chick Neural Retina Adherons. J Cell Biol 1985; 100(1):56-63.
4. Papiz MZ, Sawyer L, Eliopoulos EE et al. The structure of beta-lactoglobulin and its similarity to plasma retinol-binding protein. Nature 1986; 324(6095):383-385.
5. Morel L, Dufaure JP, Depeiges A. Lesp, an androgen-regulated lizard epididymal secretory protein family identified as a new member of the lipocalin superfamily. J Biol Chem 1993; 268(14):10274-10281.
6. Kang Y, Kulakosky PC, Vanantwerpen R et al. Sequestration of insecticyanin, a blue hemolymph protein, into the egg of the hawkmoth manduca-sexta - evidence for receptor- mediated endocytosis. Insect Biochem Mol Biol 1995; 25(4):503-510.
7. Andersen UO, Kirkeby S, BogHansen TC. Two lectin-like receptors for alpha(1)-acid glycoprotein in mouse testis. J Mol Recognit 1996; 9(5-6):364-367.
8. Boudjelal M, Sivaprasadarao A, Findlay JBC. Membrane receptor for odour-binding proteins. Biochem J 1996; 317:23-27.
9. Miller RE, Fayen JD, Chakraborty S et al. A receptor for the lipocalin placental protein 14 on human monocytes. FEBS Lett 1998; 436(3):455-460.
10. Wojnar P, Lechnar M, Merschak P et al. Molecular cloning of a novel lipocalin-1 interacting human cell membrane receptor using phage display. J Biol Chem 2001; 276(23):20206-20212.
11. Blomhoff R, Green MH, Green JB et al. Vitamin A metabolism: New perspectives on absorption, transport and storage. Physiol Rev 1991; 71:951-990.
12. Blaner WS. Retinol-binding protein: The serum transport protein for vitamin A. Endocr Rev 1989; 10(3):308-316.
13. Noy N, Slosberg E, Scarlata S. Interactions of retinol with binding-proteins - studies with retinol-binding protein and with transthyretin. Biochemistry 1992; 31(45):11118-11124.
14. Ong DE, Davis JT, Oday WT et al. Synthesis and secretion of retinol-binding protein and transthyretin by cultured retinal-pigment epithelium. Biochemistry 1994; 33(7):1835-1842.
15. Blaner WS. Radioimmunoassays for retinol-binding protein, cellular retinol-binding protein, and cellular retinoic acid-binding protein. Methods Enzymol 1990; 189:270-281.
16. Noy N, Xu ZJ. Interactions of retinol with binding-proteins - implications for the mechanism of uptake by cells. Biochemistry 1990; 29(16):3878-3883.
17. Noy N, Xu ZJ. Kinetic-parameters of the interactions of retinol with lipid bilayers. Biochemistry 1990; 29(16):3883-3888.
18. Noy N, Xu ZJ. Thermodynamic parameters of the binding of retinol to binding-proteins and to membranes. Biochemistry 1990; 29(16):3888-3892.
19. Noy N. The thermodynamic parameters of the binding of retinol to binding-proteins and to membranes. Biophysical J 1990; 57(2):A460-A460.
20. Noy N. Kinetic aspects of the interactions of vitamin-a with various retinoid-binding-proteins. FASEB J 1992; 6(1):A9-A9.
21. Blaner WS, Obunike JC, Kurlandsky SB et al. Lipoprotein lipase hydrolysis of retinyl ester. Possible implications for retinoid uptake by cells. J Biol Chem 1994; 269(24):16559-16565.
22. Vanbennekum AM, Blaner WS, Seifertbock I et al. Retinol uptake from retinol-binding protein (Rbp) by liver parenchymal-cells in vitro does not specifically depend on its binding to Rbp. Biochemistry 1993; 32(7):1727-1733.
23. Quadro L, Blaner WS, Salchow DJ et al. Impaired retinal function and vitamin A availability in mice lacking retinol-binding protein. EMBO J 1999; 18(17):4633-4644.
24. Quadro L, Blaner WS, Hamberger L et al. Muscle expression of human retinol-binding protein (RBP). Suppression of the visual defect of RBP knockout mice. J Biol Chem 2002; 277(33):30191-30197.
25. Seeliger MW, Biesalski HK, Wissinger B et al. Effects of a systemic retinol deficiency due to a hereditary loss of retinol binding protein (RBP). Invest Ophthalmol Vis Sci 1999; 40(4):1143B1151.
26. Smeland S, Bjerknes T, Malaba L et al. Tissue distribution of the receptor for plasma retinol-binding protein. Biochem J 1995; 305:419-424.

27. Burri BJ, Clifford AJ. Carotenoid and retinoid metabolism: Insights from isotope studies. Arch Biochem Biophys 2004; 430(1):110-119.
28. Davis JT, Ong DE. Synthesis and secretion of retinol-binding protein by cultured Rat sertoli cells. Biol Reprod 1992; 47(4):528-533.
29. Blomhoff R, Norum KR, Berg T. Hepatic-uptake of [H-3] retinol bound to the serum retinol binding-protein involves both parenchymal and perisinusoidal stellate cells. J Biol Chem 1985; 260(25):3571-3575.
30. Blaner WS, Dixon JL, Moriwaki H et al. Studies on the in vivo transfer of retinoids from parenchymal to stellate cells in rat liver. Eur J Biochem 1987; 164(2):301-307.
31. Davis JT, Ong DE. Retinol processing by the peritubular cell from Rat testis. Biol Reprod 1995; 52(2):356-364.
32. Yamamoto M, Drager UC, Ong DE et al. Retinoid-binding proteins in the cerebellum and choroid plexus and their relationship to regionalized retinoic acid synthesis and degradation. Eur J Biochem 1998; 257(2):344-350.
33. Bavik CO, Eriksson U, Allen RA et al. Identification and partial characterization of a retinal-pigment epithelial membrane-receptor for plasma retinol-binding protein. J Biol Chem 1991; 266(23):14978-14985.
34. Bavik CO, Levy F, Hellman U et al. The retinal-pigment epithelial membrane-receptor for plasma retinol-binding protein - Isolation and cDNA cloning of the 63- Kda protein. J Biol Chem 1993; 268(27):20540-20546.
35. Sivaprasadarao A, Findlay JBC. The interaction of retinol-binding protein with its plasma- membrane receptor. Biochem J 1988; 255(2):561-569.
36. Sivaprasadarao A, Boudjelal M, Findlay JBC. Solubilization and purification of the retinol-binding protein- Receptor from human placental membranes. Biochem J 1994; 302:245-251.
37. Vogel S, Piantedosi R, O'Byrne SM et al. Retinol-binding protein-deficient mice: Biochemical basis for impaired vision. Biochemistry 2002; 41(51):15360-15368.
38. Said HM, Ong DE, Shingleton JL. Intestinal uptake of retinol - enhancement by bovine-milk beta- Lactoglobulin. Am J Clin Nutrit 1989; 49(4):690-694.
39. Mansouri A, Gueant JL, Capiaumont J et al. Plasma membrane receptor for β-lactoglobulin and retinol-binding protein in murine hybridomas. Biofactors 1998; 7:287-298.
40. Heller J. Interactions of plasma retinol-binding protein with its receptor. Specific binding of bovine and human retinol-binding protein to pigment epithelium cells from bovine eyes. J Biol Chem 1975; 250(10):3613-3619.
41. Ward SJ, Chambon P, Ong DE et al. A retinol-binding protein receptor-mediated mechanism for uptake of vitamin A to postimplantation rat embryos. Biol Reprod 1997; 57(4):751-755.
42. Sivaprasadarao A, Findlay JBC. The mechanism of uptake of retinol by plasma-membrane vesicles. Biochem J 1988; 255(2):571-579.
43. Tosetti F, Campelli F, Levi G. Studies on the cellular uptake of retinol binding protein and retinol. Exp Cell Res 1999; 250(2):423-433.
44. Senoo H, Smeland S, Malaba L et al. Transfer of retinol-binding protein from Hepg2 human hepatoma- cells to cocultured Rat stellate cells. Proc Nat Acad Sci USA 1993; 90(8):3616-3620.
45. Matarese V, Lodish HF. Specific uptake of retinol-binding protein by variant F9 cell- lines. J Biol Chem 1993; 268(25):18859-18865.
46. Hagen E, Myhre AM, Smeland S et al. Uptake of vitamin A in macrophages from physiologic transport proteins: Role of retinol-binding protein and chylomicron remnants. J Nutrit Biochem 1999; 10(6):345-352.
47. Malaba L, Kindberg GM, Norum KR et al. Receptor-mediated endocytosis of retinol-binding protein by liver parenchymal-cells - Interference by radioactive iodination. Biochem J 1993; 291:187-191.
48. Malaba L, Smeland S, Senoo H et al. Retinol-binding protein and asialo-orosomucoid are taken up by different pathways in liver-cells. J Biol Chem 1995; 270(26):15686-15692.
49. Quadro L, Blaner WS, Hamberger L et al. The role of extrahepatic retinol binding protein in the mobilization of retinoid stores. Jo Lipid Res 2004; 45(11):1975-1982.
50. Sivaprasadarao A, Findlay JBC. Structure-function studies on human retinol-binding protein using site-directed mutagenesis. Biochem J 1994; 300:437-442.
51. Melhus H, Bavik CO, Rask L et al. Epitope mapping of a monoclonal-antibody that blocks the binding of retinol-binding protein to its receptor. Biochem Biophys Res Commun 1995; 210(1):105-112.
52. Naylor HM, Newcomer ME. The structure of human retinol-binding protein (RBP) with its carrier protein transthyretin reveals an interaction with the carboxy terminus of RBP. Biochemistry 1999; 38(9):2647-2653.

53. Monaco HL. The transthyretin-retinol-binding protein complex. Biochim Biophys Acta 2000; 1482(1-2):65-72.
54. Sundaram M, van Aalten DM, Findlay JB et al. The transfer of transthyretin and receptor-binding properties from the plasma retinol-binding protein to the epididymal retinoic acid-binding protein. Biochem J 2002; 362(Pt 2):265-271.
55. Evans J. Structure, function and engineering of CRBP [PhD]. Leeds: Biochemistry and Molecular Biology, PhD Thesis, University of Leeds, 2001.
56. Sundaram M, Sivaprasadarao A, DeSousa MM et al. The transfer of retinol from serum retinol-binding protein to cellular retinol-binding protein is mediated by a membrane receptor. J Biol Chem 1998; 273(6):3336-3342.
57. Igarashi K, Kashiwagi K. Polyamine transport in bacteria and yeast. Biochem J 1999; 344:633-642.
58. Saito A, Pietromonaco S, Loo AKC et al. Complete cloning and sequencing of Rat Gp330 megalin, a distinctive member of the low-density-lipoprotein receptor gene family. Proc Nat Acad Sci USA 1994; 91(21):9725-9729.
59. Christensen EI, Birn H. Megalin and cubilin: Multifunctional endocytic receptors. Nat Rev Mol Cell Biol 2002; 3(4):258-268A.
60. Moestrup SK, Verroust PJ. Megalin- and cubilin-mediated endocytosis of protein-bound vitamins, lipids, and hormones in polarized epithelia. Annu Rev Nutrit 2001; 21:407-428.
61. Herz J, Gotthardt M, Willnow TE. Cellular signalling by lipoprotein receptors. Curr Opin Lipidol 2000; 11(2):161-166.
62. Marino M, Pinchera A, McCluskey RT et al. Megalin in thyroid physiology and pathology. Thyroid 2001; 11(1):47-56.
63. Li YH, Cam J, Bu GJ. Low-density lipoprotein receptor family. Molecular Neurobiology 2001; 23(1):53-67.
64. McCarthy RA, Argraves WS. Megalin and the neurodevelopmental biology of sonic hedgehog and retinol. J Cell Sci 2003; 116(6):955-960.
65. Chung NS, Wasan KM. Potential role of the low-density lipoprotein receptor family as mediators of cellular drug uptake. Adv Drug Deliv Rev 2004; 56(9):1315-1334.
66. Christensen EI, Moskaug JO, Vorum H et al. Evidence for an essential role of megalin in transepithelial transport of retinol. J Am Soc Nephrol 1999; 10(4):685-695.
67. Leheste JR, Rolinski B, Vorum H et al. Megalin knockout mice as an animal model of low molecular weight proteinuria. Am J Pathol 1999; 155(4):1361-1370.
68. Flower DR. Beyond the superfamily: The lipocalin receptors. Biochim Biophys Acta 2000; 1482(1-2):327-336.
69. Redl B. Human tear lipocalin. Biochim Biophys Acta 2000; 1482(1-2):241-248.
70. Breustedt DA, Korndorfer IP, Redl B et al. The 1.8-angstrom crystal structure of human tear lipocalin reveals an extended branched cavity with capacity for multiple ligands. J Biol Chem 2005; 280(1):484-493.
71. Lechner M, Wojnar P, Redl B. Human tear lipocalin acts as an oxidative-stress-induced scavenger of potentially harmful lipid peroxidation products in a cell culture system. Biochem J 2001; 356:129-135.
72. Fluckinger M, Haas H, Merschak P et al. Human tear lipocalin exhibits antimicrobial activity by scavenging microbial siderophores. Antimicrob Agents Chemother 2004; 48(9):3367-3372.
73. Wojnar P, Lechner M, Redl B. Antisense down-regulation of lipocalin-interacting membrane receptor expression inhibits cellular internalization of lipocalin-1 in human NT2 cells. J Biol Chem 2003; 278(18):16209-16215.
74. Wojnar P, van't Hof W, Merschak P et al. The N-terminal part of recombinant human tear lipocalin/von Ebner's gland protein confers cysteine proteinase inhibition depending on the presence of the entire cystatin-like sequence motifs. Biol Chem 2001; 382(10):1515-1520.
75. Collins JF, Franck CA, Kowdley KV et al. Identification of differentially expressed genes in response to dietary iron- deprivation in Rat duodenum. Am J Physiol Gastrointest Liver Physiol 2005.
76. Goetz DH, Holmes MA, Borregaard N et al. The neutrophil lipocalin NGAL is a bacteriostatic agent that interferes with siderophore-mediated iron acquisition. Mol Cell 2002; 10(5):1033-1043.
77. Morris HR, Dell A, Easton RL et al. Gender-specific glycosylation of human glycodelin affects its contraceptive activity. J Biol Chem 1996; 271(50):32159-32167.
78. Rachmilewitz J, Borovsky Z, Riely GJ et al. Negative regulation of T cell activation by placental protein 14 is mediated by the tyrosine phosphatase receptor CD45. J Biol Chem 2003; 278(16):14059-14065.
79. Chiu PCN, Koistinen R, Koistinen H et al. Zona-binding inhibitory factor-1 from human follicular fluid is an isoform of glycodelin. Biol Reprod 2003; 69(1):365-372.

80. Chiu PCN, Koistinen R, Koistinen H et al. Binding of zona binding inhibitory factor-1 (ZIF-1) from human follicular fluid on spermatozoa. J Biol Chem 2003; 278(15):13570-13577.
81. Tse JYM, Chiu PCN, Lee KF et al. The synthesis and fate of glycodelin in human ovary during folliculogenesis (vol 8, pg 142, 2002). Mol Human Reprod 2002; 8(11):1050-1050.
82. Hanai J-I, Mammoto T, Seth P et al. Lipocalin 2 diminishes invasiveness and metastasis of ras transformed cells. J Biol Chem 2005; M413047200.
83. Gronborg M, Bunkenborg J, Kristiansen TZ et al. Comprehensive proteomic analysis of human pancreatic juice. J Proteome Res 2004; 3(5):1042-1055.
84. Missiaglia E, Blaveri E, Terris B et al. Analysis of gene expression in cancer cell lines identifies candidate markers for pancreatic tumorigenesis and metastasis. Int J Cancer 2004; 112(1):100-112.
85. Hvidberg V, Jacobsen C, Strong RK et al. The endocytic receptor megalin binds the iron transporting neutrophil-gelatinase-associated lipocalin with high affinity and mediates its cellular uptake. FEBS Lett 2005; 579(3):773-777.
86. Schlehuber S, Skerra A. Lipocalins in drug discovery: From natural ligand-binding proteins to "anticalins". Drug Discov Today 2005; 10(1):23-33.

Chapter 15

Important Mammalian Respiratory Allergens Are Lipocalins

Tuomas Virtanen* and Rauno Mäntyjärvi

Abstract

Allergy is an expanding problem in the industrialized countries. Allergenic proteins, the allergens, causing the allergic symptoms are ubiquitous materials in the environment, normally not harmful for individuals not sensitized to them. In contrast, sensitized individuals who have specific immunoglobulin E against the allergen as a result of activation of the immune system can develop immediate, even fatal allergic reactions. Sensitization to indoor allergens, for example to those derived from pets, is often associated with more severe forms of allergic diseases, such as asthma. Recent years have shown that the major mammalian allergens causing respiratory sensitization belong to the lipocalin family of proteins. Reasons for this remain to be fully elucidated but it appears that lipocalin allergens exhibit immunological features that may account for their allergenic capacity.

Introduction

Allergy is a disorder of the immune system which is manifested as inappropriate reactions to ubiquitous materials in the environment such as pollen, house dust mites, or animal dander.[1] Allergic reactions are mediated by immunoglobulin E (IgE) antibody. Proteins able to bind IgE are called allergens. They initiate allergic or type I hypersensitivity reactions by cross-linking the antibody on mast cells and basophils. This results in the release of inflammatory mediators and in the symptoms of allergy. Allergic symptoms can arise from different organs, typically from airways, digestive track, and skin. Their severity varies, e.g., from mild rhinitis and itching to a systemic anaphylactic shock. Sensitization, as indicated by the presence of IgE, can be evaluated by measuring serum IgE or by skin prick tests (SPT). Sensitization can also develop without leading to a clinical illness.

The nomenclature of allergens is based on the system introduced by the Allergen Nomenclature Sub-Committee of the International Union of Immunological Societies.[2,3] The designation comprises the first three letters of the genus of origin of the allergen, the first letter of the species name, and a running Arabic number. If two species names have identical designations they are distinguished by adding one or more letters, as necessary, to each species designation. Isoallergens sharing an amino acid identity of ≥ 67% are distinguished by a two-digit number after the sequential number of the allergen and the variants with an additional two-digit number. For example, bovine dander allergen Bos d 2.0101 is the variant 01 of the isoallergen 01 of Bos d 2.

*Corresponding Author: Tuomas Virtanen—Department of Clinical Microbiology, University of Kuopio, Kuopio, Finland. Email: tuomas.virtanen@uku.fi

Lipocalins, edited by Bo Åkerström, Niels Borregaard, Darren R. Flower and Jean-Philippe Salier. ©2006 Landes Bioscience.

Exposure and Sensitization to Animal Allergens

The major respiratory allergens of animals (Table 1), especially from mammals, belong to the lipocalin family of proteins.[4,5] While the cat allergen Fel d 1 is an important exception to the rule (see refs. 6,7), another major cat allergen, Fel d 4, was recently reported to be a lipocalin.[8] The food allergen Bos d 5 (β-lactoglobulin) also belongs to lipocalins.[5]

Household pets, cats and dogs, are frequently found in homes in the industrialized countries. As the allergens which emanate from these animals tend to stick to clothing, they are commonly found in other indoor environments, such as homes without pets, schools, day care centers, and transport vehicles. The concentrations may be high enough to sensitize and to cause symptoms in sensitized individuals.[9] Assessed by skin prick tests (SPT), dogs sensitize 7-15% of individuals in community populations.[10,11] The figure can rise up to 56% among asthmatic children.[12] Among dog-allergic subjects, sensitization to Can f 1 is detected in 50-70% of the individuals (see refs. 13-15) while the figure for Can f 2 is about 25%.[14,15] The prevalence of sensitization to cat among children with asthma is 30-50%.[16,17] The predominant cat allergen Fel d 1, not a lipocalin, sensitizes over 90% of cat-allergic individuals.[18] Sixty-three percent of cat-allergic individuals had IgE against Fel d 4, a recently discovered lipocalin allergen, typically at low levels.[8]

Exposure to rodents and to rabbit is a major occupational health problem since 20-30% of workers in laboratory animal facilities become sensitized to animal allergens.[19,20] Allergy

Table 1. Lipocalin allergens

Allergen	Animal	Comments
Bla g 4	Cockroach	An indoor allergen in the warm humid climate.
Bos d 2	Cow	An occupational allergen in the farming environment.
Bos d 5	Cow	β-lactoglobulin, a food allergen in cow's milk.
Can f 1	Dog	An indoor allergen sensitizing in homes, public places and in laboratory animal facilities.
Can f 2	Dog	Another dog allergen.
Cav p 1	Guinea pig	An indoor allergen sensitizing in homes and in laboratory animal facilities.
Cav p 2	Guinea pig	Another guinea pig allergen.
Equ c 1	Horse	An allergen related to occupational exposure and horse riding.
Equ c 2	Horse	Another horse allergen.
Fel d 4	Cat	A recently characterized feline allergen in the lipocalin family.
Mus m 1	Mouse	An indoor allergen sensitizing in homes and in laboratory animal facilities.
Ory c 1	Rabbit	An occupational allergen in laboratory animal facilities.
Ory c 2	Rabbit	Another rabbit allergen.
Rat n 1	Rat	An occupational allergen in laboratory animal facilities.
Tria p 1	Kissing bug	A salivary allergen from *Triatoma protracta* found in many regions of the Western Hemisphere.

to rabbit was observed to develop rapidly in laboratory animal workers, as its prevalence was the highest in comparison with sensitization to other animal species within one year of exposure.[19] Sixty-six percent of laboratory workers with asthma and rhinitis had IgE against rat Rat n 1.[21] In a survey of asthma patients, 35% of them had a positive SPT result with a rabbit dander extract.[22] Mice, rats, hamsters, guinea pigs, and gerbils are popular pets, but mouse can also induce allergy in subjects who live in infested apartments. The prevalence of sensitization to mouse was 18% among inner-city children with asthma in a U.S. study.[23] Keeping guinea pigs was associated with a more than threefold risk for atopic eczema in comparison with having other pets, such as dogs, cats, or hamsters.[24] Seventy to 87% of guinea pig-allergic subjects had IgE reactivity to Cav p 1 while the reactivity to Cav p 2 was about 55%.[25]

Domestic animals able to induce allergy include horse and cow. A population study reported a prevalence of about 13% for a positive SPT with horse dander.[11] Two of the characterized horse allergens, Equ c 1 and Equ c 2, are lipocalins. Seventy-six per cent of horse-allergic subjects had IgE to Equ c 1 (see ref. 26) while up to 50% had IgE against Equ c 2.[27] Cow epithelium is a potent sensitizer in the farming environment, inducing a variety of symptoms including asthma.[28] About 90% of dairy farmers with asthma of bovine origin have IgE against the major respiratory allergen of cow, Bos d 2.[29]

Lipocalin allergens have also been detected in insects. Exposure to cockroach allergens depends on socioeconomic factors including housing conditions. Sensitization to cockroach is around 40% among inner-city children with asthma.[30] Forty to sixty per cent of asthmatic patients with cockroach allergy had IgE to recombinant Bla g 4, a lipocalin allergen.[31] Another insect-derived lipocalin allergen is Tria p 1 from the "kissing bug".[32]

Immunological Features of Lipocalin Allergens

Although allergens bind IgE by definition, it is basically unknown why the inert environmental substances elicit the process of sensitization, the generation of T helper type 2 (Th2) lymphocytes, which results in the synthesis of allergen-specific IgE by B lymphocytes.[33,34] One requirement for a protein to be an allergen is that it is effectively dispersed in the environment. As discussed above, lipocalin allergens are widely encountered in human environments, obviously a result from the fact that lipocalin allergens are present in the dander and excretions of mammals. Another possible factor promoting the allergenic capacity of a protein could be its biological function. Lipocalins are known to exhibit a variety of biological functions (see ref. 35), including an enzyme activity (see ref. 36), which has been suggested to account for the allergenicity of some mite, fungal and pollen allergens.[37] The dog allergen Can f 1 has been proposed to be a cysteine protease inhibitor since it exhibits a considerable level of homology with tear lipocalin.[38] It can be speculated that like the allergens with enzyme activity (see ref. 37), lipocalin allergens might initially affect cell populations other than the T helper (Th) cells (lymphocytes), the crucial cell population in sensitization. For example, microbial pathogens can deliver signals to dendritic cells through pathogen-associated molecular patterns (PAMPs).[39] These signals can result in the polarization of T cell response, but lipocalins, or allergens in general, are not known to contain PAMPs. One endogenous lipocalin, glycodelin, acts on T cells directly. By elevating T cell receptor activation thresholds, glycodelin appears to favor the Th2 deviation of immune response.[40,41] Nevertheless, lipocalins are not known to possess a unifying physicochemical property or biological function which could explain their allergenicity.

Since Th cells play a central role in the specific immune response against exogenous substances and are therefore important in sensitization, it is possible that they are involved in determining the allergic capacity of proteins.[4,34] When we studied the cellular response against Bos d 2 we observed unexpectedly that the peripheral blood mononuclear cells (PBMC) from highly allergic cow dust-asthmatic patients with positive skin prick test reactions to Bos d 2 proliferated very weakly in vitro upon stimulation with the allergen.[42] We have recently verified that the finding was not characteristic for Bos d 2 only since recombinant dog Can f 1 and horse Equ c 1 were also weak stimulants for the PBMCs of dog and horse-allergic patients, respectively (Tuomas Virtanen, to be published). Accordingly, rat Rat n 1 was reported to be weakly stimulatory in vitro.[43] In a study with murine cells, we found that Bos d 2 was a weak immunogen for several inbred mouse strains.[44] Our recent study suggests that the weak stimulatory capacity of lipocalin allergens can result from their poor recognition by Th cells since the immunodominant T cell epitope of Bos d 2 was observed to be suboptimal for human T cells.[45] The finding is interesting because a weak stimulation through T cell receptor is known to favor Th2-type responses.[46,47] As lipocalins can exhibit considerable amino acid identity between species, it can be speculated that the presence of endogenous lipocalins could have resulted in the deletion of high-avidity lipocalin allergen-reactive T cells during thymic maturation.[4,48] Consequently, the remaining T cell population might recognize exogenous lipocalin allergens weakly, promoting the Th2 deviation of immune response.

IgE antibodies cross-reactive between different allergens can result in symptoms in an individual who encounters an allergen which is able to bind the cross-reactive antibodies but was not their initial inducer. The cross-reactivity of IgE antibodies between non-serum-derived mammalian allergens is poorly known. At the level of allergen extracts, there seems to be cross-reactivity.[49,50] It has also been suggested that the major cat and dog allergens have common IgE epitopes.[49] In a study with guinea-pig Cav p 1, a 100-fold excess of cat, mouse, and rat allergen preparations were able to cause weak inhibition, less than 10%, in IgE ELISA, pointing to a low-level cross-reactivity.[25] A further study suggested that a part of the IgE-binding epitopes of Cav p 1 and 2 are also cross-reactive.[51] Inhibition experiments with cow, dog and horse allergen extracts pointed to a possibility that cow dander contains a homolog with cat Fel d 4.[8] Most probably it is not the bovine Bos d 2 (see below) since its amino acid identity with Fel d 4 is only 29% (SBNS, Dec. 9, 2004). A monoclonal antibody raised to Bos d 5 (β-lactoglobulin), a bovine food allergen, has been reported to react against human serum retinol-binding protein.[52] The core of the epitope sequence, DTDY, is also found in glycodelin, another human lipocalin.

Mammalian Lipocalin Allergens Causing Respiratory Sensitization

Cat

The cat allergen Fel d 4 (SWISS-PROT accession number Q5VFH6) is the only feline allergen in the lipocalin family. The physicochemical properties of Fel d 4 and other mammalian lipocalin allergens causing respiratory sensitization are shown in Table 2. The allergen was cloned from the submandibular salivary gland.[8] mRNA for Fel d 4 was not detected in the other tissues examined, such as the parotid gland, liver, skin, tongue or the anal gland. Isoallergens were not isolated. Fel d 4 exhibits a 67% amino acid identity with the major horse allergen Equ c 1 and a 60% identity with boar salivary lipocalin (the SIB BLAST network service (SBNS) at the Swiss Institute of Bioinformatics, Dec. 9, 2004). The latter is an odorant and pheromone-binding protein, present in the male submaxillary salivary glands but absent in females.[53] The amino acid identity of Fel d 4 can be up to 58% with murine major urinary proteins (MUPs) and rat α_{2u}-globulins (SBNS, Dec. 9, 2004). It is 38% with a human putative MUP-like lipocalin.

Cow

Bos d 2 (SWISS-PROT Q28133) is the major allergen of cow dander and the only respiratory allergen of cow in the lipocalin family.[29,54] Bos d 2 is localized in the secretory cells of the apocrine sweat glands and in the basement membranes of the epithelium and hair follicles.[55] Three variants of Bos d 2 have been identified.[56] An immunologically related allergen is found in urine.[29] The three-dimensional structure of Bos d 2 (Protein Data Bank (PDB) code 1BJ7)

Table 2. Physicochemical characteristics of mammalian lipocalin allergens causing respiratory sensitization

Allergen	Animal	MM[1], kDa	Amino Acids	Isoelectric Point	Glycosylation	Oligomeric State	Key Reference
Bos d 2	Cow	20	156	4.2	No	M	54
Can f 1	Dog	22-25	156	5.2	Putative		14
Can f 2	Dog	22-27	162	4.9	Putative		14
Cav p 1[2]	Guinea pig	20		3.6-5.3	No	M, D[3]	25
Cav p 2[2]	Guinea pig	17		4.3-4.5	No		51
Equ c 1	Horse	22	172	3.9	Yes	D	65
Equ c 2[2]	Horse	16		3.4-3.5	No		27
Fel d 4	Cat	20	171	4.6	Putative		8
Mus m 1	Mouse	18-21	162	4.6-5.3	No	M	70
Ory c 1[2, 4]	Rabbit	17-18			Yes		76
Ory c 2[2, 4]	Rabbit	21					76
Rat n 1	Rat	17-21	162	4.2-5.5	Yes	M	70

[1]Molecular mass, [2]only N-terminus known, [3]tentative information, [4]tentatively named.

was determined with the *Pichia pastoris*-produced allergen.[57,58] It is assumed that Bos d 2 is a pheromone carrier.[55] The amino acid identity of Bos d 2 with odorant-binding proteins and other lipocalins from other species is at the level of 30-40% (SBNS, Nov. 19, 2004). The IgE-binding capacity of Bos d 2 is strongly dependent on the intact three-dimensional structure of the allergen.[59,60] Studies in cellular immunology incited us to propose a hypothesis on the allergenic capacity of Bos d 2 and other lipocalin allergens (see above).

Dog

Two of the dog allergens, Can f 1 and Can f 2, are lipocalins.[14] A third previously unidentified allergen, detected in dog epithelial extract, is probably also a lipocalin.[15] Can f 1 (SWISS-PROT O18873) is mainly found in dog saliva.[61] It is also present in dog dander but absent or in a low concentration in serum, urine, and feces.[61,62] mRNA for Can f 1 has been detected in tongue epithelial tissue but not in skin or liver.[14] The human protein which exhibits the highest amino acid identity with Can f 1, 57%, is tear lipocalin, known also as von Ebner's gland protein (SBNS, Nov. 18, 2004). Can f 2 (SWISS-PROT O18874) is present in dog dander and in saliva but urine or feces contain very little of it.[61] The mRNA of Can f 2 is predominantly found in the parotid gland. It is expressed to a lesser extent in tongue tissue and not found in skin or liver.[14] Can f 2 exhibits amino acid identities with rodent urinary proteins at a level of about 30% (SBNS, Nov. 19, 2004). Both Can f 1 and Can f 2 are available as recombinant proteins.[14,15]

Guinea Pig

Guinea pig dust contains several allergens which are present in dander, fur, urine, and saliva.[25,63] Two major allergens, Cav p 1 and Cav p 2, are probably lipocalins.[25,51] Cav p 1 (SWISS-PROT P83507), found also in urine, was isolated from hair extract.[25] Its N-terminal sequence, 14 amino acids, contains the GXW motif and it exhibits a 57% amino acid identity with Mus m 1, the mouse major urinary protein (MUP). Cav p 1 exists in isoforms and may form dimers.[51] Cav p 2 (SWISS-PROT P83508) was also isolated from hair extract.[51] It exists in isoforms. A sequence of 13 amino acids contains the characteristic GXW motif of lipocalins and shows a 69% identity with bovine Bos d 2.[51]

Horse

Two allergens, Equ c 1 and Equ c 2, out of the more than ten horse dander proteins binding IgE are lipocalins, although additional lipocalin allergens may be present in dander.[26,27] Equ c 1 (SWISS-PROT Q95182) is found in high concentrations in saliva while urine contains little of it.[64] The protein has been cloned from sublingual salivary glands. Its mRNA is expressed at about 100-fold higher level in sublingual salivary glands than in submaxillary salivary glands or liver.[65] The crystal structure of Equ c 1 (PDB 1EW3) has been resolved.[66] The allergen shows about a 50% amino acid identity with rodent major urinary proteins (see ref. 65) and identities close to 40% with a human putative MUP-like lipocalin and human lipocalin 9 (SBNS, Nov. 26, 2004). The amino acid identity with a salivary lipocalin from pig is 61% (SBNS, Nov. 26, 2004). Equ c 1 is found in several isoforms.[67] It exhibits a surfactant-like property.[68] The horse dander allergen Equ c 2 was identified as two isoforms, Equ c 2.0101 (SWISS-PROT P81216) and Equ c 2.0102 (SWISS-PROT P81217), with slightly different isoelectric points.[27] The N-terminal sequencing of a 29 amino-acid fragment showed that the allergen is a lipocalin since it contained the conserved GXW motif of lipocalins and had a 44% identity with bovine Bos d 2. The amino acid compositions of the allergens are also compatible with lipocalins[67] The nomenclature of horse allergens has been clarified by Goubran Botros et al.[68]

Mouse

Mus m 1, the mouse major urinary protein, MUP6 in the SWISS-PROT data bank (P02762), accounts for the major part of the IgE-binding capacity of the crude male urine.[69] The allergen is also present in serum and pelt.[69] MUPs are mainly produced in liver, although forms of MUPs are expressed constitutively in the exocrine glands of mice and rats.[70] Several hormones contribute to MUP synthesis, and the synthesis is influenced by androgens and is sex-dependent.[70] Mouse MUPs are encoded by numerous genes (see ref. 71), and up to 15 MUPs have been detected in male mouse urine.[72] Male urine contains the allergen in about a fourfold higher concentration than female urine.[69] The structure of mouse MUP (PDB 1MUP) has been resolved.[73] It has been produced as a recombinant protein in *Pichia pastoris*.[74] Rodent MUPs function as odorant-binding proteins which are involved in territory marking and in the endocrine priming of female conspecifics.[70] Mouse and rat MUPs have an amino acid identity of about 65%.[70] As discussed above, the homology of rodent MUPs with cat Fel d 4 and horse Equ c 1 is considerable. The amino acid identity of Mus m 1 is 40% with human putative MUP-like lipocalin (SBNS, Dec. 15, 2004).

Rabbit

The tentatively named major allergen of rabbit, Ory c 1, is a glycoprotein with a molecular mass of 17-18 kDa.[75, 76] Its aminoterminal sequence with the GXW motif and with a 72% homology with rabbit odorant-binding protein-II suggests that it is a lipocalin.[76] Ory c 1 is found in saliva, slightly less in fur, and in low amounts in dander.[75-77] It is absent from urine.[77] Another rabbit allergen, tentatively named as Ory c 2, has a molecular mass of 21 kDa.[76] Its aminoterminal sequence also shows high homology with rabbit odorant-binding protein-II.[76] It is present in several source materials.[77]

Rat

Rat n 1 (SWISS-PROT P02761), also known as rat MUP, is closely related to mouse Mus m 1, as discussed above. Like the mouse, adult male rats excrete more MUPs (about six times) in urine than females.[78] Rat urine was previously considered to contain two distinct allergens, rat urinary prealbumin and α_{2u}-globulin. As these strongly cross-reactive proteins (see ref. 21) were found to be isoforms of α_{2u}-globulin (see ref. 79), it would be better to name prealbumin as Rat n 1.01 and α_{2u}-globulin as Rat n 1.02.[48] α_{2u}-globulin has been produced in vitro.[80] The structure of the protein (PDB 2A2U) has been resolved by X-ray crystallography.[73]

Concluding Remarks

Allergenicity is a property shared by a number of animal lipocalins. Neither the molecular structure nor biologic activity has provided definite clues for the allergenicity of lipocalins. Allergenicity of a protein can be verified conclusively only immunologically, i.e., by showing its capacity to induce sensitization. We have proposed that the allergenicity of lipocalin allergens is related to the adaptation of the immune system to the presence of endogenous lipocalins. The weak capacity of lipocalin allergens to induce the proliferation of PBMCs and the suboptimal character of the immunodominant T cell epitope of Bos d 2 suggest that the recognition of a lipocalin allergen by Th cells can be involved in promoting the development of the Th2-deviated immune response. Although the mapping of T cell epitopes is a laborious process, on the practical level their identification may allow to design new modalities of allergen immunotherapy, based on natural or optimized allergen peptides.[81,82]

Acknowledgements

The work was supported by Kuopio University Hospital (project # 5021605) and the Academy of Finland (contract # 205871).

References

1. Janeway CAJ, Travers P, Walport M et al. Allergy and hypersensitivity. In: Immunobiology - The Immune System in Health and Disease. 4th ed. New York: Garland Publishing, 2001:471-500.
2. King TP, Hoffman D, Løwenstein H et al. Allergen nomenclature. Int Arch Allergy Immunol 1994; 105:224-233.
3. Chapman MD. Allergen nomenclature. In: Lockey RF, Bukantz SC, Bousquet J, eds. Allergens and Allergen Immunotherapy. 3rd ed. New York: Marcel Dekker, Inc., 2004:51-64.
4. Virtanen T, Zeiler T, Rautiainen J et al. Allergy to lipocalins: a consequence of misguided T-cell recognition of self and nonself? Immunol Today 1999; 20:398-400.
5. Åkerstrom B, Flower DR, Salier J-P. Lipocalins: unity in diversity. Biochim Biophys Acta 2000; 1482:1-8.
6. Morgenstern JP, Griffith IJ, Brauer AW et al. Amino acid sequence of Fel d I, the major allergen of the domestic cat: protein sequence analysis and cDNA cloning. Proc Natl Acad Sci USA 1991; 88:9690-9694.
7. Griffith IJ, Craig S, Pollock J et al. Expression and genomic structure of the genes encoding FdI, the major allergen from the domestic cat. Gene 1992; 113:263-268.
8. Smith W, Butler AJ, Hazell LA et al. Fel d 4, a cat lipocalin allergen. Clin Exp Allergy 2004; 34:1732-1738.
9. Munir AKM, Einarsson R, Schou C et al. Allergens in school dust.1. The amount of the major cat (Fel d I) and dog (Can f I) allergens in dust from Swedish schools is high enough to probably cause perennial symptoms in most children with asthma who are sensitized to cat and dog. J Allergy Clin Immunol 1993; 91:1067-1074.
10. Haahtela T, Björksten F, Heiskala M et al. Skin prick test reactivity to common allergens in Finnish adolescents. Allergy 1980; 35:425-431.
11. Barbee RA, Brown WG, Kaltenborn W et al. Allergen skin-test reactivity in a community population sample: correlation with age, histamine skin reactions and total serum immunoglobulin E. J Allergy Clin Immunol 1981; 68:15-19.
12. Vanto T, Koivikko A. Dog hypersensitivity in asthmatic children. A clinical study with special reference to the relationship between the exposure to dogs and the occurrence of hypersensitivity symptoms. Acta Paediatr Scand 1983; 72:571-575.
13. Schou C, Svendsen UG, Løwenstein H. Purification and characterization of the major dog allergen, Can f I. Clin Exp Allergy 1991; 21:321-328.
14. Konieczny A, Morgenstern JP, Bizinkauskas CB et al. The major dog allergens, Can f 1 and Can f 2, are salivary lipocalin proteins: cloning and immunological characterization of the recombinant forms. Immunology 1997; 92:577-586.
15. Saarelainen S, Taivainen A, Rytkönen-Nissinen M et al. Assessment of recombinant dog allergens Can f 1 and Can f 2 for the diagnosis of dog allergy. Clin Exp Allergy 2004; 34:1576-1582.
16. Nelson HS, Szefler SJ, Jacobs J et al. The relationships among environmental allergen sensitization, allergen exposure, pulmonary function, and bronchial hyperresponsiveness in the Childhood Asthma Management Program. J Allerg Clin Immunol 1999; 104:775-785.

17. Perzanowski MS, Ronmark E, Nold B et al. Relevance of allergens from cats and dogs to asthma in the northernmost province of Sweden: schools as a major site of exposure. J Allergy Clin Immunol 1999; 103:1018-1024.
18. van Ree R, van Leeuwen WA, Bulder I et al. Purified natural and recombinant Fel d 1 and cat albumin in in vitro diagnostics for cat allergy. J Allerg Clin Immunol 1999; 104:1223-1230.
19. Aoyama K, Ueda A, Manda F et al. Allergy to laboratory animals: an epidemiological study. Br J Ind Med 1992; 49:41-47.
20. Bush RK, Wood RA, Eggleston PA. Laboratory animal allergy. J Allergy Clin Immunol 1998; 102:99-112.
21. Platts-Mills TA, Longbottom J, Edwards J et al. Occupational asthma and rhinitis related to laboratory rats: serum IgG and IgE antibodies to the rat urinary allergen. J Allergy Clin Immunol 1987; 79:505-515.
22. Desjardins A, Benoit C, Ghezzo H et al. Exposure to domestic animals and risk of immunologic sensitization in subjects with asthma. J Allergy Clin Immunol 1993; 91:979-986.
23. Phipatanakul W, Eggleston PA, Wright EC et al. Mouse allergen. II. The relationship of mouse allergen exposure to mouse sensitization and asthma morbidity in inner-city children with asthma. J Allergy Clin Immunol 2000; 106:1075-1080.
24. Schäfer T, Heinrich J, Wjst M et al. Indoor risk factors for atopic eczema in school children from East Germany. Environ Res 1999; 81:151-158.
25. Fahlbusch B, Rudeschko O, Szilagyi U et al. Purification and partial characterization of the major allergen, Cav p 1, from guinea pig Cavia porcellus. Allergy 2002; 57:417-422.
26. Dandeu JP, Rabillon J, Divanovic A et al. Hydrophobic interaction chromatography for isolation and purification of Equ.c1, the horse major allergen. J Chromatogr - Biomed Appl 1993; 621:23-31.
27. Bulone V, Krogstad-Johnsen T, Smestad-Paulsen B. Separation of horse dander allergen proteins by two-dimensional electrophoresis - Molecular characterisation and identification of Equ c 2.0101 and Equ c 2.0102 as lipocalin proteins. Eur J Biochem 1998; 253:202-211.
28. Zeiler T, Taivainen A, Mäntyjärvi R et al. Threshold levels of purified natural Bos d 2 for inducing bronchial airway response in asthmatic patients. Clin Exp Allergy 2002; 32:1454-1460.
29. Ylönen J, Mäntyjärvi R, Taivainen A et al. IgG and IgE antibody responses to cow dander and urine in farmers with cow-induced asthma. Clin Exp Allergy 1992; 22:83-90.
30. Rosenstreich DL, Eggleston P, Kattan M et al. The role of cockroach allergy and exposure to cockroach allergen in causing morbidity among inner-city children with asthma. N Engl J Med 1997; 336:1356-1363.
31. Arruda LK, Vailes LD, Hayden ML et al. Cloning of cockroach allergen, Bla g 4, identifies ligand binding proteins (or calycins) as a cause of IgE antibody responses. J Biol Chem 1995; 270:31196-31201.
32. Paddock CD, McKerrow JH, Hansell E et al. Identification, cloning, and recombinant expression of procalin, a major triatomine allergen. J Immunol 2001; 167:2694-2699.
33. Blumenthal MN, Rosenberg A. Definition of an allergen (immunobiology). In: Lockey RF, Bukantz SC, Bousquet J, eds. Allergens and Allergen Immunotherapy. 3rd ed. New York: Marcel Dekker, Inc., 2004:37-50.
34. Virtanen T, Mäntyjärvi R. Mammalian allergens. In: Lockey RF, Bukantz SC, Bousquet J, eds. Allergens and Allergen Immunotherapy. 3rd ed. New York: Marcel Dekker, Inc., 2004:297-317.
35. Flower DR. The lipocalin protein family: structure and function. Biochem J 1996; 318:1-14.
36. Yusifov TN, Abduragimov AR, Gasymov OK et al. Endonuclease activity in lipocalins. Biochem J 2000; 347:815-819.
37. Reed CE, Kita H. The role of protease activation of inflammation in allergic respiratory diseases. J Allergy Clin Immunol 2004; 114:997-1008.
38. Chapman MD, Wood RA. The role and remediation of animal allergens in allergic diseases. J Allergy Clin Immunol 2001; 107:S414-421.
39. Kapsenberg ML. Dendritic-cell control of pathogen-driven T-cell polarization. Nat Rev Immunol 2003; 3:984-993.
40. Rachmilewitz J, Riely GJ, Huang JH et al. A rheostatic mechanism for T-cell inhibition based on elevation of activation thresholds. Blood 2001; 98:3727-3732.
41. Mishan-Eisenberg G, Borovsky Z, Weber MC et al. Differential regulation of TH1/TH2 cytokine responses by placental protein 14. J Immunol 2004; 173:5524-5530.
42. Zeiler T, Mäntyjärvi R, Rautiainen J et al. T cell epitopes of a lipocalin allergen colocalize with the conserved regions of the molecule. J Immunol 1999; 162:1415-1422.
43. Jeal H, Draper A, Harris J et al. Determination of the T cell epitopes of the lipocalin allergen, Rat n 1. Clin Exp Allergy 2004; 34:1919-1925.

44. Saarelainen S, Zeiler T, Rautiainen J et al. Lipocalin allergen Bos d 2 is a weak immunogen. Int Immunol 2002; 14:401-409.
45. Kinnunen T, Buhot C, Närvänen A et al. The immunodominant epitope of lipocalin allergen Bos d 2 is suboptimal for human T cells. Eur J Immunol 2003; 33:1717-1726.
46. Leitenberg D, Bottomly K. Regulation of naive T cell differentiation by varying the potency of TCR signal transduction. Semin Immunol 1999; 11:283-292.
47. Brogdon JL, Leitenberg D, Bottomly K. The potency of TCR signaling differentially regulates NFATc/p activity and early IL-4 transcription in naive CD4+ T cells. J Immunol 2002; 168:3825-3832.
48. Virtanen T, Zeiler T, Mäntyjärvi R. Important animal allergens are lipocalin proteins: Why are they allergenic? Int Arch Allergy Immunol 1999; 120:247-258.
49. Spitzauer S, Pandjaitan B, Mühl S et al. Major cat and dog allergens share IgE epitopes. J Allergy Clin Immunol 1997; 99:100-106.
50. Cabanas R, Lopez-Serrano MC, Carreira J et al. Importance of albumin in cross-reactivity among cat, dog and horse allergens. J Investig Allergol Clin Immunol 2000; 10:71-77.
51. Fahlbusch B, Rudeschko O, Schlott B et al. Further characterization of IgE-binding antigens from guinea pig hair as new members of the lipocalin family. Allergy 2003; 58:629-634.
52. Reddy BM, Karande AA, Adiga PR. A common epitope of β-lactoglobulin and serum retinol-binding proteins: elucidation of its core sequence using synthetic peptides. Mol Immunol 1992; 29:511-516.
53. Marchese S, Pes D, Scaloni A et al. Lipocalins of boar salivary glands binding odours and pheromones. Eur J Biochem 1998; 252:563-568.
54. Mäntyjärvi R, Parkkinen S, Rytkönen M et al. Complementary DNA cloning of the predominant allergen of bovine dander: a new member in the lipocalin family. J Allergy Clin Immunol 1996; 97:1297-1303.
55. Rautiainen J, Rytkönen M, Syrjänen K et al. Tissue localization of bovine dander allergen Bos d 2. J Allergy Clin Immunol 1998; 101:349-353.
56. Rautiainen J, Auriola S, Konttinen A et al. Two new variants of the lipocalin allergen Bos d 2. J Chromatogr B Biomed Sci Appl 2001; 763:91-98.
57. Rautiainen J, Auriola S, Rouvinen J et al. Molecular and crystal properties of Bos d 2, an allergenic protein of the lipocalin family. Biochem Biophys Res Commun 1998; 247:746-750.
58. Rouvinen J, Rautiainen J, Virtanen T et al. Probing the molecular basis of allergy - Three-dimensional structure of the bovine lipocalin allergen Bos d 2. J Biol Chem 1999; 274:2337-2343.
59. Zeiler T, Taivainen A, Rytkönen M et al. Recombinant allergen fragments as candidate preparations for allergen immunotherapy. J Allergy Clin Immunol 1997; 100:721-727.
60. Kauppinen J, Zeiler T, Rautiainen J et al. Mutant derivatives of the main respiratory allergen of cow are less allergenic than the intact molecule. Clin Exp Allergy 1999; 29:989-996.
61. de Groot H, Goei KG, van Swieten P et al. Affinity purification of a major and a minor allergen from dog extract: serologic activity of affinity-purified Can f I and of Can f I-depleted extract. J Allergy Clin Immunol 1991; 87:1056-1065.
62. Ford AW, Kemeny DM. The allergens of dog. II. Identification and partial purification of a major dander allergen. Clin Exp Allergy 1992; 22:793-803.
63. Walls AF, Newman Taylor AJ, Longbottom JL. Allergy to guinea pigs: I. Allergenic activities of extracts derived from the pelt, saliva, urine and other sources. Clin Allergy 1985; 15:241-251.
64. Dandeu JP, Rabillon J, Carmi-Leroy A et al. The horse major allergen and its close structural relationship to the mouse and rat urinary protein. J Allergy Clin Immunol 1995; 95:348.
65. Gregoire C, Rosinski-Chupin I, Rabillon J et al. cDNA cloning and sequencing reveal the major horse allergen Equ c 1 to be a glycoprotein member of the lipocalin superfamily. J Biol Chem 1996; 271:32951-32959.
66. Lascombe M-B, Gregoire C, Poncet P et al. Crystal structure of the allergen Equ c 1 - A dimeric lipocalin with restricted IgE-reactive epitopes. J Biol Chem 2000; 275:21572-21577.
67. Bulone V, Rademaker GJ, Pergantis S et al. Characterisation of horse dander allergen glycoproteins using amino acid and glycan structure analyses - A mass spectrometric method for glycan chain analysis of glycoproteins separated by two-dimensional electrophoresis. Int Arch Allergy Immunol 2000; 123:220-227.
68. Goubran Botros H, Poncet P, Rabillon J et al. Biochemical characterization and surfactant properties of horse allergens. Eur J Biochem 2001; 268:3126-3136.
69. Lorusso JR, Moffat S, Ohman JLJ. Immunologic and biochemical properties of the major mouse urinary allergen (Mus m I). J Allergy Clin Immunol 1986; 78:928-937.
70. Cavaggioni A, Mucignat-Caretta C. Major urinary proteins, a_{2U}-globulins and aphrodisin. Biochim Biophys Acta 2000; 1482:218-228.

71. Clark AJ, Clissold PM, Al Shawi R et al. Structure of mouse major urinary protein genes: different splicing configurations in the 3'-non-coding region. EMBO J 1984; 3:1045-1052.
72. Clissold PM, Bishop JO. Variation in mouse major urinary protein (MUP) genes and the MUP gene products within and between inbred lines. Gene 1982; 18:211-220.
73. Böcskei Z, Groom CR, Flower DR et al. Pheromone binding to two rodent urinary proteins revealed by X-ray crystallography. Nature 1992; 360:186-188.
74. Ferrari E, Lodi T, Sorbi RT et al. Expression of a lipocalin in Pichia pastoris: secretion, purification and binding activity of a recombinant mouse major urinary protein. FEBS Lett 1997; 401:73.
75. Price JA, Longbottom JL. Allergy to rabbits. II. Identification and characterization of a major rabbit allergen. Allergy 1988; 43:39-48.
76. Baker J, Berry A, Boscato LM et al. Identification of some rabbit allergens as lipocalins. Clin Exp Allergy 2001; 31:303-312.
77. Warner JA, Longbottom JL. Allergy to rabbits. III. Further identification and characterisation of rabbit allergens. Allergy 1991; 46:481-491.
78. Vandoren G, Mertens B, Heyns W et al. Different forms of a_{2u}-globulin in male and female rat urine. Eur J Biochem 1983; 134:175-181.
79. Bayard C, Holmquist L, Vesterberg O. Purification and identification of allergenic α_{2u}-globulin species of rat urine. Biochim Biophys Acta 1996; 1290:129-134.
80. Unterman RD, Lynch KR, Nakhasi HL et al. Cloning and sequence of several α_{2u}-globulin cDNAs. Proc Natl Acad Sci USA 1981; 78:3478-3482.
81. Larche M. Inhibition of human T-cell responses by allergen peptides. Immunology 2001; 104:377-382.
82. Anderton SM. Peptide-based immunotherapy of autoimmunity: a path of puzzles, paradoxes and possibilities. Immunology 2001; 104:367-376.

CHAPTER 16

Lipocalins in Clinical Medicine

Lennart Lögdberg* and Bo Åkerström

Abstract

This review highlights several possible future roles of lipocalins in human clinical medicine. Generically, due to their metabolism as low molecular weight plasma proteins, lipocalins are candidate markers of kidney functions. Clinical data strongly supporting this notion are available for α_1-microglobulin, retinol-binding protein, lipocalin prostaglandin D-synthase (also called "β-trace"), and neutrophil gelatinase-associated lipocalin (also called LCN2, siderocalin, and human neutrophil lipocalin). In addition, several lipocalins are acute phase reactants: α_1-acid glycoprotein (also called orosomucoid) and neutrophil gelatinase-associated lipocalin are particularly good markers of the acute phase response of inflammation. Finally, a group of xenogeneic lipocalins are under investigation for their possible roles in the diagnosis and treatment of allergy to animals.

Introduction

Human lipocalins have not yet won established places in clinical practice. However, an extensive literature now exists highlighting their possible roles as markers for a variety of diseases, notably in renal or inflammatory diseases (summarized in Table 1).[1-5] Moreover, based on the biological activities of these molecules, individual lipocalins may have promise as therapeutic agents, e.g., for immunomodulation,[6,7] protection against oxidative stress (c.f., Chapter 10),[8-10] amelioration of acute ischemic renal injury,[11,12] or antimicrobial siderophore sequestration (c.f., Chapter 8).[13-15]

Xenogeneic lipocalins, particularly from other mammals, but also from more divergent species, such as the *Insecta* members cockroach (*B. germanica*) and tick (*A. reflexus*), are potent allergens for atopic individuals (c.f., Chapter 15).[16,17] Thus, these lipocalins are likely to have clinical roles in future laboratory diagnostics (e.g., by serving as antigens for detecting specific IgE-antibodies in sera of allergic patients), as well as in clinical diagnostics (e.g., as antigen for prick-test or other provocation tests) and immunotherapy.

Markers for Assessing Renal Function and Disease

Markers Due to Renal Role in Metabolism of Low Molecular Weight Proteins

Members of the lipocalin superfamily are low molecular weight (LMW) proteins (~20 kD) folded into a beta-sheet dominated ("beta-barrel") structure.[18] As such, they tend to be filtered relatively freely across the glomerular membrane into the primary urine.[19] In the proximal tubules, they are reabsorbed by the tubular epithelial cells, and thus occur in the final urine in only minute amounts.[19] The reabsorption of many members of this family has

*Corresponding Author: Lennart Lögdberg—Department of Pathology and Laboratory Medicine, Emory University School of Medicine, Atlanta, Georgia, U.S.A. Email: llogdbe@emory.edu

Lipocalins, edited by Bo Åkerström, Niels Borregaard, Darren R. Flower and Jean-Philippe Salier. ©2006 Landes Bioscience.

Table 1. Lipocalins as diagnostic markers of pathological conditions

Condition	Lipocalin	Major References
I. Renal Function		
Glomerular filtration rate	α_1-microglobulin	2,24
	L-PGDS/β-trace	22,23
Proximal tubular function	α_1-microglobulin	2,3,25-27
	Retinol-binding protein	3
Acute renal failure	NGAL	28
II. Inflammation and infection		
Acute phase of inflammation	α_1-acid glycoprotein	3-5
	NGAL	3,4,55
Distinction between acute bacterial and viral infections	NGAL	3
III. Malignancy		
Lung cancer	α_1-acid glycoprotein	37,38
Hepatocellular carcinoma	α_1-microglobulin	40,41
IV. Allergy		
IgE-dependent allergy	Lipocalin allergens	17, Chapter 15

been demonstrated to be through an interaction with the endocytic receptor megalin, present on the luminal surface of the proximal tubules.[20,21]

Due to this renal processing of LMW proteins, most lipocalins could be thought to serve as markers for both glomerular filtration rate (GFR) and proximal tubular function. Thus, their serum level tends to increase with reduced GFR, and lipocalins such as α_1-microglobulin (α_1m) and lipocalin prostaglandin D synthase (L-PGDS, β-trace) have been shown to be early and sensitive indicators of reduced GFR, and exhibit elevated levels already in the so called "creatinine-blind" range.[2,22-24] Correspondingly, increased levels of LMW proteins, including lipocalins, are passed into the final urine in conditions of damage to the proximal tubules, such as occurs in poisoning by heavy metals (cadmium, lead, mercury, etc.).[2,3] Based on this principle, urinary β_2-microglobulin has long been a gold standard marker for tubular function.[2] However, other LMW proteins that are less sensitive to pH-variations or urinary density, such as α_1m (and perhaps other lipocalins) tend to perform as more consistent markers of tubular function.[2,25]

Urinary LMW protein levels are good markers for tubular function when GFR is normal. It should be noted that urinary levels of LMW proteins, including lipocalins, do increase in the absence of renal tubular dysfunction, in conditions of severely compromised GFR. Such poor GFR causes high serum levels, which is reflected in similarly high levels in the primary urine, exceeding the reabsorptive capacity in the proximal tubules.[3] Thus, in such cases, LMW proteinuria does not indicate tubular dysfunction.

In evaluating proteinuria, urinary lipocalin quantitation can be combined with other quantitative and qualitative (test strip) analyte measurements to generate computer-aided diagnostic interpretations of high accuracy. Such a computer-based "urine protein expert system" including urinary α_1m levels has been used as a decision-making tool for distinguishing between prerenal, renal, and postrenal causes for the proteinuria and has been evaluated for multi-institutional use.[26,27]

NGAL—A Specialized Role as Marker for Acute Renal Failure

Recently, the presence of the neutrophil gelatinase-associated lipocalin or NGAL (other names: lipocalin 2, LCN2; Human neutrophil lipocalin, HNL) in serum or urine at 2h post cardiopulmonary bypass was demonstrated to be a powerful independent predictor of acute renal failure (ARF).[28] Renal ischemia-reperfusion injury is the leading cause of ARF, affecting up to 30-50% of patients in intensive care units, and it is associated with high mortality and morbidity.[28]

NGAL does not only function as an early biomarker of acute ischemic renal injury, preceeding the increase in serum creatinine by 1-3 days; it also compares well with other recently proposed biomarkers for this condition.[28] This is most probably because increased NGAL levels are a consequence of two synergistic mechanisms: (1) reduced GFR and impaired proximal tubular function leading to accumulation of LMW proteins in serum and in urine, and (2) NGAL is induced by inflammatory stimuli, both generally and locally in the injured tubular epithelia.[29]

Local NGAL-expression may reflect a homeostatic response to induce renal reepithelialization, as this lipocalin appears to regulate the in vitro epithelial morphogenesis of cultured renal tubular cells[30] and also may function as an iron-transporting mechanism in nephrogenesis.[31] Consistent with these observations is the observed ameliorating effect of exogenously administered NGAL in a murine model of acute ischaemic renal injury.[11,12] Thus, in addition to being an impressive clinical marker, NGAL shows promise as a therapeutic tool in acute ischaemic renal failure.

Markers of Immunological/Inflammatory Activity

It has been noted that a subgroup of the lipocalins, denoted immunocalins and encompassing at least seven members encoded from the q32-34 region of human chromosome 9, exert a variety of immunomodulatory, anti-inflammatory, and anti-microbial effects and may be part of the innate immune system.[4] These lipocalins include α_1-microglobulin, α_1-acid glycoprotein (AGP, orosomucoid), glycodelin (Gd, PP14), tear lipocalin (TL; lipocalin 1, LCN1), NGAL, complement factor C8 γ-subunit (C8G), and L-PGDS.

Most, if not all, of these seven lipocalins are acute phase reactants, a sign that they may moderate harmful consequences of the inflammatory response, e.g., by exerting tissue protection during inflammation.[4] AGP and NGAL are classical acute phase proteins and inflammatory stimuli lead to both enhanced transcription of the corresponding genes and a subsequent substantial elevation of serum concentrations of these proteins. The other five lipocalin genes can also be upregulated by proinflammatory cytokines, by microbial stimuli, or in selective conditions, such as pregnancy, but are not necessarily associated with generally elevated serum concentrations.[4] They are nonetheless associated with a variety of protective functions.

Recent studies have illuminated the protective, anti-microbial properties of NGAL.[14,15,31-33] Briefly, endotoxin (LPS) from gram-negative bacteria stimulates a 200-fold increase of *NGAL*-message and a 20-fold increase of the NGAL serum protein concentration in vivo.[14] The liver appears to be the major source of NGAL. This activation is TLR4 (Toll-like receptor 4)-dependent, which implies an important function for NGAL in innate defense against bacterial infections.[14]

This defense mechanism of NGAL was recently disclosed. Bacteria, when growing in iron-restricted environments, such as serum, employ secreted siderophores (high-affinity, iron-sequestrating molecules) to compete for the available iron in their surrounding medium, which then enters the bacterium via receptors specific for iron-laden siderophores. It is thus intriguing that NGAL was found to specifically bind several catecholate-type siderophores, including enterochelin, with very high affinity.[15] Enterochelin-like siderophores are made by a large number of human pathogenic bacteria.

Certain *E. coli* strains can be grown to exclusively depend on catecholate-type siderophorins during growth at iron-restricted conditions. NGAL is a particularly effective anti-microbial against such strains when NGAL is stoichiometrically in excess over enterochelin. Consistent with this, the in vivo growth of such bacteria is dramatically enhanced in NGAL knock-out mice compared to wild type mice.[14] Thus, NGAL appears to be a powerful anti-microbial agent in iron-restricted environments and it acquired yet another name: siderocalin.[15]

Additional research has demonstrated that NGAL also binds soluble siderophores of mycobacteria, such as carboxymycobactins of *M. tuberculosis*, via a degenerate recognition mechanism.[33] TL, another lipocalin, also appears to exert anti-microbial effects via siderophore sequestration.[13] Its siderophore binding is less specific and of lower affinity than NGAL, and includes binding to bacterial siderophores, such as catecholate-type enterobactin and hydroxymate-type desferrioxamin, as well as to all major classes of fungal siderophores.

The inflammatory and immunomodulating properties of AGP, as well as the importance of its glycosylation pattern, were recently reviewed.[5] Increasing interest has focussed on inflammation-sensitive proteins, such as AGP, whose elevated levels in plasma or urine are associated with increased cardiovascular and overall mortality in patients with other risk factors.[34,35]

Markers in Oncology

The potential role of lipocalins as markers for malignancy was reviewed recently.[36] The known role of lipocalin ligands in cell differentiation and proliferation, and the protease-inhibitory properties of some lipocalins, suggested several pathways for interaction between lipocalins and cancer cells.

Since then, it has been demonstrated that AGP is an independent predictor of treatment response and a prognostic factor for survival in patients with non-small cell lung cancer and docetaxel chemotherapy.[37] In contrast, a high degree of fucosylation of AGP is associated with poor prognosis in patients with advanced malignancies.[38] More recently, NGAL was shown to reverse the malignant phenotype of Ras-transformed cells.[39] When added as purified NGAL or as NGAL vectors to Ras-transformed 4T1 mouse mammary tumor cells, invasiveness and metastasis were diminished.

Since the liver is the major site of synthesis of many lipocalins,these proteins may be useful as markers for hepatocyte origin and activity. Thus, elevated plasma levels of α_1m have been reported in patients with hepatocellular carcinoma (HCC).[40] Similarly, it has been suggested that the immunohistochemical demonstration of α_1m on liver sections may help distinguish between primary liver tumors vs metastases.[41]

Markers for the Diagnosis of Allergy and Possible Therapeutic Tools for Immunotherapy

Immunoglobulin E-mediated allergy to environmental antigens (present in pollen, dust, animal dander, food, etc) is an important and costly health problem in the industrialized world and its incidence/prevalence is increasing.[42-45] A subset of allergies usually linked to respiratory sensitization includes hypersensitivity to other mammals such as pets, farm animals, research animals, and their wild counterparts. Intriguingly, recent research has demonstrated that the major mammalian allergens tend to belong to the lipocalin superfamily (c.f., Chapter 15).[17]

This phenomenon likely reflects the functional properties of the lipocalin molecules such as their ligand binding and receptor interactions, and may also represent the intricate interplay between such lipocalin allergens and their endogenous human orthologues in the patients. Further research on the etiology and pathophysiology of lipocalin allergy may, therefore, reveal fundamental aspects of the normal physiology of IgE responses and of mast cell biology.

However, of immediate clinical importance is that the research identifying and characterizing lipocalin allergens has generated a new panel of purified, well-characterized natural and recombinant allergens. These are being made available gradually for the diagnosis and treatment of allergy[46] through the establishment of more sensitive detection assays for anti-lipocalin IgEs[47,48] and the biological diagnosis in prick assays.[49] With time, this is likely to be generalized to the use of purified natural and/or recombinant lipocalin allergens in diagnostic microarrays,[50] for various patient provocation testing methods,[51] and for devising innovative immunotherapy protocols.[52-54]

Concluding Remarks

Lipocalins are now positioned to find clinical roles as biomarkers for renal function, for inflammation, and for allergy diagnosis. The critical function of some lipocalins in selected pathological conditions further suggest that the future may witness the development of lipocalin-based therapeutic agents.

Acknowledgements

We thank Linda Lögdberg, Ph.D., for revising the English text.

References

1. Seppala M, Taylor RN, Koistinen H et al. Glycodelin: A major lipocalin protein of the reproductive axis with diverse actions in cell recognition and differentiation. Endocr Rev 2002; 23:401-430.
2. Penders J, Delanghe JR. Alpha 1-microglobulin: Clinical laboratory aspects and applications. Clin Chim Acta 2004; 346:107-118.
3. Xu S, Venge P. Lipocalins as biochemical markers of disease. Biochim Biophys Acta 2000; 1482:298-307.
4. Logdberg L, Wester L. Immunocalins: A lipocalin subfamily that modulates immune and inflammatory responses. Biochim Biophys Acta 2000; 1482:284-297.
5. Hochepied T, Berger FG, Baumann H et al. Alpha(1)-acid glycoprotein: An acute phase protein with inflammatory and immunomodulating properties. Cytokine Growth Factor Rev 2003; 14:25-34.
6. Muchitsch EM, Auer W, Pichler L. Effects of alpha 1-acid glycoprotein in different rodent models of shock. Fundam Clin Pharmacol 1998; 12:173-181.
7. Libert C, Brouckaert P, Fiers W. Protection by alpha 1-acid glycoprotein against tumor necrosis factor-induced lethality. J Exp Med 1994; 180:1571-1575.
8. Lechner M, Wojnar P, Redl B. Human tear lipocalin acts as an oxidative-stress-induced scavenger of potentially harmful lipid peroxidation products in a cell culture system. Biochem J 2001; 356:129-135.
9. Allhorn M, Berggard T, Nordberg J et al. Processing of the lipocalin alpha(1)-microglobulin by hemoglobin induces heme-binding and heme-degradation properties. Blood 2002; 99:1894-1901.
10. Allhorn M, Klapyta A, Åkerström B. Redox properties of the lipocalin alpha1-microglobulin: Reduction of cytochrome c, hemoglobin, and free iron. Free Radic Biol Med 2005; 38:557-567.
11. Mishra J, Mori K, Ma Q et al. Amelioration of ischemic acute renal injury by neutrophil gelatinase-associated lipocalin. J Am Soc Nephrol 2004; 15:3073-3082.
12. Mori K, Lee HT, Rapoport D et al. Endocytic delivery of lipocalin-siderophoreiron complex rescues the kidney from ischemia-reperfusion injury. J Clin Invest 2005; 115:610-621.
13. Fluckinger M, Haas H, Merschak P et al. Human tear lipocalin exhibits antimicrobial activity by scavenging microbial siderophores. Antimicrob Agents Chemother 2004; 48:3367-3372.
14. Flo TH, Smith KD, Sato S et al. Lipocalin 2 mediates an innate immune response to bacterial infection by sequestrating iron. Nature 2004; 432:917-921.
15. Goetz DH, Holmes MA, Borregaard N et al. The neutrophil lipocalin NGAL is a bacteriostatic agent that interferes with siderophoremediated iron acquisition. Mol Cell 2002; 10:1033-1043.
16. Hilger C, Bessot JC, Hutt N et al. IgE-mediated anaphylaxis caused by bites of the pigeon tick Argas reflexus: Cloning and expression of the major allergen Arg r 1. J Allergy Clin Immunol 2005; 115:617-622.
17. Mantyjarvi R, Rautiainen J, Virtanen T. Lipocalins as allergens. Biochim Biophys Acta 2000; 1482:308-317.
18. Åkerström B, Flower DR, Salier JP. Lipocalins: Unity in diversity. Biochim Biophys Acta 2000; 1482:1-8.
19. Strober W, Waldmann TA. The role of the kidney in the metabolism of plasma proteins. Nephron 1974; 13:35-66.
20. Hvidberg V, Jacobsen C, Strong RK et al. The endocytic receptor megalin binds the iron transporting neutrophil-gelatinase-associated lipocalin with high affinity and mediates its cellular uptake. FEBS Lett 2005; 579:773-777.
21. Leheste JR, Rolinski B, Vorum H et al. Megalin knockout mice as an animal model of low molecular weight proteinuria. Am J Pathol 1999; 155:1361-1370.
22. Priem F, Althaus H, Birnbaum M et al. Beta-trace protein in serum: A new marker of glomerular filtration rate in the creatinine-blind range. Clin Chem 1999; 45:567-568.
23. Filler G, Priem F, Lepage N et al. Beta-trace protein, cystatin C, beta(2)-microglobulin, and creatinine compared for detecting impaired glomerular filtration rates in children. Clin Chem 2002; 48:729-736.
24. Grubb A. Diagnostic value of analysis of cystatin C and protein HC in biological fluids. Clin Nephrol 1992; 38(Suppl 1):S20-27.

25. Ikeda M, Ezaki T, Tsukahara T et al. Critical evaluation of alpha1- and beta2-microglobulins in urine as markers of cadmium-induced tubular dysfunction. Biometals 2004; 17:539-541.
26. Ivandic M, Hofmann W, Guder WG. The use of knowledge-based systems to improve medical knowledge about urine analysis. Clin Chim Acta 2000; 297:251-260.
27. Ivandic M, Hofmann W, Guder WG. Development and evaluation of a urine protein expert system. Clin Chem 1996; 42:1214-1222.
28. Mishra J, Dent C, Tarabishi R et al. Neutrophil gelatinase-associated lipocalin (NGAL) as a biomarker for acute renal injury after cardiac surgery. Lancet 2005; 365:1231-1238.
29. Mishra J, Ma Q, Prada A et al. Identification of neutrophil gelatinase-associated lipocalin as a novel early urinary biomarker for ischemic renal injury. J Am Soc Nephrol 2003; 14:2534-2543.
30. Gwira JA, Wei F, Ishibe S et al. Expression of neutrophil gelatinase-associated lipocalin regulates epithelial morphogenesis in vitro. J Biol Chem 2005; 280:7875-7882.
31. Yang J, Goetz D, Li JY et al. An iron delivery pathway mediated by a lipocalin. Mol Cell 2002; 10:1045-1056.
32. Doneanu CE, Strong RK, Howald WN. Characterization of a noncovalent lipocalin complex by liquid chromatography/electrospray ionization mass spectrometry. J Biomol Tech 2004; 15:208-212.
33. Holmes MA, Paulsene W, Jide X et al. Siderocalin (Lcn 2) also binds carboxymycobactins, potentially defending against mycobacterial infections through iron sequestration. Structure (Camb) 2005; 13:29-41.
34. Lind P, Engstrom G, Stavenow L et al. Risk of myocardial infarction and stroke in smokers is related to plasma levels of inflammation-sensitive proteins. Arterioscler Thromb Vasc Biol 2004; 24:577-582.
35. Christiansen MS, Hommel E, Magid E et al. Orosomucoid in urine is a powerful predictor of cardiovascular mortality in normoalbuminuric patients with type 2 diabetes at five years of follow-up. Diabetologia 2005; 48:386-393.
36. Bratt T. Lipocalins and cancer. Biochim Biophys Acta 2000; 1482:318-326.
37. Bruno R, Olivares R, Berille J et al. Alpha-1-acid glycoprotein as an independent predictor for treatment effects and a prognostic factor of survival in patients with non-small cell lung cancer treated with docetaxel. Clin Cancer Res 2003; 9:1077-1082.
38. Hashimoto S, Asao T, Takahashi J et al. Alpha1-acid glycoprotein fucosylation as a marker of carcinoma progression and prognosis. Cancer 2004; 101:2825-2836.
39. Hanai J, Mammoto T, Seth P et al. Lipocalin 2 diminishes invasiveness and metastasis of Ras-transformed cells. J Biol Chem 2005; 280:13641-13647.
40. Vincent C, Kew MC, Bouic P et al. Alpha 1-microglobulin (HC protein) in human hepatocellular carcinoma. Br J Cancer 1989; 59:415-416.
41. Badve S, Tanaka K, Steinberg JJ et al. Alpha-1 microglobulin as an immunohistochemical marker for evaluation of liver tumors. Modern Pathology 1998; 11:150A.
42. Sullivan SD, Weiss KB. Health economics of asthma and rhinitis. II. Assessing the value of interventions. J Allergy Clin Immunol 2001; 107:203-210.
43. Weiss KB, Sullivan SD. The health economics of asthma and rhinitis. I. Assessing the economic impact. J Allergy Clin Immunol 2001; 107:3-8.
44. Bresciani M, Parisi C, Manghi G et al. The hygiene hypothesis: Does it function worldwide? Curr Opin Allergy Clin Immunol 2005; 5:147-151.
45. Strachan DP. Hay fever, hygiene, and household size. Bmj 1989; 299:1259-1260.
46. Chapman MD, Smith AM, Vailes LD et al. Recombinant allergens for diagnosis and therapy of allergic disease. J Allergy Clin Immunol 2000; 106:409-418.
47. Korpi A, Mantyjarvi R, Rautiainen J et al. Detection of mouse and rat urinary aeroallergens with an improved ELISA. J Allergy Clin Immunol 2004; 113:677-682.
48. Ferrari E, Tsay A, Eggleston PA et al. Environmental detection of mouse allergen by means of immunoassay for recombinant Mus m 1. J Allergy Clin Immunol 2004; 114:341-346.
49. Saarelainen S, Taivainen A, Rytkonen-Nissinen M et al. Assessment of recombinant dog allergens Can f 1 and Can f 2 for the diagnosis of dog allergy. Clin Exp Allergy 2004; 34:1576-1582.
50. Hiller R, Laffer S, Harwanegg C et al. Microarrayed allergen molecules: Diagnostic gatekeepers for allergy treatment. FASEB J 2002; 16:414-416.
51. van Hage-Hamsten M, Pauli G. Provocation testing with recombinant allergens. Methods 2004; 32:281-291.
52. Niederberger V, Horak F, Vrtala S et al. Vaccination with genetically engineered allergens prevents progression of allergic disease. Proc Natl Acad Sci USA 2004; 101(Suppl 2):14677-14682.
53. Zhu D, Kepley CL, Zhang K et al. A chimeric human-cat fusion protein blocks cat-induced allergy. Nat Med 2005; 11:446-449.
54. Larche M, Wraith DC. Peptide-based therapeutic vaccines for allergic and autoimmune diseases. Nat Med 2005; 11:S69-76.
55. Kjeldsen L, Cowland JB, Borregaard N. Human neutrophil gelatinase-associated lipocalin and homologous proteins in rat and mouse. Biochim Biophys Acta 2000; 1482:272-283.

Chapter 17

The Lipocalin Protein Family:

Perspectives for Future Research

Darren R. Flower* and Arne Skerra

Introduction

Lipocalinology, as a discipline, has been with us for more or less twenty years. After an initial period of exciting, if capricious, growth, study of the lipocalin protein family has now entered a period of solid and significant maturity. The modern era of lipocalin research is marked by the astonishing and burgeoning diversity of function currently becoming apparent among members of this remarkable family. The lipocalins are, perhaps, best known to those outside the field for two notable properties. First, the simplicity and aesthetic appeal of their highly symmetrical β-barrel fold. Secondly, and more significantly, is the seeming paradox of their low sequence conservation, which stands in marked contrast to their high structural propinquity.[1]

While it is now clear that neither property is unique to the lipocalins, nonetheless the family remains noteworthy for several reasons. Three kinds of intriguing molecular recognition events have long been seen to characterize the lipocalins: small molecule binding within the intra-calyx cavity, macromolecular complex formation, and receptor binding. Deep insights have been gained into these properties, and it is clear that we still understand structural and biophysical aspects of the lipocalins better than we grasp their physiological role and thus their place in biology. While it is true that lipocalinists have now catalogued many important physiological functions, much of biological significance remains to be discovered.

However, gone are the days when the family could be dismissed as a small group of obscure transporters of hydrophobic molecules. Lipocalins are important proteins doing important things, not only in the delivery of essential metabolic compounds but also in cellular signalling or as part of the innate immune response, to name just a few examples. However, the link between their molecular recognition properties and their function is not always obvious. Biological function, itself, can remain cryptic. The underlying coherence of the apparently separate biochemical activities of individual lipocalins can appear obscure. Nonetheless, in the time since the first definitive review of the family was written,[1] our view of the lipocalin family has changed dramatically. Important strides are being made and will continue to be made, and it is the future perspectives for lipocalin research that forms the focus of this chapter. Specifically, we will endeavour to throw light upon the way that certain recent developments will shape the future of lipocalin research.

*Corresponding Author: Darren R. Flower—Edward Jenner Institute for Vaccine Research, Compton, Berkshire, RG20 7NN, U.K. Email: darren.flower@jenner.ac.uk

Lipocalins, edited by Bo Åkerström, Niels Borregaard, Darren R. Flower and Jean-Philippe Salier. ©2006 Landes Bioscience.

The Lipocalin Protein Family: Their Changing Functional Roles

Amongst other things, the lipocalin protein family undertakes a general role in the storage and transport of poorly soluble or chemically sensitive compounds like vitamins, steroids, and metabolic products. Other lipocalins have a more specialized function in vertebrates: in olfaction, in the mediation of cell homeostasis and immune regulation, and even as an enzyme catalyzing the formation of prostaglandin D_2.

A subset, with functional rather than structural similarity, of the lipocalin protein family is involved in mediating protective immunoregulation and antibacterial effects. In particular, the role of lipocalins as mediators of defense against oxidative stress has recently been highlighted. For example, the archetypal plasma lipocalin α_1-microglobulin acts as a radical scavenger during extravascular hemolysis and chronic ulcer inflammation. Furthermore, in the context of innate immunity, a physiological role of NGAL is to act as a tight binder of bacterial virulence factors called siderophores, whose function is to sequester iron from the human body. Another important lipocalin implicated in immunology is Glycodelin. Coincidentally, many lipocalins from bloodsucking insects act as virulence factors, releasing heme-bound nitric oxide or sequestering histamine at wound sites. Thus, lipocalins have evolved to act both as virulence factors and to provide a host defense against them.

As is clear from other chapters in this book, the study of the lipocalin family has revealed a number of previously hidden functions far beyond simple transport of hydrophobic molecules, important though that is. It is likely that as research into the lipocalins continues, new and even more interesting biological activities will become evident.

The Lipocalin Protein Family: A Paradigm for Molecular Recognition

Rightly, there has been much interest of late in the emergent functional properties of the lipocalins, yet there is much still to be learnt about their more fundamental properties. Sequence and structural analysis of lipocalin genes and proteins will reveal ever more about the evolutionary origin and development of the family. Structural insight will also grow in terms of the relationship between the lipocalin family and an ever more extensive and diverse set of related proteins. The molecular interactions made by lipocalins with their receptor proteins may, in time, also be revealed, but this will first require a deeper characterization of lipocalin receptors themselves.

Much is still to be learned about the different ways in which small molecules can form complexes with lipocalins. Binding of ligands to members of the lipocalin protein family is now well characterized. Some lipocalins possess high ligand specificity, e.g., RBP for retinol, ApoD for progesterone and arachidonic acid, or NGAL for FeIII•enterobactin, while others, such as human tear lipocalin, bovine β-lactoglobulin, odorant binding proteins, and even crustacyanin, demonstrate considerable promiscuity in their binding patterns.

The affinity of lipocalins for small molecules is typically intermediate, with K_d values around 1 μM, which is consistent with their functioning as physiological buffers for their bound molecules. However, there are notable exceptions to this, reflecting, presumably, a particular biological role: e.g., NGAL (K_d of 0.4 nM for enterobactin) and tick histamine-binding protein (K_d of 1.7 nM for histamine). As so many lipocalins are "nonspecific" binders, forming complexes with a variety of small or medium sized hydrophobic molecules, it will become increasingly important to properly classify true small molecule ligands from those which can be bound without physiological meaning.

This is particularly true of synthetic exogenous molecules, for example rifampin in conjunction with Tlc, where binding may be coincidental and without functional implication. Hence, identifying binding of a particular molecule to a particular lipocalin, while interesting, does not necessarily imply a natural function of the protein. We will see progressively more sophisticated instrumental approaches, for example Isothermal Titration Calorimetry, displacing conventional assays, such as equilibrium dialysis or competition ELISA, as the principal tool for investigating lipocalin-small molecule interactions, and we can be confident that much exciting data remains to be uncovered.

Lipocalins are proteins whose properties challenge and inspire the creativity of the chemist, pharmacist, biotechnologist, and biophysicist. Long ago, crustacyanin was primarily of interest as a surrogate for bathochromic shifts observed in rhodopsin. More recently, Matile and co-workers have been motivated by the eight-stranded structure of lipocalins to develop novel host-guest chemistry.[2] They designed and synthesized a surrogate supramolecular β-barrel model: oligo(p-phenylene)s were assembled with complementary Lys-Leu-Lys and Glu-Leu-Glu side chains to form water-soluble rigid-rod pseudo-β-barrels of fixed length but with flexible diameter. The presence of a hydrophobic interior was confirmed by guest encapsulation. Complexes with β-carotene were formed within tetrameric barrels, and the carotenoid astaxanthin was complexed by dodecameric barrels. Astaxanthin is the natural ligand of crustacyanin, a carotenoid-protein complex from the calcified outer layer of the lobster carapace. Crustacyanin is the main lobster pigment and gives them their characteristic blue colour: the blue-to-red change on boiling results from the liberation of free astaxanthin upon heat denaturation of the lipocalin protein.

Drugs, Drug Discovery, and Drug Delivery: The Role of Lipocalins and Lipocalin Receptors

In principle, some lipocalins may deliver their ligand from the extracellular medium via a cell membrane receptor, possibly involving endocytosis of the whole complex, whereas certain lipocalins could even function in cellular signalling by themselves. A number of receptors for the lipocalins are now known,[3] but examples properly characterized at the molecular level have only emerged recently. Evidence exists for the specific interaction of RBP with a cellular receptor and also for the uptake of NGAL by a receptor on kidney cells. Megalin, a very large membrane receptor has been shown to bind several lipocalins, but in a nonspecific manner. Glycodelin binds to CD45, in a lectin-like manner, with an ensuing negative regulation of T-cell signaling. Lipocalin-1 interacting membrane receptor (LIMR), a 57 kD transmembrane protein, is the first specifically cloned lipocalin receptor. It is responsible for the receptor-mediated internalization of tear lipocalin. This area is one of the most exciting in lipocalin research because much remains to be discovered about lipocalin receptors and their physiological relevance.

The observation that lipocalins participate in a variety of key physiological mechanisms, together with their molecular properties as robust globular ligand-binding proteins, suggests potential in diverse medical applications. For instance, certain human lipocalins are associated with pathological disorders and may thus provide interesting targets for pharmaceutical intervention.[4] Moreover, because of their ability to complex small hydrophobic molecules while simultaneously being bound by cell surface receptors, lipocalins were long ago suggested as potential systems for drug delivery. Certain lipocalins, such as bovine β-lactoglobulin, are acid stable and could thus survive a journey through the stomach before being taken up by receptors in the intestine. Were it possible to manipulate the specificity of β-lactoglobulin or other acid-stable lipocalins, they could thus become excellent transporters of acid-labile small molecule drugs or pro-drugs. Other members of the lipocalin protein family, i.e., those originating in arthropods, possess scavenging functions for inflammatory mediators, such as histamine. Potentially, these lipocalins could themselves be useful as therapeutic protein drugs.

It is also possible to engineer artificial lipocalins with specifically tailored binding properties. Along with other protein scaffolds with biotechnological applications that have recently emerged, artificial lipocalins could fulfil similar medical applications as do therapeutic antibodies now.[5] Although many lipocalins display high ligand specificity, many, such as odorant binding proteins, exhibit a catholic taste for different ligands. So the possibility exists of tailoring lipocalin structures to bind one or several molecules over a considerable range of molecular affinities. Beyond ligand binding, lipocalins also offer other benefits for protein engineering and raise the expectation that they will find important applications in biotechnology and medicine as therapies, diagnostics, and reagents. For example, many lipocalins

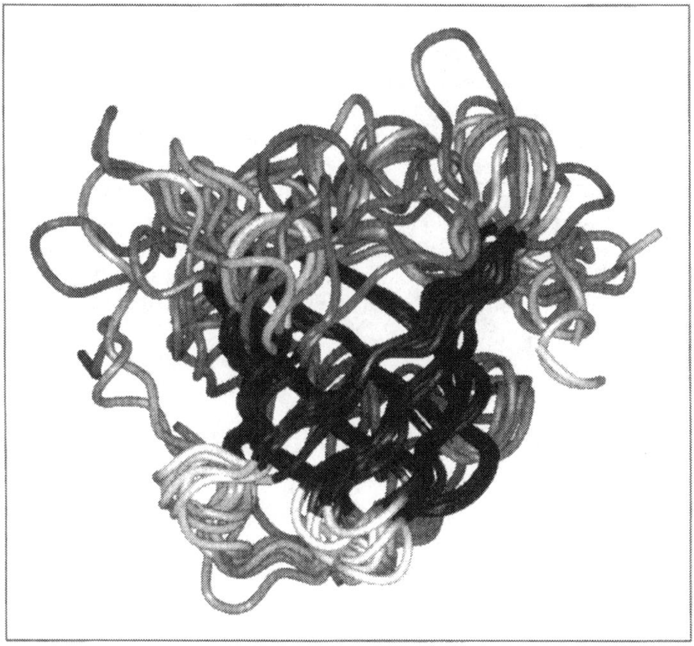

Figure 1. Structural plasticity of the lipocalin scaffold: potential for reshaping the binding site. The coordinates of six different lipocalins were superimposed (black: β-barrel; colour: hypervariable loop region at the open end of the β-barrel; grey: extraneous structural elements; PDB entries 1RBP, 1BBP, 1MUP, 1EPA, 1BEB, 1BJ7).

are remarkably thermostable and constitute extracellular, monomeric globular proteins with low immunogenic potential. Their small size and lack of glycosylation render them highly appropriate for mass production as recombinant proteins using bacterial expression systems.

Using a combinatorial approach to protein mutagenesis, Skerra and colleagues have done just this. By reshaping the natural binding site of lipocalins formed by the four loops at the open end of the β-barrel fold (Fig. 1), they constructed novel proteins able to bind new small molecules and, more recently, also to recognize protein targets.[4] Initially, they have described engineered lipocalins able to bind fluorescein and digoxigenin. Digoxigenin is the steroid component of cardioactive glycosides of theurapeutic interest. Its chemically reactive analogues also make the rather hydrophilic digoxigenin group an important reagent for the nonradioactive labeling of biological molecules. Starting with the structure of the butterfly bilin-binding protein, libraries of mutant lipocalins were generated by random mutagenesis. Phage display and colony screening then led to the identification of proteins showing high affinity for the prescribed ligands (Fig. 2).

These artificial lipocalins have been likened to antibodies, and the phrase "Anticalins" was coined to describe them.[4] Nature has exploited the antibody structure to generate a vast array of biomolecules able to bind other proteins or haptens strongly. "Anticalins" can perform a similar, but engineered, function for molecular recognition.[5] The way is open to develop a whole new repertoire of protein reagents able to bind small molecules. It is also possible in a similar way to engineer their binding to soluble macromolecules and cell surface receptors.[4] The potential exploitation of lipocalins in chemistry, biotechnology, and medicine is just beginning. One day the diversity and importance of engineered lipocalin reagents may come to rival the variety and importance of their natural biological functions. A combination of their

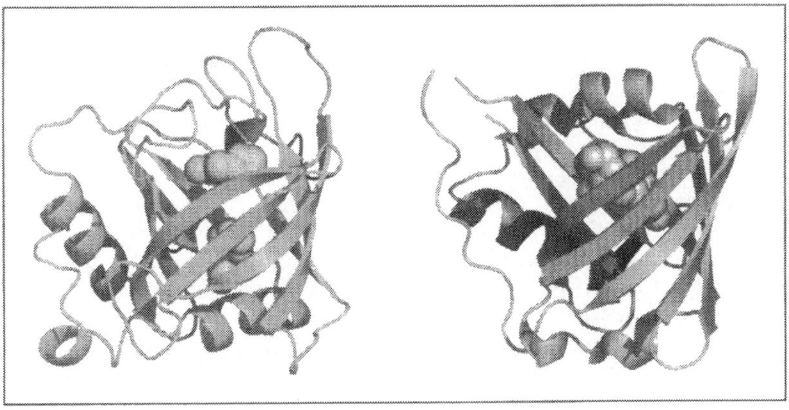

Figure 2. A natural and an engineered lipocalin with therapeutic potential: Histacalin vs. Digical®. Ribbon diagram of the crystal structures of HBP2 (PDB entry 1QFT), left, and of DigA16 (PDB entry 1LKE), right. The insect lipocalin HBP2 binds two histamine ligands, one at the open end (top) and one buried within the lower part of the β-barrel. The anticalin DigA16 complexes the prescribed digoxigenin steroid at the open end of the β-barrel, in the region of the four structurally variable loops, as is typical for most natural lipocalins. Due to its high affinity it can be used as an effecctive scavenger for the cardioactive drug.

structural flexibility, their capacity to be designed, and their receptor mediated function means that they may also, in time, deliver blockbuster small molecule drugs and yield protein therapeutics of true significance.

The Limitless Horizons of Lipocalin Research

The molecular recognition properties displayed by the lipocalins manifest themselves in remarkably different ways and this is mirrored in their highly divergent sequences, their wide phyletic spread across species, and, as far as is known, in their differing apparent functions. As our knowledge of the family, and its inherent complexity, grows it becomes ever more evident how little we understand about the lipocalins. Even simple questions remain unanswered. How many of them are there in nature? The answer to this should, in the genomic era, be relatively straightforward, at least at the conceptual level, but is it? The current, putative size of the human genome has been revised down from figures in excess of 100,000 to estimates around 40,000 genes. Most recently, a number closer to 20,000 has been suggested.[6] Clearly, the size of the human genome and the number of lipocalins within it remain just that: estimates. Both are likely to change, although we can be confident that, with a number around ten, most lipocalin genes have now been identified, at least for humans. For other genomes, the situation still lags some way behind in terms of annotation and lipocalin identification. Although lipocalins are present in phyla from bacteria to man, family members have, surprisingly, yet to be found in some common and well studied species, such as yeast.

Genomic lipocalin identification is, however, the beginning rather than the end. This is much more obvious at the protein level than it is at the nucleic acid level. One gene may imply many proteins: through splice variants or contingent post-translational modifications, etc. These distinct proteins may have different properties and thus new functions. As is well known, the proteome is much larger and more complex than the genome. Some estimates place the number of proteins encoded by the human genome orders of magnitude higher than the number of genes. Moreover, the proteome is also much more dynamic than the genome. Identifying, cataloguing and characterising the lipocalin complement within the human proteome will thus prove significantly more challenging than annotation of the genome.

As the overall number of genes, apparent in the human genome, has been seen to decrease, the relative importance of each gene has demonstrated a concomitant increase. The fundamental augmentation we see in the importance of each lipocalin is likewise reflected in the growing relevance of its functions. Our present knowledge of biology is chiefly phenomenological, though in the post-genomic era we are slowly gaining a much deeper physical comprehension of individual molecular events. Our appreciation of life is altering from a top-down to an integrated, bottom-up view. The lipocalins illustrate this view well: protein multi-functionalism was once perceived as the exception, yet it is fast becoming the rule. The diversity exhibited by the biological functioning of lipocalins is doubtlessly matched by the diversity of interacting partners mediating their effects. Recently, evidence has emerged of lipocalin receptors of both known structure, such as CCR5,[7] and those for which a definite structure has yet to be determined, such as the mouse spermatozoan cell-surface receptor for 24p3, the murine homologue of NGAL.[8] We can thus be confident that several more lipocalin receptors await discovery.

While many lipocalin functions remain to be discovered, extant work suggests that they fulfil many critical roles and the accurate characterization of lipocalin receptors will foment the hunt for novel agonists and antagonists of lipocalin function. However, trying to foresee the future is no easy business and it would be foolish to guess beyond what is reasonable in this area. The lipocalins have always been full of surprises, always ready to subvert our expectations. Nonetheless, exciting times lie ahead not only for the study of natural lipocalins but also for creating artificial lipocalins with novel functions. We may hope that their role as diagnostic markers, laboratory reagents, therapeutic agents, as well as drug discovery targets in their own right, together with the pharmacology of their receptors, will yet raise the profile of the lipocalins to the preeminent position that it so richly deserves.

Acknowledgements

D.R. Flower would like to thank Prof. A.C.T. North for introducing him to the lipocalins and to Prof. T.K. Attwood for long term collaboration. A. Skerra wishes to thank Prof. Alwyn Jones and Prof. Robert Huber for attracting his attention towards the molecular structures of RBP and BBP, respectively.

References

1. Flower DR. The lipocalin protein family: Structure and function. Biochem J 1996; 318:1-14.
2. Matile S. Bioorganic chemistry a la baguette: Studies on molecular recognition in biological systems using rigid-rod molecules. Chem Rec 2001; 1:162-172.
3. Flower DR. Beyond the superfamily: The lipocalin receptors. Biochim Biophys Acta 2000; 1482:327-336.
4. Schlehuber S, Skerra A. Lipocalins in drug discovery: From natural ligand-binding proteins to 'anticalins'. Drug Discov Today 2005; 10:23-33.
5. Skerra A. Imitating the humoral immune response. Curr Opin Chem Biol 2003; 7:683-693.
6. International Human Genome Sequencing Consortium. Finishing the euchromatic sequence of the human genome. Nature 2004; 431:931-945.
7. Atemezem A, Mbemba E, Vassy R et al. Human α_1-acid glycoprotein binds to CCR5 expressed on the plasma membrane of human primary macrophages. Biochem J 2001; 356:121-128.
8. Elangovan N, Lee YC, Tzeng WF et al. Delivery of ferric ion to mouse spermatozoa is mediated by lipocalin internalization. Biochem Biophys Res Commun 2004; 319:1096-1104.

Index

C

D

E

F

G

H

I

K

L

U

V

W

X

Y

Z